VOLUME SIX HUNDRED AND THIRTY SIX

# METHODS IN ENZYMOLOGY

## Tumor Immunology and Immunotherapy – Integrated Methods Part B

# METHODS IN ENZYMOLOGY

VOLUME SIX HUNDRED AND THIRTY SIX

# METHODS IN ENZYMOLOGY

## Tumor Immunology and Immunotherapy – Integrated Methods Part B

Edited by


**LORENZO GALLUZZI**

*Department of Radiation Oncology,*
*Weill Cornell Medical College; Sandra and Edward Meyer Cancer Center, New York, NY;*
*Department of Dermatology, Yale School of Medicine, New Haven, CT, United States;*
*Université de Paris, Paris, France*

**NILS-PETTER RUDQVIST**

*Department of Radiation Oncology,*
*Weill Cornell Medicine, New York, NY, United States*



ELSEVIER

ACADEMIC PRESS
An imprint of Elsevier

Academic Press is an imprint of Elsevier
50 Hampshire Street, 5th Floor, Cambridge, MA 02139, United States
525 B Street, Suite 1650, San Diego, CA 92101, United States
The Boulevard, Langford Lane, Kidlington, Oxford OX5 1GB, United Kingdom
125 London Wall, London, EC2Y 5AS, United Kingdom

First edition 2020



**Notices**
Knowledge and best practice in this field are constantly changing. As new research and experience broaden our understanding, changes in research methods, professional practices, or medical treatment may become necessary.

Practitioners and researchers must always rely on their own experience and knowledge in evaluating and using any information, methods, compounds, or experiments described herein. In using such information or methods they should be mindful of their own safety and the safety of others, including parties for whom they have a professional responsibility.

To the fullest extent of the law, neither the Publisher nor the authors, contributors, or editors, assume any liability for any injury and/or damage to persons or property as a matter of products liability, negligence or otherwise, or from any use or operation of any methods, products, instructions, or ideas contained in the material herein.

ISBN: 978-0-12-820667-6
ISSN: 0076-6879

For information on all Academic Press publications
visit our website at https://www.elsevier.com/books-and-journals

*Publisher:* Zoe Kruze
*Acquisitions Editor:* Ashlie M. Jackman
*Editorial Project Manager:* Shellie Bryant
*Production Project Manager:* James Selvam
*Cover Designer:* Greg Harris

Typeset by SPi Global, India

# Contents

# Contributors

**Patrizia Agostinis**
Department for Cellular and Molecular Medicine, Cell Death Research & Therapy (CDRT) Unit, KU Leuven; Center for Cancer Biology (CCB), VIB, Leuven, Belgium

**Fernando Aranda**
Immunoreceptors of the Innate and Adaptative System, Institut d'Investigacions Biomèdiques August Pi i Sunyer, Barcelona, Spain

**Giuseppe Astarita**
Department of Biochemistry and Molecular & Cellular Biology, Georgetown University, Washington, DC, United States

**Ann-Marie Baker**
Barts Cancer Institute, Barts and the London School of Medicine and Dentistry, Queen Mary University of London, London, United Kingdom

**John W. Barrett**
Department of Otolaryngology, Head & Neck Surgery, The University of Western Ontario, London, ON, Canada

**Bhavneet Bhinder**
Caryl and Israel Englander Institute for Precision Medicine; Department of Physiology and Biophysics, Weill Cornell Medicine, New York, NY, United States

**Sabrina Carpentier**
Innate Pharma Research Laboratories, Innate Pharma, Marseille, France

**Daniel D. De Carvalho**
UHN Princess Margaret Cancer Centre; Department of Medical Biophysics, University of Toronto, Toronto, ON, Canada

**Ankur Chakravarthy**
UHN Princess Margaret Cancer Centre, Toronto, ON, Canada

**Robert Clarke**
Lombardi Comprehensive Cancer Center, Georgetown University, Washington, DC, United States

**Suraj Dhungana**
Waters Corporation, Milford, MA, United States

**Marianna Di Scala**
VIVEbiotech, San Sebastián, Spain

**Javier Dotor**
Facultad de Ciencias Biosanitarias, Universidad Francisco de Vitoria, Madrid, Spain

**Olivier Elemento**
Caryl and Israel Englander Institute for Precision Medicine; Department of Physiology and Biophysics, Weill Cornell Medicine, New York, NY, United States

**Shentong Fang**
Wihuri Research Institute and Translational Cancer Medicine, University of Helsinki, Helsinki, Finland

**Francesca Finotello**
Division for Bioinformatics, Biocenter, Medical University of Innsbruck, Innsbruck, Austria

**Michael Friedrich**
Institute for Medical Immunology, Faculty of Medicine, Martin Luther University Halle-Wittenberg, Halle, Germany

**Gabriel Gallo-Oller**
Group of Translational Research in Child and Adolescent Cancer, Vall d'Hebron Research Institute, Barcelona, Spain

**Jérôme Galon**
INSERM, Laboratory of Integrative Cancer Immunology, Équipe Labellisée Ligue Contre le Cancer, Sorbonne Université, Sorbonne Paris Cité, Université Paris Descartes, Université Paris Diderot; Centre de Recherche des Cordeliers, Paris, France

**Steven F. Gameiro**
Department of Microbiology and Immunology, The University of Western Ontario, London, ON, Canada

**Abhishek D. Garg**
Department for Cellular and Molecular Medicine, Cell Death Research & Therapy (CDRT) Unit, KU Leuven, Leuven, Belgium

**Farhad Ghasemi**
Department of Otolaryngology, Head & Neck Surgery, The University of Western Ontario, London, ON, Canada

**Erica A. Golemis**
Program in Molecular Therapeutics, Fox Chase Cancer Center, Philadelphia, PA, United States

**Trevor A. Graham**
Barts Cancer Institute, Barts and the London School of Medicine and Dentistry, Queen Mary University of London, London, United Kingdom

**Nadine Heimer**
Institute for Medical Immunology, Faculty of Medicine, Martin Luther University Halle-Wittenberg, Halle, Germany

**Sandra A. Jablonski**
Lombardi Comprehensive Cancer Center, Georgetown University, Washington, DC, United States

**Simon Jasinski-Bergner**
Institute for Medical Immunology, Faculty of Medicine, Martin Luther University Halle-Wittenberg, Halle, Germany

**Michael Johnson**
Lombardi Comprehensive Cancer Center, Georgetown University, Washington, DC, United States

**James Koropatnick**
Department of Microbiology and Immunology; Department of Oncology; Department of Pathology; Department of Physiology and Pharmacology, The University of Western Ontario; London Regional Cancer Program, Lawson Health Research Institute, London, ON, Canada

**Lucie Lafontaine**
INSERM, Laboratory of Integrative Cancer Immunology, Équipe Labellisée Ligue Contre le Cancer, Sorbonne Université, Sorbonne Paris Cité, Université Paris Descartes, Université Paris Diderot; Centre de Recherche des Cordeliers, Paris, France

**Evagelia C. Laiakis**
Department of Biochemistry and Molecular & Cellular Biology; Department of Oncology, Lombardi Comprehensive Cancer Center, Georgetown University, Washington, DC, United States

**Maria-Filothei Lazaridou**
Institute for Medical Immunology, Faculty of Medicine, Martin Luther University Halle-Wittenberg, Halle, Germany

**Saman Maleki Vareki**
Department of Oncology, The University of Western Ontario; London Regional Cancer Program, Lawson Health Research Institute, London, ON, Canada

**Florentina Marches**
The Jackson Laboratory for Genomic Medicine, Farmington, CT, United States

**Florence Marliot**
INSERM, Laboratory of Integrative Cancer Immunology, Équipe Labellisée Ligue Contre le Cancer, Sorbonne Université, Sorbonne Paris Cité, Université Paris Descartes, Université Paris Diderot; Centre de Recherche des Cordeliers; Laboratory of Immunology, AP-HP, European Hospital Georges Pompidou, Paris, France

**Jan Martinek**
The Jackson Laboratory for Genomic Medicine, Farmington, CT, United States

**Joseph Murray**
Lombardi Comprehensive Cancer Center, Georgetown University, Washington, DC, United States

**Joe S. Mymryk**
Department of Microbiology and Immunology; Department of Otolaryngology, Head & Neck Surgery; Department of Oncology, The University of Western Ontario; London Regional Cancer Program, Lawson Health Research Institute, London, ON, Canada

**Rochelle E. Nasto**
Lombardi Comprehensive Cancer Center, Georgetown University, Washington, DC; Program in Molecular Therapeutics, Fox Chase Cancer Center; School of Biomedical Engineering, Science and Health Systems, Drexel University, Philadelphia, PA, United States

**Anthony C. Nichols**
Department of Otolaryngology, Head & Neck Surgery; Department of Oncology, The University of Western Ontario; London Regional Cancer Program, Lawson Health Research Institute, London, ON, Canada

**Karolina Palucka**
The Jackson Laboratory for Genomic Medicine, Farmington, CT, United States

**Christina Plattner**
Division for Bioinformatics, Biocenter, Medical University of Innsbruck, Innsbruck, Austria

**Jette Rahn**
Institute for Medical Immunology, Faculty of Medicine, Martin Luther University Halle-Wittenberg, Halle, Germany

**Dietmar Rieder**
Division for Bioinformatics, Biocenter, Medical University of Innsbruck, Innsbruck, Austria

**Rebecca B. Riggins**
Lombardi Comprehensive Cancer Center, Georgetown University, Washington, DC, United States

**François Romagné**
MI-mAbs, Aix-Marseille Univ, Marseille, France

**Nils-Petter Rudqvist**
Department of Radiation Oncology, Weill Cornell Medicine, New York, NY, United States

**Petri Salven**
Department of Pathology, Haartman Institute, University of Helsinki, Helsinki, Finland

**Barbara Seliger**
Institute for Medical Immunology, Faculty of Medicine, Martin Luther University Halle-Wittenberg, Halle, Germany

**Ilya G. Serebriiskii**
Program in Molecular Therapeutics, Fox Chase Cancer Center, Philadelphia, PA, United States; Kazan Federal University, Kazan, Russia

**Bindesh Shrestha**
Waters Corporation, Milford, MA, United States

**Sheila Spada**
Department of Radiation Oncology, Weill Cornell Medicine, New York, NY, United States

**Rishi Surana**
Lombardi Comprehensive Cancer Center, Georgetown University, Washington, DC, United States

**Christoforos Vaxevanis**
Institute for Medical Immunology, Faculty of Medicine, Martin Luther University Halle-Wittenberg, Halle, Germany

**Eric Vivier**
Innate Pharma Research Laboratories, Innate Pharma; Aix-Marseille Univ, APHM, CNRS, INSERM, CIML, Hôpital de la Timone, Marseille-Immunopole, Marseille, France

**Louis M. Weiner**
Lombardi Comprehensive Cancer Center, Georgetown University, Washington, DC, United States

**Erik Wennerberg**
Department of Radiation Oncology, Weill Cornell Medicine, New York, NY, United States

**Te-Chia Wu**
The Jackson Laboratory for Genomic Medicine, Farmington, CT, United States

**Chun I. Yu**
The Jackson Laboratory for Genomic Medicine, Farmington, CT, United States

**Yong-Wei Zhang**
Lombardi Comprehensive Cancer Center, Georgetown University, Washington, DC, United States

# Preface: More than two decades of modern tumor immunology

**Lorenzo Galluzzi**[a,b,c,d,e,*], **Nils-Petter Rudqvist**[a,*]
[a]Department of Radiation Oncology, Weill Cornell Medical College, New York, NY, United States
[b]Sandra and Edward Meyer Cancer Center, New York, NY, United States
[c]Caryl and Israel Englander Institute for Precision Medicine, New York, NY, United States
[d]Department of Dermatology, Yale University School of Medicine, New Haven, CT, United States
[e]Université de Paris, Paris, France
*Corresponding authors: e-mail address: deadoc80@gmail.com; nils.rudqvist@gmail.com

The roots of modern tumor immunology extend back to the pioneering work of William B. Coley, an American surgeon who—at the end of the 19th century—was the first to successfully use a partially inactivated bacterial preparation (the so-called "Coley toxin") for treating soft tissue sarcoma (Coley, 1893), and the revolutionary proposals of Paul Ehlrich, a German-Jewish physician who—at the beginning of the 20th century—introduced for the first time in history the concept of cancer immunosurveillance (Ehrlich, 1909; Rao, Gharib, & Han, 2019). However, with the exception of punctual episodes including the establishment of the Cancer Research Institute of New York (in 1957), whose specific purpose was the development of immunological approaches to treat cancer (Decker et al., 2017), and the publication of the first cancer vaccine study (in 1959) (Graham & Graham, 1959), the field remained virtually unborn until the 1990s. Indeed, the 1990s witnessed not only the discovery that cytotoxic T lymphocyte associated protein 4 (CTLA4) mediates robust immunosuppressive functions, hence constituting a potential target for cancer therapy (Krummel & Allison, 1995; Leach, Krummel, & Allison, 1996), but also the realization that oncogenesis occurs in the context of a bidirectional crosstalk between malignant cells and the immune system of the host (Dighe, Richards, Old, & Schreiber, 1994; Kaplan et al., 1998), laying the foundation for the modern cancer immunosurveillance and immunoediting theories (Dunn, Bruce, Ikeda, Old, & Schreiber, 2002; Dunn, Old, & Schreiber, 2004). In the 2000s, the field gathered additional momentum from preclinical and clinical findings elucidating multiple mechanisms that underlie the ability of the immune system to control developing tumors and the capacity of progressing neoplasms to evade immunosurveillance (Bargou et al., 2008; Dudley et al., 2002; Galon et al., 2006;

Hildner et al., 2008; Jaiswal et al., 2009; Koebel et al., 2007; Majeti et al., 2009; Morgan et al., 2006; Savage et al., 2008; Shankaran et al., 2001), as well as from the discovery that the efficacy of multiple conventional anticancer agents relies (at least in part) on the (re-)activation of tumor-targeting immune responses as a consequence of immunogenic cell death (ICD) induction (Bezu et al., 2018; Casares et al., 2005; Galluzzi, Bravo-San Pedro, Demaria, Formenti, & Kroemer, 2017; Galluzzi, Buque, Kepp, Zitvogel, & Kroemer, 2015, 2017; Galluzzi, Kepp, Chan, & Kroemer, 2017; Galluzzi, Vitale, et al., 2018; Galluzzi, Zitvogel, & Kroemer, 2016; Obeid et al., 2007; Pol et al., 2015). These data paved the way to the cancer immunotherapy revolution of the 2010s (Kelly, 2018).

The 2010s represent indeed the decade in which modern tumor immunology revolutionized the clinical management of multiple tumor types upon the approval of no less than 12 different immunotherapeutic agents for use in cancer patients by the US Food and Drug Administration (FDA) and various equivalent regulatory agencies worldwide (Galluzzi, Chan, Kroemer, Wolchok, & Lopez-Soto, 2018). In January 2019, when this Preface was being redacted, these agents included: (1) sipuleucel T, a myeloid cell-derived product originally approved for the treatment of asymptomatic or minimally symptomatic metastatic castration resistant (hormone refractory) prostate cancer in April 2010 (Hagihara et al., 2019; Kantoff et al., 2010); (2) ipilimumab, a CTLA4-targeting immune checkpoint inhibitor (ICI) (Aranda et al., 2014) initially licensed for use in patients with unresectable or metastatic melanoma who received at least one prior systemic treatment in March 2011 (Hodi et al., 2010; McDermott, Haanen, Chen, Lorigan, & O'Day, 2013; Robert, Schadendorf, Messina, Hodi, & O'Day, 2013), and currently approved for no less than two additional indications (Coens et al., 2017; Eggermont et al., 2015; Eggermont et al., 2016; Geoerger et al., 2017; Merchant et al., 2016); (3) nivolumab, an ICI targeting programmed cell death 1 (PDCD1, best known as PD-1) first licensed for the management of patients with unresectable or metastatic melanoma and disease progression following ipilimumab and, if $BRAF^{V600}$ mutation positive, a BRAF inhibitor in December 2014 (Weber et al., 2015), and currently available for no less than 10 additional indications (Ansell et al., 2015; Borghaei et al., 2015; Cella et al., 2016; El-Khoueiry et al., 2017; Escudier, Motzer, et al., 2017; Escudier, Sharma, et al., 2017; Ferris et al., 2016; Harrington et al., 2017; Horn et al., 2017; Kazandjian et al., 2016; Kiyota et al., 2017; Motzer et al., 2015; Overman et al., 2017; Sharma et al., 2017; Vokes et al., 2018; Weber et al., 2017; Xu et al., 2017; Younes et al., 2016);

(4) pembrolizumab, a PD-1-targeting ICI originally licensed for the treatment of advanced or unresectable melanoma patients who are no longer responding to ipilimumab and BRAF inhibitors (when indicated) in September 2014 (Hodi et al., 2016; Joseph et al., 2018; Patnaik et al., 2015; Ribas et al., 2016; Robert et al., 2018, 2014), and currently approved for no less than 10 additional indications (Alley et al., 2017; Balar, Castellano, et al., 2017; Brahmer et al., 2017; Brogden et al., 2018; Chatterjee et al., 2016; Chen et al., 2017; Chow et al., 2016; Doi et al., 2018; Frenel et al., 2017; Fuchs et al., 2018; Gandhi et al., 2018; Garon et al., 2015; Hsu et al., 2017; Hui et al., 2017; Le et al., 2017, 2015; Muro et al., 2016; Nanda et al., 2016; O'Neil et al., 2017; Ott, Elez, et al., 2017; Ott, Piha-Paul, et al., 2017; Patnaik et al., 2015; Petrella et al., 2017; Plimack et al., 2017; Reck et al., 2016; Robert et al., 2015; Schachter et al., 2017; Seiwert et al., 2016; Shaverdian et al., 2017; Tahara et al., 2018; Zinzani et al., 2017); (5) cemiplimab, yet another ICI specific for PD-1 originally approved for the treatment of metastatic or locally advanced cutaneous squamous cell carcinoma in September 2018 (Migden et al., 2018); (6) durvalumab, an ICI specific for CD274 (best known as PD-L1) originally licensed for the treatment of locally advanced or metastatic urothelial carcinoma progressing during or following platinum-containing chemotherapy in May 2017 (Levy, Massard, Soria, & Deutsch, 2016; Powles et al., 2017), and currently also available for the management of patients with unresectable nonsmall cell lung carcinoma (Antonia et al., 2017); (7) avelumab, a PD-L1-targeting ICI originally approved for the treatment of adult and pediatric patients with metastatic Merkel cell carcinoma in March 2017 (D'Angelo et al., 2018; Kaufman et al., 2016, 2018), and currently licensed also for use in patients with locally advanced or metastatic urothelial carcinoma progressing during or following platinum-containing chemotherapy (Apolo et al., 2017; Heery et al., 2017); (8) atezolizumab, yet another PD-L1 inhibitor originally approved for the treatment of locally advanced or metastatic urothelial carcinoma progressing during or following platinum-containing chemotherapy in May 2016 (Balar, Galsky, et al., 2017; Necchi et al., 2017; Perez-Gracia et al., 2018; Rosenberg et al., 2016), and currently approved for no less than two additional indications (Fehrenbacher et al., 2016; Hida et al., 2018; Rittmeyer et al., 2017); (9) PEGylated interferon alpha 2B, a recombinant cytokine (Garcia-Martinez et al., 2018) originally approved for the treatment of stage III melanoma patients after surgery in March 2011 (Bottomley et al., 2009; Bouwhuis et al., 2010; Di Trolio et al., 2012; Eggermont et al., 2008; Eggermont et al., 2010; Fusi et al., 2009);

(10) talimogene laherparepvec, an oncolytic virus originally approved for the treatment of metastatic, unresectable melanoma patients in October 2015 (Andtbacka et al., 2015; Pol et al., 2018; Pol, Kroemer, & Galluzzi, 2016; Vacchelli et al., 2013), (11) tisagenlecleucel, a preparation of autologous T cells engineered to express a chimeric antigen receptor (CAR) (Borrie & Maleki Vareki, 2018; Raikar et al., 2018) that recognizes CD19, originally approved for the treatment of pediatric and young adults with acute lymphocytic leukemia who relapsed or are refractory to recommended prior therapy in August 2017 (Maude et al., 2018), and now also licensed for the therapy of subjects with various B-cell lymphomas relapsing on or refractory to recommended prior therapy (Schuster et al., 2019); and (12) axicabtagene ciloleucel, another variant of CAR-expressing T cells specific for CD19 (Kellner, Peipp, Gramatzki, Schrappe, & Schewe, 2018), originally approved for the treatment of several variants of refractory or relapsing B-cell lymphoma in October 2017 (Neelapu et al., 2017).

Alongside, considerable translational advances have been made in the 2010s, including (but not limited to): (1) the realization of the major impact of the immunological tumor microenvironment on disease outcome among cancer patients treated with a variety of therapeutic regimens beyond immunotherapy (Atherton et al., 2018; Bantug, Galluzzi, Kroemer, & Hess, 2018; Chevrier et al., 2017; De Henau et al., 2016; Herbst et al., 2014; Humbert & Hugues, 2019; Lavin et al., 2017; Li et al., 2019; Mariathasan et al., 2018; Riaz et al., 2017; Spranger, Bao, & Gajewski, 2015; Straussman et al., 2012; Tauriello et al., 2018; Zheng et al., 2017); (2) the identification of the gut microbiota as a major determinant of the sensitivity of cancer patients to multiple treatments including ICIs (Bhutiani et al., 2018; Gopalakrishnan et al., 2018; Iida et al., 2013; Routy et al., 2018; Vetizou et al., 2015; Viaud et al., 2013); (3) the discovery that novel antigens expressed by cancer cells as a consequence of nonsynonymous mutations play a key role in tumor progression and response to treatment (Gubin et al., 2014; Janikovits et al., 2018; Ozcan, Janikovits, von Knebel Doeberitz, & Kloor, 2018; Rizvi et al., 2015; Schumacher & Schreiber, 2015; Vitale et al., 2019); (4) the elucidation of innate immune mechanisms that provide a major contribution to both natural and therapy-elicited immunosurveillance (Andre et al., 2018; Dou et al., 2017; Ferrari de Andrade et al., 2018; Galluzzi, Vanpouille-Box, Bakhoum, & Demaria, 2018; Harding et al., 2017; Lopez-Soto, Gonzalez, Smyth, & Galluzzi, 2017; Mackenzie et al., 2017; Oyer, Gitto, Altomare, & Copik, 2018; Ruscetti et al., 2018;

van Montfoort et al., 2018; Vanpouille-Box et al., 2017; Vanpouille-Box, Demaria, Formenti, & Galluzzi, 2018); and (5) the identification of exosomes as a prominent vehicle for malignant cells to condition distant anatomical sites and interface with the host immune system (Chen et al., 2018; Diamond et al., 2018; Gabrusiewicz et al., 2018; Hoshino et al., 2015; Mirzaei et al., 2018; Yaddanapudi et al., 2019; Zhang et al., 2015). Finally, major technological advances have been achieved in the 2010s, including (but not limited to): (1) the development of reliable and cost-effective platforms for single cell DNA or RNA sequencing (Alemany, Florescu, Baron, Peterson-Maduro, & van Oudenaarden, 2018; Mulqueen et al., 2018; Zhang et al., 2018); (2) the design of robust algorithms enabling not only the analysis of large single cell omic datasets (Becht et al., 2019; Grobner et al., 2018; Wu et al., 2014), but also various other *in silico* functions that are widely employed in modern tumor immunology, such as neoantigen and TCR motif prediction (Balachandran et al., 2017; Brown & Holt, 2019; Dash et al., 2017; Formenti et al., 2018; Glanville et al., 2017; Hundal et al., 2019; Karasaki et al., 2017); (3) the refinement of the CRISPR/Cas system for gene editing (Halperin et al., 2018; Hille et al., 2018; Kalhor et al., 2018; Nishimasu et al., 2018; Shen et al., 2018; Wang et al., 2018; Yan et al., 2019); and (4) the introduction of mass cytometry (Chevrier et al., 2017; Galluzzi, Yamazaki, & Demaria, 2017; Lavin et al., 2017; Spitzer & Nolan, 2016; Wierz, Janji, Berchem, Moussay, & Paggetti, 2018). Beyond any reasonable doubt, the preclinical, clinical and technological progress we witnessed in the 2010s will pave the way to an exciting decade of achievements in the fields of tumor immunology and immunotherapy as the 2020s are on their way.

In *Tumor Immunology and Immunotherapy* (a group of volumes of the *Methods in Enzymology* series), world leaders in the field gathered to provide the first comprehensive compendium of methods for the study of cancer immunology in both its fundamental and translational components. Each of the chapters of *Tumor Immunology and Immunotherapy* provides a detailed description of one or a few protocols that are suitable for the characterization of (components of) the tumor–host interaction, *in vitro, in vivo* or *in silico*. The book consists of five volumes: “Molecular Methods,” “Cellular Methods—Part A,” “Cellular Methods—Part B,” “Integrated Methods—Part A,” and “Integrated Methods—Part B,” and is equally addressed to expert researchers who may want to extend their technical portfolio, and to beginners in this exciting and rapidly growing field of investigation.

## Acknowledgments

L.G. is supported by a Breakthrough Level 2 grant from the US Department of Defense (DoD), Breast Cancer Research Program (BRCP) [#BC180476P1], by a startup grant from the Department of Radiation Oncology at Weill Cornell Medicine (New York, USA), by industrial collaborations with Lytix (Oslo, Norway) and Phosplatin (New York, USA), and by donations from Phosplatin (New York, USA), the Luke Heller TECPR2 Foundation (Boston, USA) and Sotio a.s. (Prague, Czech Republic).

## Disclosures

L.G. provides remunerated consulting to OmniSEQ (Buffalo, NY, USA), Astra Zeneca (Gaithersburg, MD, USA), Boehringer Ingelheim (Vienna, Austria), VL47 (New York, NY, USA), and the Luke Heller TECPR2 Foundation (Boston, MA, USA), and is member of the Scientific Advisory Committee of OmniSEQ (Buffalo, NY, USA).

## References

Alemany, A., Florescu, M., Baron, C. S., Peterson-Maduro, J., & van Oudenaarden, A. (2018). Whole-organism clone tracing using single-cell sequencing. *Nature*, *556*(7699), 108–112. https://doi.org/10.1038/nature25969.

Alley, E. W., Lopez, J., Santoro, A., Morosky, A., Saraf, S., Piperdi, B., et al. (2017). Clinical safety and activity of pembrolizumab in patients with malignant pleural mesothelioma (KEYNOTE-028): Preliminary results from a non-randomised, open-label, phase 1b trial. *The Lancet Oncology*, *18*(5), 623–630. https://doi.org/10.1016/s1470-2045(17)30169-9.

Andre, P., Denis, C., Soulas, C., Bourbon-Caillet, C., Lopez, J., Arnoux, T., et al. (2018). Anti-NKG2A mAb is a checkpoint inhibitor that promotes anti-tumor immunity by unleashing both T and NK cells. *Cell*, *175*(7). 1731–1743.e1713. https://doi.org/10.1016/j.cell.2018.10.014.

Andtbacka, R. H., Kaufman, H. L., Collichio, F., Amatruda, T., Senzer, N., Chesney, J., et al. (2015). Talimogene laherparepvec improves durable response rate in patients with advanced melanoma. *Journal of Clinical Oncology*, *33*(25), 2780–2788. https://doi.org/10.1200/jco.2014.58.3377.

Ansell, S. M., Lesokhin, A. M., Borrello, I., Halwani, A., Scott, E. C., Gutierrez, M., et al. (2015). PD-1 blockade with nivolumab in relapsed or refractory Hodgkin's lymphoma. *The New England Journal of Medicine*, *372*(4), 311–319. https://doi.org/10.1056/NEJMoa1411087.

Antonia, S. J., Villegas, A., Daniel, D., Vicente, D., Murakami, S., Hui, R., et al. (2017). Durvalumab after chemoradiotherapy in stage III non-small-cell lung cancer. *The New England Journal of Medicine*, *377*(20), 1919–1929. https://doi.org/10.1056/NEJMoa1709937.

Apolo, A. B., Infante, J. R., Balmanoukian, A., Patel, M. R., Wang, D., Kelly, K., et al. (2017). Avelumab, an anti-programmed death-ligand 1 antibody, in patients with refractory metastatic urothelial carcinoma: Results from a multicenter, phase Ib study. *Journal of Clinical Oncology*, *35*(19), 2117–2124. https://doi.org/10.1200/jco.2016.71.6795.

Aranda, F., Vacchelli, E., Eggermont, A., Galon, J., Fridman, W. H., Zitvogel, L., et al. (2014). Trial watch: Immunostimulatory monoclonal antibodies in cancer therapy. *Oncoimmunology*, *3*(1), e27297. https://doi.org/10.4161/onci.27297.

Atherton, M. J., Stephenson, K. B., Tzelepis, F., Bakhshinyan, D., Nikota, J. K., Son, H. H., et al. (2018). Transforming the prostatic tumor microenvironment with oncolytic virotherapy. *Oncoimmunology*, 7(7), e1445459. https://doi.org/10.1080/2162402X.2018.1445459.

Balachandran, V. P., Luksza, M., Zhao, J. N., Makarov, V., Moral, J. A., Remark, R., et al. (2017). Identification of unique neoantigen qualities in long-term survivors of pancreatic cancer. *Nature*, *551*(7681), 512–516. https://doi.org/10.1038/nature24462.

Balar, A. V., Castellano, D., O'Donnell, P. H., Grivas, P., Vuky, J., Powles, T., et al. (2017). First-line pembrolizumab in cisplatin-ineligible patients with locally advanced and unresectable or metastatic urothelial cancer (KEYNOTE-052): A multicentre, single-arm, phase 2 study. *The Lancet Oncology*, *18*(11), 1483–1492. https://doi.org/10.1016/s1470-2045(17)30616-2.

Balar, A. V., Galsky, M. D., Rosenberg, J. E., Powles, T., Petrylak, D. P., Bellmunt, J., et al. (2017). Atezolizumab as first-line treatment in cisplatin-ineligible patients with locally advanced and metastatic urothelial carcinoma: A single-arm, multicentre, phase 2 trial. *Lancet*, *389*(10064), 67–76. https://doi.org/10.1016/s0140-6736(16)32455-2.

Bantug, G. R., Galluzzi, L., Kroemer, G., & Hess, C. (2018). The spectrum of T cell metabolism in health and disease. *Nature Reviews. Immunology*, *18*(1), 19–34. https://doi.org/10.1038/nri.2017.99.

Bargou, R., Leo, E., Zugmaier, G., Klinger, M., Goebeler, M., Knop, S., et al. (2008). Tumor regression in cancer patients by very low doses of a T cell-engaging antibody. *Science*, *321*(5891), 974–977. https://doi.org/10.1126/science.1158545.

Becht, E., McInnes, L., Healy, J., Dutertre, C. A., Kwok, I. W. H., Ng, L. G., et al. (2019). Dimensionality reduction for visualizing single-cell data using UMAP. *Nature Biotechnology*, *37*, 38–44. https://doi.org/10.1038/nbt.4314.

Bezu, L., Sauvat, A., Humeau, J., Leduc, M., Kepp, O., & Kroemer, G. (2018). eIF2alpha phosphorylation: A hallmark of immunogenic cell death. *Oncoimmunology*, 7(6). e1431089. https://doi.org/10.1080/2162402X.2018.1431089.

Bhutiani, N., Li, Q., Anderson, C. D., Gallagher, H. C., De Jesus, M., Singh, R., et al. (2018). Enhanced gut barrier integrity sensitizes colon cancer to immune therapy. *Oncoimmunology*, 7(11), e1498438. https://doi.org/10.1080/2162402X.2018.1498438.

Borghaei, H., Paz-Ares, L., Horn, L., Spigel, D. R., Steins, M., Ready, N. E., et al. (2015). Nivolumab versus docetaxel in advanced nonsquamous non-small-cell lung cancer. *The New England Journal of Medicine*, *373*(17), 1627–1639. https://doi.org/10.1056/NEJMoa1507643.

Borrie, A. E., & Maleki Vareki, S. (2018). T lymphocyte-based cancer immunotherapeutics. *International Review of Cell and Molecular Biology*, *341*, 201–276. https://doi.org/10.1016/bs.ircmb.2018.05.010.

Bottomley, A., Coens, C., Suciu, S., Santinami, M., Kruit, W., Testori, A., et al. (2009). Adjuvant therapy with pegylated interferon alfa-2b versus observation in resected stage III melanoma: A phase III randomized controlled trial of health-related quality of life and symptoms by the European Organisation for Research and Treatment of Cancer Melanoma Group. *Journal of Clinical Oncology*, *27*(18), 2916–2923. https://doi.org/10.1200/jco.2008.20.2069.

Bouwhuis, M. G., Suciu, S., Testori, A., Kruit, W. H., Sales, F., Patel, P., et al. (2010). Phase III trial comparing adjuvant treatment with pegylated interferon Alfa-2b versus observation: Prognostic significance of autoantibodies—EORTC 18991. *Journal of Clinical Oncology*, *28*(14), 2460–2466. https://doi.org/10.1200/jco.2009.24.6264.

Brahmer, J. R., Rodriguez-Abreu, D., Robinson, A. G., Hui, R., Csoszi, T., Fulop, A., et al. (2017). Health-related quality-of-life results for pembrolizumab versus chemotherapy in advanced, PD-L1-positive NSCLC (KEYNOTE-024): A multicentre, international, randomised, open-label phase 3 trial. *The Lancet Oncology*, *18*(12), 1600–1609. https://doi.org/10.1016/s1470-2045(17)30690-3.

Brogden, K. A., Parashar, D., Hallier, A. R., Braun, T., Qian, F., Rizvi, N. A., et al. (2018). Genomics of NSCLC patients both affirm PD-L1 expression and predict their clinical responses to anti-PD-1 immunotherapy. *BMC Cancer, 18*(1), 225. https://doi.org/10.1186/s12885-018-4134-y.

Brown, S. D., & Holt, R. A. (2019). Neoantigen characteristics in the context of the complete predicted MHC class I self-immunopeptidome. *Oncoimmunology, 8*(3), 1556080. https://doi.org/10.1080/2162402X.2018.1556080.

Casares, N., Pequignot, M. O., Tesniere, A., Ghiringhelli, F., Roux, S., Chaput, N., et al. (2005). Caspase-dependent immunogenicity of doxorubicin-induced tumor cell death. *The Journal of Experimental Medicine, 202*(12), 1691–1701. https://doi.org/10.1084/jem.20050915.

Cella, D., Grunwald, V., Nathan, P., Doan, J., Dastani, H., Taylor, F., et al. (2016). Quality of life in patients with advanced renal cell carcinoma given nivolumab versus everolimus in CheckMate 025: A randomised, open-label, phase 3 trial. *The Lancet Oncology, 17*(7), 994–1003. https://doi.org/10.1016/s1470-2045(16)30125-5.

Chatterjee, M., Turner, D. C., Felip, E., Lena, H., Cappuzzo, F., Horn, L., et al. (2016). Systematic evaluation of pembrolizumab dosing in patients with advanced non-small-cell lung cancer. *Annals of Oncology, 27*(7), 1291–1298. https://doi.org/10.1093/annonc/mdw174.

Chen, G., Huang, A. C., Zhang, W., Zhang, G., Wu, M., Xu, W., et al. (2018). Exosomal PD-L1 contributes to immunosuppression and is associated with anti-PD-1 response. *Nature, 560*(7718), 382–386. https://doi.org/10.1038/s41586-018-0392-8.

Chen, R., Zinzani, P. L., Fanale, M. A., Armand, P., Johnson, N. A., Brice, P., et al. (2017). Phase II study of the efficacy and safety of pembrolizumab for relapsed/refractory classic Hodgkin lymphoma. *Journal of Clinical Oncology, 35*(19), 2125–2132. https://doi.org/10.1200/jco.2016.72.1316.

Chevrier, S., Levine, J. H., Zanotelli, V. R. T., Silina, K., Schulz, D., Bacac, M., et al. (2017). An immune atlas of clear cell renal cell carcinoma. *Cell, 169*(4), 736–749.e718. https://doi.org/10.1016/j.cell.2017.04.016.

Chow, L. Q. M., Haddad, R., Gupta, S., Mahipal, A., Mehra, R., Tahara, M., et al. (2016). Antitumor activity of pembrolizumab in biomarker-unselected patients with recurrent and/or metastatic head and neck squamous cell carcinoma: Results from the Phase Ib KEYNOTE-012 expansion cohort. *Journal of Clinical Oncology, 34*(32), 3838–3845. https://doi.org/10.1200/jco.2016.68.1478.

Coens, C., Suciu, S., Chiarion-Sileni, V., Grob, J. J., Dummer, R., Wolchok, J. D., et al. (2017). Health-related quality of life with adjuvant ipilimumab versus placebo after complete resection of high-risk stage III melanoma (EORTC 18071): Secondary outcomes of a multinational, randomised, double-blind, phase 3 trial. *The Lancet Oncology, 18*(3), 393–403. https://doi.org/10.1016/s1470-2045(17)30015-3.

Coley, W. B. (1893). The treatment of malignant tumors by repeated inoculations of erysipelas. With a report of ten original cases. *The American Journal of the Medical Sciences, 105*, 487–511.

D'Angelo, S. P., Russell, J., Lebbe, C., Chmielowski, B., Gambichler, T., Grob, J. J., et al. (2018). Efficacy and safety of first-line avelumab treatment in patients with Stage IV metastatic Merkel cell carcinoma: A preplanned interim analysis of a clinical trial. *JAMA Oncology, 4*(9). e180077. https://doi.org/10.1001/jamaoncol.2018.0077.

Dash, P., Fiore-Gartland, A. J., Hertz, T., Wang, G. C., Sharma, S., Souquette, A., et al. (2017). Quantifiable predictive features define epitope-specific T cell receptor repertoires. *Nature, 547*(7661), 89–93. https://doi.org/10.1038/nature22383.

De Henau, O., Rausch, M., Winkler, D., Campesato, L. F., Liu, C., Cymerman, D. H., et al. (2016). Overcoming resistance to checkpoint blockade therapy by targeting PI3Kgamma in myeloid cells. *Nature, 539*(7629), 443–447. https://doi.org/10.1038/nature20554.

Decker, W. K., da Silva, R. F., Sanabria, M. H., Angelo, L. S., Guimaraes, F., Burt, B. M., et al. (2017). Cancer immunotherapy: Historical perspective of a clinical revolution and emerging preclinical animal models. *Frontiers in Immunology*, *8*, 829. https://doi.org/10.3389/fimmu.2017.00829.

Di Trolio, R., Simeone, E., Di Lorenzo, G., Grimaldi, A. M., Romano, A., Ayala, F., et al. (2012). Update on PEG-interferon alpha-2b as adjuvant therapy in melanoma. *Anticancer Research*, *32*(9), 3901–3909.

Diamond, J. M., Vanpouille-Box, C., Spada, S., Rudqvist, N. P., Chapman, J. R., Ueberheide, B. M., et al. (2018). Exosomes shuttle TREX1-sensitive IFN-stimulatory dsDNA from irradiated cancer cells to DCs. *Cancer Immunology Research*, *6*(8), 910–920. https://doi.org/10.1158/2326-6066.cir-17-0581.

Dighe, A. S., Richards, E., Old, L. J., & Schreiber, R. D. (1994). Enhanced in vivo growth and resistance to rejection of tumor cells expressing dominant negative IFN gamma receptors. *Immunity*, *1*(6), 447–456.

Doi, T., Piha-Paul, S. A., Jalal, S. I., Saraf, S., Lunceford, J., Koshiji, M., et al. (2018). Safety and antitumor activity of the anti-programmed death-1 antibody pembrolizumab in patients with advanced esophageal carcinoma. *Journal of Clinical Oncology*, *36*(1), 61–67. https://doi.org/10.1200/jco.2017.74.9846.

Dou, Z., Ghosh, K., Vizioli, M. G., Zhu, J., Sen, P., Wangensteen, K. J., et al. (2017). Cytoplasmic chromatin triggers inflammation in senescence and cancer. *Nature*, *550*(7676), 402–406. https://doi.org/10.1038/nature24050.

Dudley, M. E., Wunderlich, J. R., Robbins, P. F., Yang, J. C., Hwu, P., Schwartzentruber, D. J., et al. (2002). Cancer regression and autoimmunity in patients after clonal repopulation with antitumor lymphocytes. *Science*, *298*(5594), 850–854. https://doi.org/10.1126/science.1076514.

Dunn, G. P., Bruce, A. T., Ikeda, H., Old, L. J., & Schreiber, R. D. (2002). Cancer immunoediting: From immunosurveillance to tumor escape. *Nature Immunology*, *3*(11), 991–998. https://doi.org/10.1038/ni1102-991.

Dunn, G. P., Old, L. J., & Schreiber, R. D. (2004). The three Es of cancer immunoediting. *Annual Review of Immunology*, *22*, 329–360. https://doi.org/10.1146/annurev.immunol.22.012703.104803.

Eggermont, A. M., Bouwhuis, M. G., Kruit, W. H., Testori, A., ten Hagen, T., Yver, A., et al. (2010). Serum concentrations of pegylated interferon alpha-2b in patients with resected stage III melanoma receiving adjuvant pegylated interferon alpha-2b in a randomized phase III trial (EORTC 18991). *Cancer Chemotherapy and Pharmacology*, *65*(4), 671–677. https://doi.org/10.1007/s00280-009-1072-z.

Eggermont, A. M., Chiarion-Sileni, V., Grob, J. J., Dummer, R., Wolchok, J. D., Schmidt, H., et al. (2015). Adjuvant ipilimumab versus placebo after complete resection of high-risk stage III melanoma (EORTC 18071): A randomised, double-blind, phase 3 trial. *The Lancet Oncology*, *16*(5), 522–530. https://doi.org/10.1016/s1470-2045(15)70122-1.

Eggermont, A. M., Chiarion-Sileni, V., Grob, J. J., Dummer, R., Wolchok, J. D., Schmidt, H., et al. (2016). Prolonged survival in stage III melanoma with ipilimumab adjuvant therapy. *The New England Journal of Medicine*, *375*(19), 1845–1855. https://doi.org/10.1056/NEJMoa1611299.

Eggermont, A. M., Suciu, S., Santinami, M., Testori, A., Kruit, W. H., Marsden, J., et al. (2008). Adjuvant therapy with pegylated interferon alfa-2b versus observation alone in resected stage III melanoma: Final results of EORTC 18991, a randomised phase III trial. *Lancet*, *372*(9633), 117–126. https://doi.org/10.1016/s0140-6736(08)61033-8.

Ehrlich, P. (1909). Über den jetzigen stand der karzinomforschung. *Nederlands Tijdschrift voor Geneeskunde*, *5*, 273–290.

El-Khoueiry, A. B., Sangro, B., Yau, T., Crocenzi, T. S., Kudo, M., Hsu, C., et al. (2017). Nivolumab in patients with advanced hepatocellular carcinoma (CheckMate 040): An open-label, non-comparative, phase 1/2 dose escalation and expansion trial. *Lancet, 389*(10088), 2492–2502. https://doi.org/10.1016/s0140-6736(17)31046-2.

Escudier, B., Motzer, R. J., Sharma, P., Wagstaff, J., Plimack, E. R., Hammers, H. J., et al. (2017). Treatment beyond progression in patients with advanced renal cell carcinoma treated with nivolumab in CheckMate 025. *European Urology, 72*(3), 368–376. https://doi.org/10.1016/j.eururo.2017.03.037.

Escudier, B., Sharma, P., McDermott, D. F., George, S., Hammers, H. J., Srinivas, S., et al. (2017). CheckMate 025 randomized phase 3 study: Outcomes by key baseline factors and prior therapy for nivolumab versus everolimus in advanced renal cell carcinoma. *European Urology, 72*(6), 962–971. https://doi.org/10.1016/j.eururo.2017.02.010.

Fehrenbacher, L., Spira, A., Ballinger, M., Kowanetz, M., Vansteenkiste, J., Mazieres, J., et al. (2016). Atezolizumab versus docetaxel for patients with previously treated non-small-cell lung cancer (POPLAR): A multicentre, open-label, phase 2 randomised controlled trial. *Lancet, 387*(10030), 1837–1846. https://doi.org/10.1016/s0140-6736(16)00587-0.

Ferrari de Andrade, L., Tay, R. E., Pan, D., Luoma, A. M., Ito, Y., Badrinath, S., et al. (2018). Antibody-mediated inhibition of MICA and MICB shedding promotes NK cell-driven tumor immunity. *Science, 359*(6383), 1537–1542. https://doi.org/10.1126/science.aao0505.

Ferris, R. L., Blumenschein, G., Jr., Fayette, J., Guigay, J., Colevas, A. D., Licitra, L., et al. (2016). Nivolumab for recurrent squamous-cell carcinoma of the head and neck. *The New England Journal of Medicine, 375*(19), 1856–1867. https://doi.org/10.1056/NEJMoa1602252.

Formenti, S. C., Rudqvist, N. P., Golden, E., Cooper, B., Wennerberg, E., Lhuillier, C., et al. (2018). Radiotherapy induces responses of lung cancer to CTLA-4 blockade. *Nature Medicine, 24*(12), 1845–1851. https://doi.org/10.1038/s41591-018-0232-2.

Frenel, J. S., Le Tourneau, C., O'Neil, B., Ott, P. A., Piha-Paul, S. A., Gomez-Roca, C., et al. (2017). Safety and efficacy of pembrolizumab in advanced, programmed death ligand 1-positive cervical cancer: Results from the phase Ib KEYNOTE-028 trial. *Journal of Clinical Oncology, 35*(36), 4035–4041. https://doi.org/10.1200/jco.2017.74.5471.

Fuchs, C. S., Doi, T., Jang, R. W., Muro, K., Satoh, T., Machado, M., et al. (2018). Safety and efficacy of pembrolizumab monotherapy in patients with previously treated advanced gastric and gastroesophageal junction cancer: Phase 2 clinical KEYNOTE-059 trial. *JAMA Oncology, 4*(5). e180013. https://doi.org/10.1001/jamaoncol.2018.0013.

Fusi, A., Collette, S., Busse, A., Suciu, S., Rietz, A., Santinami, M., et al. (2009). Circulating melanoma cells and distant metastasis-free survival in stage III melanoma patients with or without adjuvant interferon treatment (EORTC 18991 side study). *European Journal of Cancer, 45*(18), 3189–3197. https://doi.org/10.1016/j.ejca.2009.09.004.

Gabrusiewicz, K., Li, X., Wei, J., Hashimoto, Y., Marisetty, A. L., Ott, M., et al. (2018). Glioblastoma stem cell-derived exosomes induce M2 macrophages and PD-L1 expression on human monocytes. *Oncoimmunology*, 7(4), e1412909. https://doi.org/10.1080/2162402X.2017.1412909.

Galluzzi, L., Bravo-San Pedro, J. M., Demaria, S., Formenti, S. C., & Kroemer, G. (2017). Activating autophagy to potentiate immunogenic chemotherapy and radiation therapy. *Nature Reviews. Clinical Oncology, 14*(4), 247–258. https://doi.org/10.1038/nrclinonc.2016.183.

Galluzzi, L., Buque, A., Kepp, O., Zitvogel, L., & Kroemer, G. (2015). Immunological effects of conventional chemotherapy and targeted anticancer agents. *Cancer Cell, 28*(6), 690–714. https://doi.org/10.1016/j.ccell.2015.10.012.

Galluzzi, L., Buque, A., Kepp, O., Zitvogel, L., & Kroemer, G. (2017). Immunogenic cell death in cancer and infectious disease. *Nature Reviews. Immunology, 17*(2), 97–111. https://doi.org/10.1038/nri.2016.107.

Galluzzi, L., Chan, T. A., Kroemer, G., Wolchok, J. D., & Lopez-Soto, A. (2018). The hallmarks of successful anticancer immunotherapy. *Science Translational Medicine*, *10*(459), eaat7807. https://doi.org/10.1126/scitranslmed.aat7807.

Galluzzi, L., Kepp, O., Chan, F. K., & Kroemer, G. (2017). Necroptosis: Mechanisms and relevance to disease. *Annual Review of Pathology*, *12*, 103–130. https://doi.org/10.1146/annurev-pathol-052016-100247.

Galluzzi, L., Vanpouille-Box, C., Bakhoum, S. F., & Demaria, S. (2018). SnapShot: CGAS-STING signaling. *Cell*, *173*(1), 276–276.e271. https://doi.org/10.1016/j.cell.2018.03.015.

Galluzzi, L., Vitale, I., Aaronson, S. A., Abrams, J. M., Adam, D., Agostinis, P., et al. (2018). Molecular mechanisms of cell death: Recommendations of the Nomenclature Committee on Cell Death 2018. *Cell Death and Differentiation*, *25*(3), 486–541. https://doi.org/10.1038/s41418-017-0012-4.

Galluzzi, L., Yamazaki, T., & Demaria, S. (2017). Heavy metal to rock the immune infiltrate. *Trends in Immunology*, *38*(8), 539–541. https://doi.org/10.1016/j.it.2017.05.007.

Galluzzi, L., Zitvogel, L., & Kroemer, G. (2016). Immunological mechanisms underneath the efficacy of cancer therapy. *Cancer Immunology Research*, *4*(11), 895–902. https://doi.org/10.1158/2326-6066.CIR-16-0197.

Galon, J., Costes, A., Sanchez-Cabo, F., Kirilovsky, A., Mlecnik, B., Lagorce-Pages, C., et al. (2006). Type, density, and location of immune cells within human colorectal tumors predict clinical outcome. *Science*, *313*(5795), 1960–1964. https://doi.org/10.1126/science.1129139.

Gandhi, L., Rodriguez-Abreu, D., Gadgeel, S., Esteban, E., Felip, E., De Angelis, F., et al. (2018). Pembrolizumab plus chemotherapy in metastatic non-small-cell lung cancer. *The New England Journal of Medicine*, *378*(22), 2078–2092. https://doi.org/10.1056/NEJMoa1801005.

Garcia-Martinez, E., Smith, M., Buque, A., Aranda, F., de la Pena, F. A., Ivars, A., et al. (2018). Trial watch: Immunostimulation with recombinant cytokines for cancer therapy. *Oncoimmunology*, 7(6), e1433982. https://doi.org/10.1080/2162402X.2018.1433982.

Garon, E. B., Rizvi, N. A., Hui, R., Leighl, N., Balmanoukian, A. S., Eder, J. P., et al. (2015). Pembrolizumab for the treatment of non-small-cell lung cancer. *The New England Journal of Medicine*, *372*(21), 2018–2028. https://doi.org/10.1056/NEJMoa1501824.

Geoerger, B., Bergeron, C., Gore, L., Sender, L., Dunkel, I. J., Herzog, C., et al. (2017). Phase II study of ipilimumab in adolescents with unresectable stage III or IV malignant melanoma. *European Journal of Cancer*, *86*, 358–363. https://doi.org/10.1016/j.ejca.2017.09.032.

Glanville, J., Huang, H., Nau, A., Hatton, O., Wagar, L. E., Rubelt, F., et al. (2017). Identifying specificity groups in the T cell receptor repertoire. *Nature*, *547*(7661), 94–98. https://doi.org/10.1038/nature22976.

Gopalakrishnan, V., Spencer, C. N., Nezi, L., Reuben, A., Andrews, M. C., Karpinets, T. V., et al. (2018). Gut microbiome modulates response to anti-PD-1 immunotherapy in melanoma patients. *Science*, *359*(6371), 97–103. https://doi.org/10.1126/science.aan4236.

Graham, J. B., & Graham, R. M. (1959). The effect of vaccine on cancer patients. *Surgery, Gynecology & Obstetrics*, *109*(2), 131–138.

Grobner, S. N., Worst, B. C., Weischenfeldt, J., Buchhalter, I., Kleinheinz, K., Rudneva, V. A., et al. (2018). The landscape of genomic alterations across childhood cancers. *Nature*, *555*(7696), 321–327. https://doi.org/10.1038/nature25480.

Gubin, M. M., Zhang, X., Schuster, H., Caron, E., Ward, J. P., Noguchi, T., et al. (2014). Checkpoint blockade cancer immunotherapy targets tumour-specific mutant antigens. *Nature*, *515*(7528), 577–581. https://doi.org/10.1038/nature13988.

Hagihara, K., Chan, S., Zhang, L., Oh, D. Y., Wei, X. X., Simko, J., et al. (2019). Neoadjuvant sipuleucel-T induces both Th1 activation and immune regulation in localized prostate cancer. *Oncoimmunology*, *8*(1), e1486953. https://doi.org/10.1080/2162402X.2018.1486953.

Halperin, S. O., Tou, C. J., Wong, E. B., Modavi, C., Schaffer, D. V., & Dueber, J. E. (2018). CRISPR-guided DNA polymerases enable diversification of all nucleotides in a tunable window. *Nature*, *560*(7717), 248–252. https://doi.org/10.1038/s41586-018-0384-8.

Harding, S. M., Benci, J. L., Irianto, J., Discher, D. E., Minn, A. J., & Greenberg, R. A. (2017). Mitotic progression following DNA damage enables pattern recognition within micronuclei. *Nature*, *548*(7668), 466–470. https://doi.org/10.1038/nature23470.

Harrington, K. J., Ferris, R. L., Blumenschein, G., Jr., Colevas, A. D., Fayette, J., Licitra, L., et al. (2017). Nivolumab versus standard, single-agent therapy of investigator's choice in recurrent or metastatic squamous cell carcinoma of the head and neck (CheckMate 141): Health-related quality-of-life results from a randomised, phase 3 trial. *The Lancet Oncology*, *18*(8), 1104–1115. https://doi.org/10.1016/s1470-2045(17)30421-7.

Heery, C. R., O'Sullivan-Coyne, G., Madan, R. A., Cordes, L., Rajan, A., Rauckhorst, M., et al. (2017). Avelumab for metastatic or locally advanced previously treated solid tumours (JAVELIN Solid Tumor): A phase 1a, multicohort, dose-escalation trial. *The Lancet Oncology*, *18*(5), 587–598. https://doi.org/10.1016/s1470-2045(17)30239-5.

Herbst, R. S., Soria, J. C., Kowanetz, M., Fine, G. D., Hamid, O., Gordon, M. S., et al. (2014). Predictive correlates of response to the anti-PD-L1 antibody MPDL3280A in cancer patients. *Nature*, *515*(7528), 563–567. https://doi.org/10.1038/nature14011.

Hida, T., Kaji, R., Satouchi, M., Ikeda, N., Horiike, A., Nokihara, H., et al. (2018). Atezolizumab in Japanese patients with previously treated advanced non-small-cell lung cancer: A subgroup analysis of the phase 3 OAK study. *Clinical Lung Cancer*, *19*(4), e405–e415. https://doi.org/10.1016/j.cllc.2018.01.004.

Hildner, K., Edelson, B. T., Purtha, W. E., Diamond, M., Matsushita, H., Kohyama, M., et al. (2008). Batf3 deficiency reveals a critical role for CD8alpha+ dendritic cells in cytotoxic T cell immunity. *Science*, *322*(5904), 1097–1100. https://doi.org/10.1126/science.1164206.

Hille, F., Richter, H., Wong, S. P., Bratovic, M., Ressel, S., & Charpentier, E. (2018). The biology of CRISPR-Cas: Backward and forward. *Cell*, *172*(6), 1239–1259. https://doi.org/10.1016/j.cell.2017.11.032.

Hodi, F. S., Hwu, W. J., Kefford, R., Weber, J. S., Daud, A., Hamid, O., et al. (2016). Evaluation of immune-related response criteria and RECIST v1.1 in patients with advanced melanoma treated with pembrolizumab. *Journal of Clinical Oncology*, *34*(13), 1510–1517. https://doi.org/10.1200/jco.2015.64.0391.

Hodi, F. S., O'Day, S. J., McDermott, D. F., Weber, R. W., Sosman, J. A., Haanen, J. B., et al. (2010). Improved survival with ipilimumab in patients with metastatic melanoma. *The New England Journal of Medicine*, *363*(8), 711–723. https://doi.org/10.1056/NEJMoa1003466.

Horn, L., Spigel, D. R., Vokes, E. E., Holgado, E., Ready, N., Steins, M., et al. (2017). Nivolumab versus docetaxel in previously treated patients with advanced non-small-cell lung cancer: Two-year outcomes from two randomized, open-label, phase III trials (CheckMate 017 and CheckMate 057). *Journal of Clinical Oncology*, *35*(35), 3924–3933. https://doi.org/10.1200/jco.2017.74.3062.

Hoshino, A., Costa-Silva, B., Shen, T. L., Rodrigues, G., Hashimoto, A., Tesic Mark, M., et al. (2015). Tumour exosome integrins determine organotropic metastasis. *Nature*, *527*(7578), 329–335. https://doi.org/10.1038/nature15756.

Hsu, C., Lee, S. H., Ejadi, S., Even, C., Cohen, R. B., Le Tourneau, C., et al. (2017). Safety and antitumor activity of pembrolizumab in patients with programmed death-ligand 1-positive nasopharyngeal carcinoma: results of the KEYNOTE-028 study. *Journal of Clinical Oncology*, *35*(36), 4050–4056. https://doi.org/10.1200/jco.2017.73.3675.

Hui, R., Garon, E. B., Goldman, J. W., Leighl, N. B., Hellmann, M. D., Patnaik, A., et al. (2017). Pembrolizumab as first-line therapy for patients with PD-L1-positive advanced non-small cell lung cancer: A phase 1 trial. *Annals of Oncology*, *28*(4), 874–881. https://doi.org/10.1093/annonc/mdx008.

Humbert, M., & Hugues, S. (2019). Warming up the tumor microenvironment in order to enhance immunogenicity. *Oncoimmunology*, *8*(1), e1510710. https://doi.org/10.1080/2162402X.2018.1510710.

Hundal, J., Kiwala, S., Feng, Y. Y., Liu, C. J., Govindan, R., Chapman, W. C., et al. (2019). Accounting for proximal variants improves neoantigen prediction. *Nature Genetics*, *51*(1), 175–179. https://doi.org/10.1038/s41588-018-0283-9.

Iida, N., Dzutsev, A., Stewart, C. A., Smith, L., Bouladoux, N., Weingarten, R. A., et al. (2013). Commensal bacteria control cancer response to therapy by modulating the tumor microenvironment. *Science*, *342*(6161), 967–970. https://doi.org/10.1126/science.1240527.

Jaiswal, S., Jamieson, C. H., Pang, W. W., Park, C. Y., Chao, M. P., Majeti, R., et al. (2009). CD47 is upregulated on circulating hematopoietic stem cells and leukemia cells to avoid phagocytosis. *Cell*, *138*(2), 271–285. https://doi.org/10.1016/j.cell.2009.05.046.

Janikovits, J., Muller, M., Krzykalla, J., Korner, S., Echterdiek, F., Lahrmann, B., et al. (2018). High numbers of PDCD1 (PD-1)-positive T cells and B2M mutations in microsatellite-unstable colorectal cancer. *Oncoimmunology*, 7(2), e1390640. https://doi.org/10.1080/2162402X.2017.1390640.

Joseph, R. W., Elassaiss-Schaap, J., Kefford, R., Hwu, W. J., Wolchok, J. D., Joshua, A. M., et al. (2018). Baseline tumor size is an independent prognostic factor for overall survival in patients with melanoma treated with pembrolizumab. *Clinical Cancer Research*, *24*(20), 4960–4967. https://doi.org/10.1158/1078-0432.ccr-17-2386.

Kalhor, R., Kalhor, K., Mejia, L., Leeper, K., Graveline, A., Mali, P., et al. (2018). Developmental barcoding of whole mouse via homing CRISPR. *Science*, *361*(6405), eaat9804. https://doi.org/10.1126/science.aat9804.

Kantoff, P. W., Higano, C. S., Shore, N. D., Berger, E. R., Small, E. J., Penson, D. F., et al. (2010). Sipuleucel-T immunotherapy for castration-resistant prostate cancer. *The New England Journal of Medicine*, *363*(5), 411–422. https://doi.org/10.1056/NEJMoa1001294.

Kaplan, D. H., Shankaran, V., Dighe, A. S., Stockert, E., Aguet, M., Old, L. J., et al. (1998). Demonstration of an interferon gamma-dependent tumor surveillance system in immunocompetent mice. *Proceedings of the National Academy of Sciences of the United States of America*, *95*(13), 7556–7561.

Karasaki, T., Nagayama, K., Kuwano, H., Nitadori, J. I., Sato, M., Anraku, M., et al. (2017). Prediction and prioritization of neoantigens: Integration of RNA sequencing data with whole-exome sequencing. *Cancer Science*, *108*(2), 170–177. https://doi.org/10.1111/cas.13131.

Kaufman, H. L., Russell, J., Hamid, O., Bhatia, S., Terheyden, P., D'Angelo, S. P., et al. (2016). Avelumab in patients with chemotherapy-refractory metastatic Merkel cell carcinoma: A multicentre, single-group, open-label, phase 2 trial. *The Lancet Oncology*, *17*(10), 1374–1385. https://doi.org/10.1016/s1470-2045(16)30364-3.

Kaufman, H. L., Russell, J. S., Hamid, O., Bhatia, S., Terheyden, P., D'Angelo, S. P., et al. (2018). Updated efficacy of avelumab in patients with previously treated metastatic Merkel cell carcinoma after >/=1 year of follow-up: JAVELIN Merkel 200, a phase 2 clinical trial. *Journal for Immunotherapy of Cancer*, *6*(1), 7. https://doi.org/10.1186/s40425-017-0310-x.

Kazandjian, D., Suzman, D. L., Blumenthal, G., Mushti, S., He, K., Libeg, M., et al. (2016). FDA approval summary: Nivolumab for the treatment of metastatic non-small cell lung cancer with progression on or after platinum-based chemotherapy. *The Oncologist, 21*(5), 634–642. https://doi.org/10.1634/theoncologist.2015-0507.

Kellner, C., Peipp, M., Gramatzki, M., Schrappe, M., & Schewe, D. M. (2018). Perspectives of Fc engineered antibodies in CD19 targeting immunotherapies in pediatric B-cell precursor acute lymphoblastic leukemia. *Oncoimmunology*, 7(8), e1448331. https://doi.org/10.1080/2162402X.2018.1448331.

Kelly, P. N. (2018). The cancer immunotherapy revolution. *Science, 359*(6382), 1344–1345. https://doi.org/10.1126/science.359.6382.1344.

Kiyota, N., Hasegawa, Y., Takahashi, S., Yokota, T., Yen, C. J., Iwae, S., et al. (2017). A randomized, open-label, phase III clinical trial of nivolumab vs. therapy of investigator's choice in recurrent squamous cell carcinoma of the head and neck: A subanalysis of Asian patients versus the global population in checkmate 141. *Oral Oncology, 73*, 138–146. https://doi.org/10.1016/j.oraloncology.2017.07.023.

Koebel, C. M., Vermi, W., Swann, J. B., Zerafa, N., Rodig, S. J., Old, L. J., et al. (2007). Adaptive immunity maintains occult cancer in an equilibrium state. *Nature, 450*(7171), 903–907. https://doi.org/10.1038/nature06309.

Krummel, M. F., & Allison, J. P. (1995). CD28 and CTLA-4 have opposing effects on the response of T cells to stimulation. *The Journal of Experimental Medicine, 182*(2), 459–465.

Lavin, Y., Kobayashi, S., Leader, A., Amir, E. D., Elefant, N., Bigenwald, C., et al. (2017). Innate immune landscape in early lung adenocarcinoma by paired single-cell analyses. *Cell, 169*(4), 750–765.e717. https://doi.org/10.1016/j.cell.2017.04.014.

Le, D. T., Durham, J. N., Smith, K. N., Wang, H., Bartlett, B. R., Aulakh, L. K., et al. (2017). Mismatch repair deficiency predicts response of solid tumors to PD-1 blockade. *Science, 357*(6349), 409–413. https://doi.org/10.1126/science.aan6733.

Le, D. T., Uram, J. N., Wang, H., Bartlett, B. R., Kemberling, H., Eyring, A. D., et al. (2015). PD-1 blockade in tumors with mismatch-repair deficiency. *The New England Journal of Medicine, 372*(26), 2509–2520. https://doi.org/10.1056/NEJMoa1500596.

Leach, D. R., Krummel, M. F., & Allison, J. P. (1996). Enhancement of antitumor immunity by CTLA-4 blockade. *Science, 271*(5256), 1734–1736.

Levy, A., Massard, C., Soria, J. C., & Deutsch, E. (2016). Concurrent irradiation with the anti-programmed cell death ligand-1 immune checkpoint blocker durvalumab: Single centre subset analysis from a phase 1/2 trial. *European Journal of Cancer, 68*, 156–162. https://doi.org/10.1016/j.ejca.2016.09.013.

Li, H., van der Leun, A. M., Yofe, I., Lubling, Y., Gelbard-Solodkin, D., van Akkooi, A. C. J., … Amit, I. (2019). Dysfunctional CD8 T cells form a proliferative, dynamically regulated compartment within human melanoma. Cell, 176(4), 775-789 e718. doi:https://doi.org/10.1016/j.cell.2018.11.043

Lopez-Soto, A., Gonzalez, S., Smyth, M. J., & Galluzzi, L. (2017). Control of metastasis by NK cells. *Cancer Cell, 32*(2), 135–154. https://doi.org/10.1016/j.ccell.2017.06.009.

Mackenzie, K. J., Carroll, P., Martin, C. A., Murina, O., Fluteau, A., Simpson, D. J., et al. (2017). cGAS surveillance of micronuclei links genome instability to innate immunity. *Nature, 548*(7668), 461–465. https://doi.org/10.1038/nature23449.

Majeti, R., Chao, M. P., Alizadeh, A. A., Pang, W. W., Jaiswal, S., Gibbs, K. D., Jr., et al. (2009). CD47 is an adverse prognostic factor and therapeutic antibody target on human acute myeloid leukemia stem cells. *Cell, 138*(2), 286–299. https://doi.org/10.1016/j.cell.2009.05.045.

Mariathasan, S., Turley, S. J., Nickles, D., Castiglioni, A., Yuen, K., Wang, Y., et al. (2018). TGFbeta attenuates tumour response to PD-L1 blockade by contributing to exclusion of T cells. *Nature, 554*(7693), 544–548. https://doi.org/10.1038/nature25501.

Maude, S. L., Laetsch, T. W., Buechner, J., Rives, S., Boyer, M., Bittencourt, H., et al. (2018). Tisagenlecleucel in children and young adults with B-cell lymphoblastic leukemia. *The New England Journal of Medicine*, *378*(5), 439–448. https://doi.org/10.1056/NEJMoa1709866.

McDermott, D., Haanen, J., Chen, T. T., Lorigan, P., & O'Day, S. (2013). Efficacy and safety of ipilimumab in metastatic melanoma patients surviving more than 2 years following treatment in a phase III trial (MDX010-20). *Annals of Oncology*, *24*(10), 2694–2698. https://doi.org/10.1093/annonc/mdt291.

Merchant, M. S., Wright, M., Baird, K., Wexler, L. H., Rodriguez-Galindo, C., Bernstein, D., et al. (2016). Phase I clinical trial of ipilimumab in pediatric patients with advanced solid tumors. *Clinical Cancer Research*, *22*(6), 1364–1370. https://doi.org/10.1158/1078-0432.ccr-15-0491.

Migden, M. R., Rischin, D., Schmults, C. D., Guminski, A., Hauschild, A., Lewis, K. D., et al. (2018). PD-1 blockade with cemiplimab in advanced cutaneous squamous-cell carcinoma. *The New England Journal of Medicine*, *379*(4), 341–351. https://doi.org/10.1056/NEJMoa1805131.

Mirzaei, R., Sarkar, S., Dzikowski, L., Rawji, K. S., Khan, L., Faissner, A., et al. (2018). Brain tumor-initiating cells export tenascin-C associated with exosomes to suppress T cell activity. *Oncoimmunology*, 7(10), e1478647. https://doi.org/10.1080/2162402X.2018.1478647.

Morgan, R. A., Dudley, M. E., Wunderlich, J. R., Hughes, M. S., Yang, J. C., Sherry, R. M., et al. (2006). Cancer regression in patients after transfer of genetically engineered lymphocytes. *Science*, *314*(5796), 126–129. https://doi.org/10.1126/science.1129003.

Motzer, R. J., Escudier, B., McDermott, D. F., George, S., Hammers, H. J., Srinivas, S., et al. (2015). Nivolumab versus everolimus in advanced renal-cell carcinoma. *The New England Journal of Medicine*, *373*(19), 1803–1813. https://doi.org/10.1056/NEJMoa1510665.

Mulqueen, R. M., Pokholok, D., Norberg, S. J., Torkenczy, K. A., Fields, A. J., Sun, D., et al. (2018). Highly scalable generation of DNA methylation profiles in single cells. *Nature Biotechnology*, *36*(5), 428–431. https://doi.org/10.1038/nbt.4112.

Muro, K., Chung, H. C., Shankaran, V., Geva, R., Catenacci, D., Gupta, S., et al. (2016). Pembrolizumab for patients with PD-L1-positive advanced gastric cancer (KEYNOTE-012): A multicentre, open-label, phase 1b trial. *The Lancet Oncology*, *17*(6), 717–726. https://doi.org/10.1016/s1470-2045(16)00175-3.

Nanda, R., Chow, L. Q., Dees, E. C., Berger, R., Gupta, S., Geva, R., et al. (2016). Pembrolizumab in patients with advanced triple-negative breast cancer: Phase Ib KEYNOTE-012 study. *Journal of Clinical Oncology*, *34*(21), 2460–2467. https://doi.org/10.1200/jco.2015.64.8931.

Necchi, A., Joseph, R. W., Loriot, Y., Hoffman-Censits, J., Perez-Gracia, J. L., Petrylak, D. P., et al. (2017). Atezolizumab in platinum-treated locally advanced or metastatic urothelial carcinoma: Post-progression outcomes from the phase II IMvigor210 study. *Annals of Oncology*, *28*(12), 3044–3050. https://doi.org/10.1093/annonc/mdx518.

Neelapu, S. S., Locke, F. L., Bartlett, N. L., Lekakis, L. J., Miklos, D. B., Jacobson, C. A., et al. (2017). Axicabtagene ciloleucel CAR T-cell therapy in refractory large B-cell lymphoma. *The New England Journal of Medicine*, 377(26), 2531–2544. https://doi.org/10.1056/NEJMoa1707447.

Nishimasu, H., Shi, X., Ishiguro, S., Gao, L., Hirano, S., Okazaki, S., et al. (2018). Engineered CRISPR-Cas9 nuclease with expanded targeting space. *Science*, *361*(6408), 1259–1262. https://doi.org/10.1126/science.aas9129.

Obeid, M., Tesniere, A., Ghiringhelli, F., Fimia, G. M., Apetoh, L., Perfettini, J. L., et al. (2007). Calreticulin exposure dictates the immunogenicity of cancer cell death. *Nature Medicine*, *13*(1), 54–61. https://doi.org/10.1038/nm1523.

O'Neil, B. H., Wallmark, J. M., Lorente, D., Elez, E., Raimbourg, J., Gomez-Roca, C., et al. (2017). Safety and antitumor activity of the anti-PD-1 antibody pembrolizumab in patients with advanced colorectal carcinoma. *PLoS One*, *12*(12), e0189848. https://doi.org/10.1371/journal.pone.0189848.

Ott, P. A., Elez, E., Hiret, S., Kim, D. W., Morosky, A., Saraf, S., et al. (2017). Pembrolizumab in patients with extensive-stage small-cell lung cancer: Results from the phase Ib KEYNOTE-028 study. *Journal of Clinical Oncology*, *35*(34), 3823–3829. https://doi.org/10.1200/jco.2017.72.5069.

Ott, P. A., Piha-Paul, S. A., Munster, P., Pishvaian, M. J., van Brummelen, E. M. J., Cohen, R. B., et al. (2017). Safety and antitumor activity of the anti-PD-1 antibody pembrolizumab in patients with recurrent carcinoma of the anal canal. *Annals of Oncology*, *28*(5), 1036–1041. https://doi.org/10.1093/annonc/mdx029.

Overman, M. J., McDermott, R., Leach, J. L., Lonardi, S., Lenz, H. J., Morse, M. A., et al. (2017). Nivolumab in patients with metastatic DNA mismatch repair-deficient or microsatellite instability-high colorectal cancer (CheckMate 142): An open-label, multicentre, phase 2 study. *The Lancet Oncology*, *18*(9), 1182–1191. https://doi.org/10.1016/s1470-2045(17)30422-9.

Oyer, J. L., Gitto, S. B., Altomare, D. A., & Copik, A. J. (2018). PD-L1 blockade enhances anti-tumor efficacy of NK cells. *Oncoimmunology*, 7(11), e1509819. https://doi.org/10.1080/2162402X.2018.1509819.

Ozcan, M., Janikovits, J., von Knebel Doeberitz, M., & Kloor, M. (2018). Complex pattern of immune evasion in MSI colorectal cancer. *Oncoimmunology*, 7(7), e1445453. https://doi.org/10.1080/2162402X.2018.1445453.

Patnaik, A., Kang, S. P., Rasco, D., Papadopoulos, K. P., Elassaiss-Schaap, J., Beeram, M., et al. (2015). Phase I study of pembrolizumab (MK-3475; Anti-PD-1 monoclonal antibody) in patients with advanced solid tumors. *Clinical Cancer Research*, *21*(19), 4286–4293. https://doi.org/10.1158/1078-0432.ccr-14-2607.

Perez-Gracia, J. L., Loriot, Y., Rosenberg, J. E., Powles, T., Necchi, A., Hussain, S. A., et al. (2018). Atezolizumab in platinum-treated locally advanced or metastatic urothelial carcinoma: Outcomes by prior number of regimens. *European Urology*, *73*, 462–468. https://doi.org/10.1016/j.eururo.2017.11.023.

Petrella, T. M., Robert, C., Richtig, E., Miller, W. H., Jr., Masucci, G. V., Walpole, E., et al. (2017). Patient-reported outcomes in KEYNOTE-006, a randomised study of pembrolizumab versus ipilimumab in patients with advanced melanoma. *European Journal of Cancer*, *86*, 115–124. https://doi.org/10.1016/j.ejca.2017.08.032.

Plimack, E. R., Bellmunt, J., Gupta, S., Berger, R., Chow, L. Q., Juco, J., et al. (2017). Safety and activity of pembrolizumab in patients with locally advanced or metastatic urothelial cancer (KEYNOTE-012): A non-randomised, open-label, phase 1b study. *The Lancet Oncology*, *18*(2), 212–220. https://doi.org/10.1016/S1470-2045(17)30007-4.

Pol, J., Kroemer, G., & Galluzzi, L. (2016). First oncolytic virus approved for melanoma immunotherapy. *Oncoimmunology*, *5*(1), e1115641. https://doi.org/10.1080/2162402X.2015.1115641.

Pol, J., Levesque, S., Workenhe, S. T., Gujar, S., Le Boeuf, F., D, R. C., et al. (2018). Trial watch: Oncolytic viro-immunotherapy of hematologic and solid tumors. *Oncoimmunology*, 7(12), e1503032. https://doi.org/10.1080/2162402X.2018.1503032.

Pol, J., Vacchelli, E., Aranda, F., Castoldi, F., Eggermont, A., Cremer, I., et al. (2015). Trial watch: Immunogenic cell death inducers for anticancer chemotherapy. *Oncoimmunology*, *4*(4). e1008866. https://doi.org/10.1080/2162402X.2015.1008866.

Powles, T., O'Donnell, P. H., Massard, C., Arkenau, H. T., Friedlander, T. W., Hoimes, C. J., et al. (2017). Efficacy and safety of durvalumab in locally advanced or metastatic urothelial carcinoma: Updated results from a Phase 1/2 open-label study. *JAMA Oncology*, *3*(9), e172411. https://doi.org/10.1001/jamaoncol.2017.2411.

Raikar, S. S., Fleischer, L. C., Moot, R., Fedanov, A., Paik, N. Y., Knight, K. A., et al. (2018). Development of chimeric antigen receptors targeting T-cell malignancies using two structurally different anti-CD5 antigen binding domains in NK and CRISPR-edited T cell lines. *Oncoimmunology*, 7(3), e1407898. https://doi.org/10.1080/2162402X.2017.1407898.

Rao, S., Gharib, K., & Han, A. (2019). Cancer immunosurveillance by T cells. *International Review of Cell and Molecular Biology*, *342*, 149–173. https://doi.org/10.1016/bs.ircmb.2018.08.001.

Reck, M., Rodriguez-Abreu, D., Robinson, A. G., Hui, R., Csoszi, T., Fulop, A., et al. (2016). Pembrolizumab versus chemotherapy for PD-L1-positive non-small-cell lung cancer. *The New England Journal of Medicine*, *375*(19), 1823–1833. https://doi.org/10.1056/NEJMoa1606774.

Riaz, N., Havel, J. J., Makarov, V., Desrichard, A., Urba, W. J., Sims, J. S., et al. (2017). Tumor and microenvironment evolution during immunotherapy with nivolumab. *Cell*, *171*(4), 934–949.e916. https://doi.org/10.1016/j.cell.2017.09.028.

Ribas, A., Hamid, O., Daud, A., Hodi, F. S., Wolchok, J. D., Kefford, R., et al. (2016). Association of pembrolizumab with tumor response and survival among patients with advanced melanoma. *JAMA*, *315*(15), 1600–1609. https://doi.org/10.1001/jama.2016.4059.

Rittmeyer, A., Barlesi, F., Waterkamp, D., Park, K., Ciardiello, F., von Pawel, J., et al. (2017). Atezolizumab versus docetaxel in patients with previously treated non-small-cell lung cancer (OAK): A phase 3, open-label, multicentre randomised controlled trial. *Lancet*, *389*(10066), 255–265. https://doi.org/10.1016/s0140-6736(16)32517-x.

Rizvi, N. A., Hellmann, M. D., Snyder, A., Kvistborg, P., Makarov, V., Havel, J. J., et al. (2015). Cancer immunology. Mutational landscape determines sensitivity to PD-1 blockade in non-small cell lung cancer. *Science*, *348*(6230), 124–128. https://doi.org/10.1126/science.aaa1348.

Robert, C., Ribas, A., Hamid, O., Daud, A., Wolchok, J. D., Joshua, A. M., et al. (2018). Durable complete response after discontinuation of pembrolizumab in patients with metastatic melanoma. *Journal of Clinical Oncology*, *36*(17), 1668–1674. https://doi.org/10.1200/jco.2017.75.6270.

Robert, C., Ribas, A., Wolchok, J. D., Hodi, F. S., Hamid, O., Kefford, R., et al. (2014). Anti-programmed-death-receptor-1 treatment with pembrolizumab in ipilimumab refractory advanced melanoma: A randomised dose-comparison cohort of a phase 1 trial. *Lancet*, *384*(9948), 1109–1117. https://doi.org/10.1016/s0140-6736(14)60958-2.

Robert, C., Schachter, J., Long, G. V., Arance, A., Grob, J. J., Mortier, L., et al. (2015). Pembrolizumab versus ipilimumab in advanced melanoma. *The New England Journal of Medicine*, *372*(26), 2521–2532. https://doi.org/10.1056/NEJMoa1503093.

Robert, C., Schadendorf, D., Messina, M., Hodi, F. S., & O'Day, S. (2013). Efficacy and safety of retreatment with ipilimumab in patients with pretreated advanced melanoma who progressed after initially achieving disease control. *Clinical Cancer Research*, *19*(8), 2232–2239. https://doi.org/10.1158/1078-0432.ccr-12-3080.

Rosenberg, J. E., Hoffman-Censits, J., Powles, T., van der Heijden, M. S., Balar, A. V., Necchi, A., et al. (2016). Atezolizumab in patients with locally advanced and metastatic urothelial carcinoma who have progressed following treatment with platinum-based chemotherapy: A single-arm, multicentre, phase 2 trial. *Lancet*, *387*(10031), 1909–1920. https://doi.org/10.1016/s0140-6736(16)00561-4.

Routy, B., Le Chatelier, E., Derosa, L., Duong, C. P. M., Alou, M. T., Daillere, R., et al. (2018). Gut microbiome influences efficacy of PD-1-based immunotherapy against epithelial tumors. *Science*, *359*(6371), 91–97. https://doi.org/10.1126/science.aan3706.

Ruscetti, M., Leibold, J., Bott, M. J., Fennell, M., Kulick, A., Salgado, N. R., et al. (2018). NK cell-mediated cytotoxicity contributes to tumor control by a cytostatic drug combination. *Science*, *362*(6421), 1416–1422. https://doi.org/10.1126/science.aas9090.

Savage, P. A., Vosseller, K., Kang, C., Larimore, K., Riedel, E., Wojnoonski, K., et al. (2008). Recognition of a ubiquitous self antigen by prostate cancer-infiltrating CD8+ T lymphocytes. *Science*, *319*(5860), 215–220. https://doi.org/10.1126/science.1148886.

Schachter, J., Ribas, A., Long, G. V., Arance, A., Grob, J. J., Mortier, L., et al. (2017). Pembrolizumab versus ipilimumab for advanced melanoma: final overall survival results of a multicentre, randomised, open-label phase 3 study (KEYNOTE-006). *Lancet*, *390*(10105), 1853–1862. https://doi.org/10.1016/s0140-6736(17)31601-x.

Schumacher, T. N., & Schreiber, R. D. (2015). Neoantigens in cancer immunotherapy. *Science*, *348*(6230), 69–74. https://doi.org/10.1126/science.aaa4971.

Schuster, S. J., Bishop, M. R., Tam, C. S., Waller, E. K., Borchmann, P., McGuirk, J. P., et al. (2019). Tisagenlecleucel in adult relapsed or refractory diffuse large B-cell lymphoma. *The New England Journal of Medicine*, *380*(1), 45–56. https://doi.org/10.1056/NEJMoa1804980.

Seiwert, T. Y., Burtness, B., Mehra, R., Weiss, J., Berger, R., Eder, J. P., et al. (2016). Safety and clinical activity of pembrolizumab for treatment of recurrent or metastatic squamous cell carcinoma of the head and neck (KEYNOTE-012): An open-label, multicentre, phase 1b trial. *The Lancet Oncology*, *17*(7), 956–965. https://doi.org/10.1016/s1470-2045(16)30066-3.

Shankaran, V., Ikeda, H., Bruce, A. T., White, J. M., Swanson, P. E., Old, L. J., et al. (2001). IFNgamma and lymphocytes prevent primary tumour development and shape tumour immunogenicity. *Nature*, *410*(6832), 1107–1111. https://doi.org/10.1038/35074122.

Sharma, P., Retz, M., Siefker-Radtke, A., Baron, A., Necchi, A., Bedke, J., et al. (2017). Nivolumab in metastatic urothelial carcinoma after platinum therapy (CheckMate 275): A multicentre, single-arm, phase 2 trial. *The Lancet Oncology*, *18*(3), 312–322. https://doi.org/10.1016/s1470-2045(17)30065-7.

Shaverdian, N., Lisberg, A. E., Bornazyan, K., Veruttipong, D., Goldman, J. W., Formenti, S. C., et al. (2017). Previous radiotherapy and the clinical activity and toxicity of pembrolizumab in the treatment of non-small-cell lung cancer: A secondary analysis of the KEYNOTE-001 phase 1 trial. *The Lancet Oncology*, *18*(7), 895–903. https://doi.org/10.1016/s1470-2045(17)30380-7.

Shen, M. W., Arbab, M., Hsu, J. Y., Worstell, D., Culbertson, S. J., Krabbe, O., et al. (2018). Predictable and precise template-free CRISPR editing of pathogenic variants. *Nature*, *563*(7733), 646–651. https://doi.org/10.1038/s41586-018-0686-x.

Spitzer, M. H., & Nolan, G. P. (2016). Mass cytometry: Single cells, many features. *Cell*, *165*(4), 780–791. https://doi.org/10.1016/j.cell.2016.04.019.

Spranger, S., Bao, R., & Gajewski, T. F. (2015). Melanoma-intrinsic beta-catenin signalling prevents anti-tumour immunity. *Nature*, *523*(7559), 231–235. https://doi.org/10.1038/nature14404.

Straussman, R., Morikawa, T., Shee, K., Barzily-Rokni, M., Qian, Z. R., Du, J., et al. (2012). Tumour micro-environment elicits innate resistance to RAF inhibitors through HGF secretion. *Nature*, *487*(7408), 500–504. https://doi.org/10.1038/nature11183.

Tahara, M., Muro, K., Hasegawa, Y., Chung, H. C., Lin, C. C., Keam, B., et al. (2018). Pembrolizumab in Asia-Pacific patients with advanced head and neck squamous cell carcinoma: Analyses from KEYNOTE-012. *Cancer Science*, *109*(3), 771–776. https://doi.org/10.1111/cas.13480.

Tauriello, D. V. F., Palomo-Ponce, S., Stork, D., Berenguer-Llergo, A., Badia-Ramentol, J., Iglesias, M., et al. (2018). TGFbeta drives immune evasion in genetically reconstituted colon cancer metastasis. *Nature*, *554*(7693), 538–543. https://doi.org/10.1038/nature25492.

Vacchelli, E., Eggermont, A., Sautes-Fridman, C., Galon, J., Zitvogel, L., Kroemer, G., et al. (2013). Trial watch: Oncolytic viruses for cancer therapy. *Oncoimmunology*, *2*(6). e24612. https://doi.org/10.4161/onci.24612.

van Montfoort, N., Borst, L., Korrer, M. J., Sluijter, M., Marijt, K. A., Santegoets, S. J., et al. (2018). NKG2A blockade potentiates CD8 T cell immunity induced by cancer vaccines. *Cell*, *175*(7), 1744–1755.e1715. https://doi.org/10.1016/j.cell.2018.10.028.

Vanpouille-Box, C., Alard, A., Aryankalayil, M. J., Sarfraz, Y., Diamond, J. M., Schneider, R. J., et al. (2017). DNA exonuclease Trex1 regulates radiotherapy-induced tumour immunogenicity. *Nature Communications*, *8*, 15618. https://doi.org/10.1038/ncomms15618.

Vanpouille-Box, C., Demaria, S., Formenti, S. C., & Galluzzi, L. (2018). Cytosolic DNA sensing in organismal tumor control. *Cancer Cell*, *34*(3), 361–378. https://doi.org/10.1016/j.ccell.2018.05.013.

Vetizou, M., Pitt, J. M., Daillere, R., Lepage, P., Waldschmitt, N., Flament, C., et al. (2015). Anticancer immunotherapy by CTLA-4 blockade relies on the gut microbiota. *Science*, *350*(6264), 1079–1084. https://doi.org/10.1126/science.aad1329.

Viaud, S., Saccheri, F., Mignot, G., Yamazaki, T., Daillere, R., Hannani, D., et al. (2013). The intestinal microbiota modulates the anticancer immune effects of cyclophosphamide. *Science*, *342*(6161), 971–976. https://doi.org/10.1126/science.1240537.

Vitale, I., Sistigu, A., Manic, G., Rudqvist, N. P., Trajanoski, Z., & Galluzzi, L. (2019). Mutational and antigenic landscape in tumor progression and cancer immunotherapy. *Trends in Cell Biology*, *29*, 395–416. https://doi.org/10.1016/j.tcb.2019.01.003.

Vokes, E. E., Ready, N., Felip, E., Horn, L., Burgio, M. A., Antonia, S. J., et al. (2018). Nivolumab versus docetaxel in previously treated advanced non-small-cell lung cancer (CheckMate 017 and CheckMate 057): 3-year update and outcomes in patients with liver metastases. *Annals of Oncology*, *29*(4), 959–965. https://doi.org/10.1093/annonc/mdy041.

Wang, H., Xu, X., Nguyen, C. M., Liu, Y., Gao, Y., Lin, X., et al. (2018). CRISPR-mediated programmable 3D genome positioning and nuclear organization. *Cell*, *175*(5), 1405–1417.e1414. https://doi.org/10.1016/j.cell.2018.09.013.

Weber, J. S., D'Angelo, S. P., Minor, D., Hodi, F. S., Gutzmer, R., Neyns, B., et al. (2015). Nivolumab versus chemotherapy in patients with advanced melanoma who progressed after anti-CTLA-4 treatment (CheckMate 037): A randomised, controlled, open-label, phase 3 trial. *The Lancet Oncology*, *16*(4), 375–384. https://doi.org/10.1016/s1470-2045(15)70076-8.

Weber, J., Mandala, M., Del Vecchio, M., Gogas, H. J., Arance, A. M., Cowey, C. L., et al. (2017). Adjuvant nivolumab versus ipilimumab in resected stage III or IV melanoma. *The New England Journal of Medicine*, *377*(19), 1824–1835. https://doi.org/10.1056/NEJMoa1709030.

Wierz, M., Janji, B., Berchem, G., Moussay, E., & Paggetti, J. (2018). High-dimensional mass cytometry analysis revealed microenvironment complexity in chronic lymphocytic leukemia. *Oncoimmunology*, *7*(8), e1465167. https://doi.org/10.1080/2162402X.2018.1465167.

Wu, Y., Williams, E. G., Dubuis, S., Mottis, A., Jovaisaite, V., Houten, S. M., et al. (2014). Multilayered genetic and omics dissection of mitochondrial activity in a mouse reference population. *Cell*, *158*(6), 1415–1430. https://doi.org/10.1016/j.cell.2014.07.039.

Xu, J. X., Maher, V. E., Zhang, L., Tang, S., Sridhara, R., Ibrahim, A., et al. (2017). FDA approval summary: nivolumab in advanced renal cell carcinoma after anti-angiogenic therapy and exploratory predictive biomarker analysis. *The Oncologist*, *22*(3), 311–317. https://doi.org/10.1634/theoncologist.2016-0476.

Yaddanapudi, K., Meng, S., Whitt, A. G., Al Rayyan, N., Richie, J., Tu, A., et al. (2019). Exosomes from GM-CSF expressing embryonic stem cells are an effective prophylactic vaccine for cancer prevention. *Oncoimmunology*, *8*(3), 1561119. https://doi.org/10.1080/2162402X.2018.1561119.

Yan, W. X., Hunnewell, P., Alfonse, L. E., Carte, J. M., Keston-Smith, E., Sothiselvam, S., et al. (2019). Functionally diverse type V CRISPR-Cas systems. *Science*, *363*(6422), 88–91. https://doi.org/10.1126/science.aav7271.

Younes, A., Santoro, A., Shipp, M., Zinzani, P. L., Timmerman, J. M., Ansell, S., et al. (2016). Nivolumab for classical Hodgkin's lymphoma after failure of both autologous stem-cell transplantation and brentuximab vedotin: A multicentre, multicohort, single-arm phase 2 trial. *The Lancet Oncology*, *17*(9), 1283–1294. https://doi.org/10.1016/s1470-2045(16)30167-x.

Zhang, S. Q., Ma, K. Y., Schonnesen, A. A., Zhang, M., He, C., Sun, E., et al. (2018). High-throughput determination of the antigen specificities of T cell receptors in single cells. *Nature Biotechnology*. https://doi.org/10.1038/nbt.4282.

Zhang, L., Zhang, S., Yao, J., Lowery, F. J., Zhang, Q., Huang, W. C., et al. (2015). Microenvironment-induced PTEN loss by exosomal microRNA primes brain metastasis outgrowth. *Nature*, *527*(7576), 100–104. https://doi.org/10.1038/nature15376.

Zheng, C., Zheng, L., Yoo, J. K., Guo, H., Zhang, Y., Guo, X., et al. (2017). Landscape of infiltrating T cells in liver cancer revealed by single-cell sequencing. *Cell*, *169*(7), 1342–1356.e1316. https://doi.org/10.1016/j.cell.2017.05.035.

Zinzani, P. L., Ribrag, V., Moskowitz, C. H., Michot, J. M., Kuruvilla, J., Balakumaran, A., et al. (2017). Safety and tolerability of pembrolizumab in patients with relapsed/refractory primary mediastinal large B-cell lymphoma. *Blood*, *130*(3), 267–270. https://doi.org/10.1182/blood-2016-12-758383.

CHAPTER ONE

# A comprehensive approach to gene expression profiling in immune cells

**Sabrina Carpentier[a], François Romagné[b], Eric Vivier[a,c,*]**
[a]Innate Pharma Research Laboratories, Innate Pharma, Marseille, France
[b]MI-mAbs, Aix-Marseille Univ, Marseille, France
[c]Aix-Marseille Univ, APHM, CNRS, INSERM, CIML, Hôpital de la Timone, Marseille-Immunopole, Marseille, France
*Corresponding author: e-mail address: vivier@ciml.univ-mrs.fr

## Contents



## Abstract

With the advent of whole-transcriptome studies and the growing need for public repositories, it has become essential to combine multiple heterogeneous datasets for immune cells. In this chapter, we describe the implementation of a compendium of 10,833 genes for 975 samples, corresponding to 52 resting immune cell types. We begin by describing the datasets, and their selection, in particular. We then explain the methodology implemented to create a qualified compendium: the processing of each array (quality control, normalization and bias correction), integration (merging rules, global normalization and batch removal) and validation. Finally some examples of use will be detailed. The utility and limitations of the compendium are also discussed, as an introduction to the next version.

*Methods in Enzymology*, Volume 636
ISSN 0076-6879
https://doi.org/10.1016/bs.mie.2019.07.005

## 1. Introduction

Transcriptome profiling by microarray analysis, and, more recently, next-generation sequencing, have been massively used to capture information at the system scale for diseases. These approaches have proved highly efficient in cell biology. As the number of studies based on these technologies increased, many journals insisted on free access to the data generated, to make it possible for anyone to validate the findings. As a result, huge amounts of transcriptome data are freely available from public databases, such as the NCBI Gene Expression Omnibus (GEO) and EMBL-EBI ArrayExpress databases.

A large number of transcriptional experiments have been performed to improve the characterization of immune cells and their functional classification in conditions of health and disease. The Immunological Genome Project (ImmGen), which was designed to address the question of immune cell characterization, was also intended to be a public resource. However, this project initially focused on mice and did not aim to collect together all free-access datasets.

One way to make use of the large amount of collective data available to obtain a more complete picture of the immune system is to combine several public datasets into a single compendium. This project, essential to work in this field, was challenging due to the tremendous heterogeneity of the data obtained by the different research groups in terms of sample preparation, hybridization protocols and, of course, the diverse types and versions of the arrays used.

We initially focused on resting immune cells, successfully merging 54 public datasets obtained with 13 microarray platforms from three manufacturers. The result is a compendium of 10,833 genes for 975 samples, corresponding to 52 resting immune cell types.

In this chapter, we aim to describe the datasets, including their selection in particular, and to explain the methodology implemented to create a qualified compendium: the processing of each array (quality control, normalization and bias correction), integration (merging rules, global normalization and batch removal) and validation. Finally, we describe some examples of its use and discuss its utility and limitations, to provide an introduction to the next version.

## 2. Resuts See Fig. 1

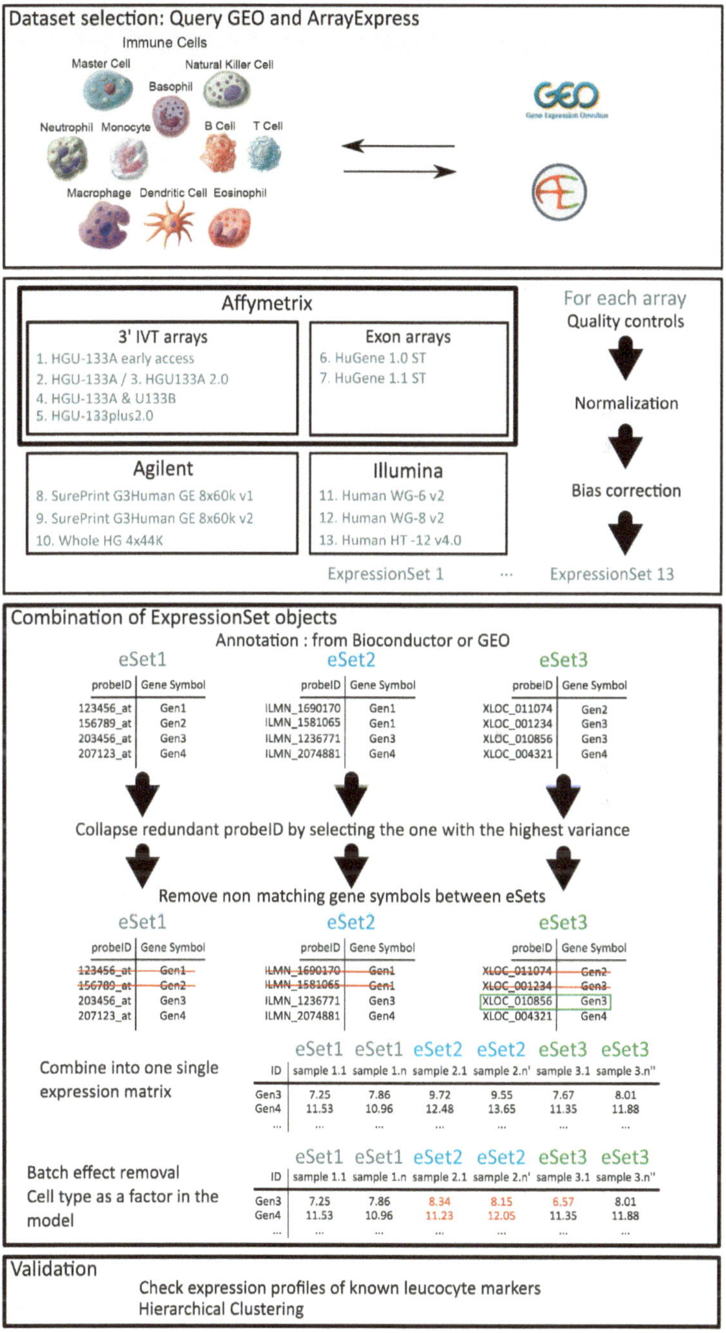


**Fig. 1** Scheme of implementation steps. Four main steps were required to build the curated compendium. These steps are described in detail in the implementation section.

## 2.1 Step 1: Dataset selection

Potentially relevant datasets deposited in GEO (Edgar, Domrachev, & Lash, 2002) or ArrayExpress (Parkinson, Sarkans, Kolesnikov, Abeygunawardena, et al., 2011) were identified by querying each repository for each type of immune cell. Multiple search queries were designed for each immune cell type, to identify as many datasets as possible. For example, CD8 T-cell datasets were identified by searching for "T cells" or "T lymphocytes" or "CD8 T cells" or "CD8 T lymphocytes" or "cytotoxic T cells". Datasets relating to dendritic cells and their subsets were identified by searching for "dendritic cells," "DC," "classical DC," "conventional DC," "cDC," "BDCA1 DC," "BDCA3 DC," "mDC," "myeloid dendritic cells," "pDC," "plasmacytoid dendritic cells," "DC-LC," "LC, "Langerhans cells," "iDC," or "immature DC."

Our queries focused on immune cells at steady state in healthy individuals. The relevance of each of the datasets identified was evaluated individually, by reading the description and annotations of each sample. Only samples fulfilling the following criteria were retained: immune cell types well-characterized according to the sorting protocol, individuals healthy and immune cells at steady state. Where possible, we preferentially selected datasets generated with whole-genome microarrays from Affymetrix, Agilent or Illumina.

The 54 public datasets constituting the final compendium are listed in Table 1. All datasets were downloaded in raw data format (CEL files for Affymetrix arrays and TXT files for Agilent and Illumina arrays) (Fig. 2).

## 2.2 Step 2: Processing of datasets by array

Datasets were first organized by array (i.e., the version of the array, for the arrays of each manufacturer). Thirteen different arrays were processed: seven from Affymetrix, three from Illumina and three from Agilent. First steps of processing were normalization and quality controls. These methods aim to make samples of good quality comparable. Normalization and quality control were manufacturer-specific, except for the Affymetrix arrays, for which these procedures differed between 3′ IVT Arrays and Exon arrays. All these analyses were processed with Bioconductor (Huber, Carey, Gentleman, Anders, et al., 2015) in the R statistical environment. At the end of this step, an object ExpressionSet with a $\log_2$-normalized gene expression matrix was available for each platform (i.e., array).

**Table 1** List of the 54 public datasets constituting the final compendium.

| Author | Dataset.ID | Platform | Tissue | Cell types | N |
|---|---|---|---|---|---|
| Jourdan et al. (2009) | *E*-MEXP-2360 | Affymetrix HG U133 Plus 2.0 Array (GPL570) | BM, blood | Memory B, plasma cells | 9 |
| Nodland, Berkowska, Bajer, Shah, et al. (2011) | E-MEXP-384 | Affymetrix HG U133A Array/ Affymetrix HG U133B Array (GPL96/GPL97) | BM | Naive B, preB, proB | 12 |
| Rasheed, Rahn, Sallusto, Lipp, et al. (2006) | E-MEXP-750 | Affymetrix HG U133A Array/ Affymetrix HG U133B Array (GPL96/GPL97) | Blood, tonsil | CD4 T (FHT, TCM, TEM) | 21 |
| Kassambara, Rème, Jourdan, Fest, et al. (2015) | E-MTAB-1771 | Affymetrix HG U133 Plus 2.0 Array (GPL570) | | Naive B | 5 |
| Lindstedt, Lundberg, and Borrebaeck (2005) | E-TABM-34 | Affymetrix HG U133 Plus 2.0 Array (GPL570) | Blood, tonsil | DC (e.g., cDC, pDC) | 21 |
| Raghavachari, Xu, Harris, Villagra, et al. (2007) | GSE11524 | Affymetrix HG U133 Plus 2.0 Array (GPL570) | Blood | Platelets | 4 |
| Emonts et al. (2008) | GSE11755 | Affymetrix HG U133 Plus 2.0 Array (GPL570) | Blood | Monocytes | 3 |
| Fujisaki, Kakuda, Shimasaki, Imai, et al. (2009) | GSE12198 | Affymetrix HG U133 Plus 2.0 Array (GPL570) | Blood | NK cells | 5 |
| Good, Avery, and Tangye (2009) | GSE13411 | Affymetrix HG U133A Array (GPL96) | Plasma, spleen | Memory B, naive B, plasma cells | 7 |
| Zhang, Ohyashiki, Shimizu, and Ohyashiki (2010) | GSE13906 | Affymetrix HG U133 Plus 2.0 Array (GPL570) | | γδ T cells | 2 |

*Continued*

**Table 1** List of the 54 public datasets constituting the final compendium.—cont'd

| Author | Dataset.ID | Platform | Tissue | Cell types | N |
|---|---|---|---|---|---|
| Miyara, Yoshioka, Kitoh, Shima, et al. (2009) | GSE15659 | Affymetrix HG U133 Plus 2.0 Array (GPL570) | Blood | CD4 T cells | 4 |
| Vinh, Patel, Uzel, Anderson, et al. (2010) | GSE16020 | Affymetrix HG U133 Plus 2.0 Array (GPL570) | Blood | PMN | 12 |
| Ancuta, Liu, Misra, Wacleche, et al. (2009) | GSE16836 | Affymetrix HG U133 Plus 2.0 Array (GPL570) | Blood | $CD14^+CD16^-$ monocytes, $CD14^+CD16^+$ monocytes, | 6 |
| Cassani, Montini, Maruggi, Ambrosi, et al. (2009) | GSE17354 | Affymetrix HG U133A Array (GPL96) | Blood | CD4 T, CD8 T cells | 8 |
| Baram, Dekel, Mekori, and Sagi-Eisenberg (2010) | GSE19888 | Affymetrix HG U133A 2.0 Array (GPL571) | | Mast cells | 2 |
| Juelke, Killig, Luetke-Eversloh, Parente, et al. (2010) | GSE21774 | Affymetrix HG U133 Plus 2.0 Array (GPL570) | Blood | NK cells | 9 |
| Merryweather-Clarke, Atzberger, Soneji, Gray, et al. (2011) | GSE22552 | Affymetrix HG U133 Plus 2.0 Array (GPL570) | Blood | Erythroblasts | 15 |
| Abbas, Baldwin, Ma, Ouyang, et al. (2005) | GSE22886 | Affymetrix HG U133A Array/ Affymetrix HG U133B Array (GPL96/GPL97) | BM, blood | CD4 T, CD8 T, NK, naive B, memory B, plasma cells, MoDc, monocytes, neutrophils | 59 |
| Széles, Póliska, Nagy, Szatmari, et al. (2010) | GSE23618 | Affymetrix HG U133 Plus 2.0 Array (GPL570) | Blood, skin | DC (MoDC, DC-LC, cDC) | 12 |

| | | | | | |
|---|---|---|---|---|---|
| Novershtern, Subramanian, Lawton, Mak, et al. (2011) | GSE24759 | Affymetrix GeneChip HT-HG_U133A Early Access Array (GPL4685) | Umbilical cord blood | CD4 T (TCM, TEM), CD8 T(TCM, TEM), NKT, NK, B (early, proB, naive, mature, memory), DC (mDc, pDC), monocytes, neutrophils, eosinophils, basophils, erythroblasts | 155 |
| Wong, Tai, Wong, Han, et al. (2011) | GSE25913 | Illumina human-6 v2.0 expression beadchip (GPL6102) | Blood | $CD14^{+}CD16^{-}$ monocytes, $CD14^{+}CD16^{+}$ monocytes, $CD14^{dim}CD16^{+}$ monocytes | 24 |
| Garg, Szmania, Khan, Hoering, et al. (2012) | GSE27838 | Affymetrix HG U133 Plus 2.0 Array (GPL570) | Blood | NK cells | 8 |
| Allantaz, Cheng, Bergauer, Ravindran, et al. (2012) | GSE28490 | Affymetrix HG U133 Plus 2.0 Array (GPL570) | Blood | B, CD4 T, CD8 T, NK cells, DC (cDC, pDC), monocytes, neutrophils, eosinophils | 46 |
| Allantaz et al. (2012) | GSE28491 | Affymetrix HG U133 Plus 2.0 Array (GPL570) | Blood | B, CD4 T, CD8 T, NK cells, monocytes, neutrophils, eosinophils | 33 |
| Sirois, Robitaille, Allary, Shah, et al. (2011) | GSE30536 | Affymetrix HG U133 Plus 2.0 Array (GPL570) | | Macrophages (monocyte-derived macrophages) | 5 |
| Santarlasci, Maggi, Capone, Querci, et al. (2012) | GSE30664 | Agilent-014850 Whole HG Microarray 4x44K G4112F (GPL4133) | | Th1, Th17 | 8 |
| Wang, Shalek, Lawrence, Ding, et al. (2014) | GSE31048 | Affymetrix HG U133 Plus 2.0 Array (GPL570) | Blood | B | 24 |

*Continued*

**Table 1** List of the 54 public datasets constituting the final compendium.—cont'd

| Author | Dataset.ID | Platform | Tissue | Cell types | N |
|---|---|---|---|---|---|
| Zhang, Ferguson, Ng, Hui, et al. (2012) | GSE32901 | Affymetrix Human Gene 1.0 ST Array (GPL6244) | Blood | CD4 T cells (e.g., Th1, Th17) | 13 |
| Hutter, Kauer, Simonitsch-Klupp, Jug, et al. (2012) | GSE35340 | Affymetrix HG U133 Plus 2.0 Array (GPL570) | Blood, skin | DC (cDC, pDC, DC-LC) | 7 |
| Haniffa, Shin, Bigley, McGovern, et al. (2012) | GSE35457 | Illumina HumanHT-12 V4.0 expression beadchip (GPL10558) | Blood, skin | DC, monocytes | 49 |
| Le Gallou, Caron, Delaloy, Rossille, et al. (2012) | GSE36975 | Illumina HumanHT-12 V4.0 expression beadchip (GPL10558) | Blood | Naive B | 3 |
| Segura, Touzot, Bohineust, Cappuccio, et al. (2013) | GSE40484 | Affymetrix Human Gene 1.1 ST Array (GPL11532) | | DC, monocytes | 12 |
| Covens, Verbinnen, Geukens, Meyts, et al. (2013) | GSE42724 | Affymetrix Human Gene 1.0 ST Array (GPL6244) | Blood | Naive B, memory B, preplasmablasts, plasmablasts | 12 |
| Zhang, Nestor, Zhao, Lentini, et al. (2013) | GSE43005 | Agilent-028004 SurePrint G3 Human GE 8x60K Microarray (GPL14550) | Blood | CD4 T cells (e.g., Th1, Th17, Th2, Treg) | 24 |
| Yu, Becker, Wang, Marches, et al. (2013) | GSE43184 | Illumina HumanHT-12 V4.0 expression beadchip (GPL10558) | Blood, lung | cDC | 10 |
| Glatzer, Killig, Meisig, Ommert, et al. (2013) | GSE43409 | Affymetrix HG U133 Plus 2.0 Array (GPL570) | Tonsil | ILC3 | 2 |
| Streicher, Morehouse, Groves, Rajan, et al. (2014) | GSE45535 | Affymetrix HG U133 Plus 2.0 Array (GPL570) | Blood | B cells, plasma cells, T cells, monocytes, granulocytes | 10 |
| López-Corral, Corchete, Sarasquete, Mateos, et al. (2014) | GSE47552 | Affymetrix Human Gene 1.0 ST Array (GPL6244) | BM | Plasma cells | 5 |

| | | | | | |
|---|---|---|---|---|---|
| Mabbott, Baillie, Brown, Freeman, et al. (2013) | GSE49910 | Affymetrix HG U133 Plus 2.0 Array (GPL570) | | B cells, CD4 T, CD8 T cells, monocytes, neutrophils, erythroblasts | 45 |
| Bernier, Cleret-Buhot, Zhang, Goulet, et al. (2013) | GSE50175 | Affymetrix HG U133 Plus 2.0 Array (GPL570) | Blood | Th1 | 4 |
| Lee, Ye, Villani, Raj, et al. (2014) | GSE53166 | Affymetrix Human Gene 1.0 ST Array (GPL6244) | | MoDC | 38 |
| Touzot, Grandclaudon, Cappuccio, Satoh, et al. (2014) | GSE54627 | Affymetrix Human Gene 1.1 ST Array (GPL11532) | Blood | CD4 T cells (Th0, Th1, Th2, Th17) | 48 |
| Laudanski, Miller-Graziano, Xiao, Mindrinos, et al. (2006) | GSE5580 | Affymetrix HG U133A Array (GPL96) | Blood | T cells, monocytes | 14 |
| Balan, Ollion, Colletti, Chelbi, et al. (2014) | GSE57671 | Affymetrix Human Gene 1.0 ST Array (GPL6244) | Blood | DC (cDC, pDC) | 5 |
| Dürig, Bug, Klein-Hitpass, Boes, et al. (2007) | GSE5788 | Affymetrix HG U133A Array (GPL96) | Blood | T cells | 8 |
| Ono, Yu, Kasahara, Kikuchi, et al. (2014) | GSE59295 | Illumina human-6 v2.0 expression beadchip (GPL6102) | Blood | CD4 T cells (Th1, Th2) | 24 |
| Montaldo, Teixeira-Alves, Glatzer, Durek, et al. (2014) | GSE63197 | Agilent-039494 SurePrint G3 Human GE v2 8x60K Microarray 039381 (GPL17077) | Tonsil | ILC3, NK cells | 14 |
| Chng, Kumar, Vanwier, Ahmann, et al. (2007) | GSE6477 | Affymetrix HG U133A Array (GPL96) | BM | Plasma cells | 12 |
| Gutiérrez, Ocio, de Las Rivas, Maiso, et al. (2007) | GSE6691 | Affymetrix HG U133A Array (GPL96) | Blood | B cells | 7 |

*Continued*

**Table 1** List of the 54 public datasets constituting the final compendium.—cont'd

| Author | Dataset.ID | Platform | Tissue | Cell types | N |
|---|---|---|---|---|---|
| Hyrcza, Kovacs, Loutfy, Halpenny, et al. (2007) | GSE6740 | Affymetrix HG U133A Array (GPL96) | Blood | CD4 T, CD8 T cells | 10 |
| Mezger, Wozniok, Blockhaus, Kurzai, et al. (2008) | GSE6965 | Affymetrix HG U133 Plus 2.0 Array (GPL570) | Blood | iDC | 2 |
| Dybkaer, Iqbal, Zhou, Geng, et al. (2007) | GSE8059 | Affymetrix HG U133 Plus 2.0 Array (GPL570) | Blood | CD8 T, NK cells | 2 |
| Görgün, Holderried, Zahrieh, Neuberg, et al. (2005) | GSE8835 | Affymetrix HG U133A Array (GPL96) | Blood | CD4 T, CD8 T cells | 24 |
| Schirmer, Fledderus, and van der Laan (2009) | GSE9820 | Illumina humanRef-8 v2.0 expression beadchip (GPL6255) | | T cells, monocytes, macrophages, stem cells | 52 |

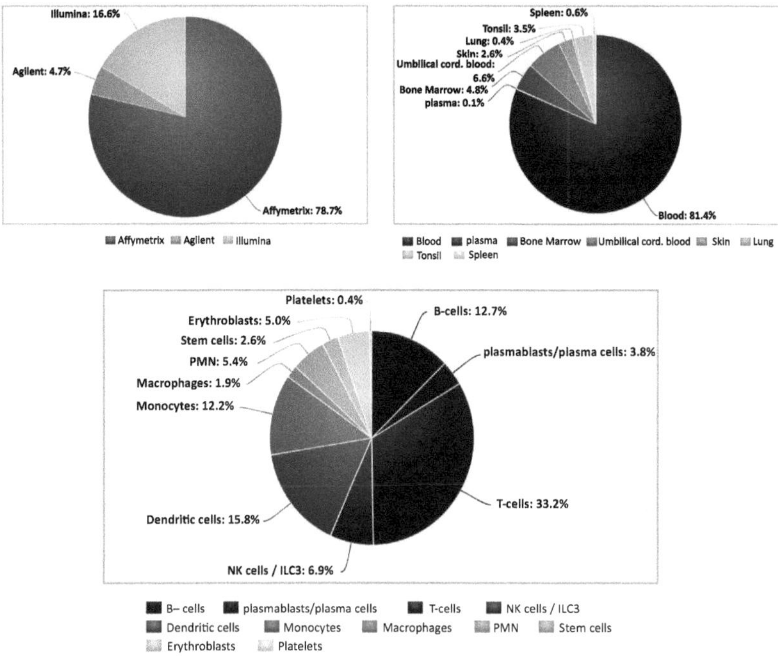


**Fig. 2** Pie chart of dataset collection, by manufacturer, tissue and cell type.

### *2.2.1 Affymetrix array processing*

All samples processed on Affymetrix HG-U133 plus 2.0 arrays were placed in a single Affybatch object with the affy package (Gauthier, Cope, Bolstad, & Irizarry, 2004). Quality controls were performed for array hybridization (NUSE plot (Bolstad, 2004) and Affymetrix quality metrics) with the affyPLM and yaqcaffy packages (Bolstad, 2004; Gatto, 2018). Raw expression data were normalized and summarized by Robust Multichip Analysis (RMA; Irizarry et al., 2003). Some samples/datasets previously identified as potentially relevant were removed at this step because the NUSE plot revealed that they were not comparable to the others.(Fig. 3).

Other 3′ IVT arrays (Affymetrix HG U133A Array, Affymetrix HG U133A 2.0 Array and Affymetrix HG U133A Early Access Array) were processed in the same way.

Samples from three datasets (E-MEXP-384, E-MEXP-750 and GSE22886) were processed on Affymetrix HG U133A and U133B arrays. These two arrays are complementary and have common probes and can together be

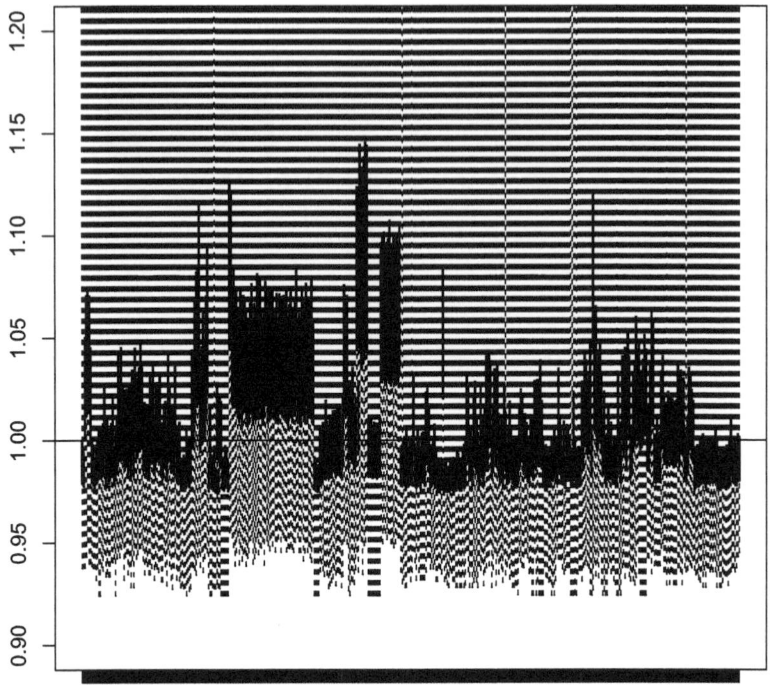


**Fig. 3** NUSE plot for samples processed on Affymetrix HG U133 Plus 2.0 arrays. The boxplots can be used to check two variables, to control for poor quality. All distributions must be centered on or close to 1. Typically an array with a boxplot centered on 1.1 is considered to be of poor quality. Moreover, the spreads of NUSE distribution must be comparable. If one array has a, NUSE distribution with a greater spread than others, this may also be a sign of poor quality.

considered to constitute an HG U133 Plus 2.0 array. The raw expression data for U133A samples were normalized and summarized with RMA. RMA was applied independently to the raw expression data for U133B samples. The set of probes common to U133A and U133B was used to estimate parameters from the prediction model (weighted linear regression). This weighted linear regression, thus defined, is used to predict the probe set signals of U133B. Then U133B probesets are concatenated to the U133A matrix.

Affymetrix Exon arrays (Affymetrix Human Gene 1.0 ST Array and Human Gene 1.1 ST Array) were processed in a similar manner, with NUSE plots for quality control and RMA normalization, but the oligo package was required for the processing of these arrays.

### 2.2.2 Agilent array processing

Agilent arrays (SurePrint G3 Human GE 8x60k (v1 and v2) and Whole HG 4x44k) were processed with the limma package (Ritchie et al., 2015): the signals were first background-corrected by the "normexp" method (Ritchie et al., 2007). They were then quantile-normalized (Bolstad, Irizarry, Astrand, & Speed, 2003) and within-array replicates were replaced by their means provided that the sequences of multiple probe names were strictly identical.

### 2.2.3 Ilumina array processing

Quantile-normalized gene expression signals from GSE9820 were downloaded directly from GEO. The other datasets were merged by array version (Illumina Human WG-6 v2, WG-8 v2 and Human HT-12 v4.0) and quantile-normalized with the lumiN function of the lumi package (Du, Kibbe, & Lin, 2008).

### 2.2.4 Common steps

Principal Component Analysis (PCA) was performed with the ade4 package (Dray & Dufour, 2007) on each normalized gene-expression matrix obtained as described above, to evaluate the effect of dataset. (Fig. 4) A dataset effect was indeed evident in the first components of PCA for each platform, but could be corrected only for the U133 set matrix. For the others, the dataset effect was not corrected because no cell population is in common. Having common cell types across different datasets is essential to estimate correctly the part of the signals induced by the dataset factor alone. If no cell population is in common between datasets, then cell types and datasets are confounded factors. In this case, correcting the datasets effects can remove actual biological differences between cell populations: the biological differences we are looking for.

The dataset effect for the U133 set was corrected with the Combat (Johnson, Li, & Rabinovic, 2007) function of sva package (Leek et al., 2019), using the model including cell type as a factor (Fig. 5).

## 2.3 Step 3: Combination of ExpressionSet objects

After quality control and normalization for each platform, the 13 expressionSet objects were combined by a method inspired by the virtualArray package (Heider & Alt, 2013).

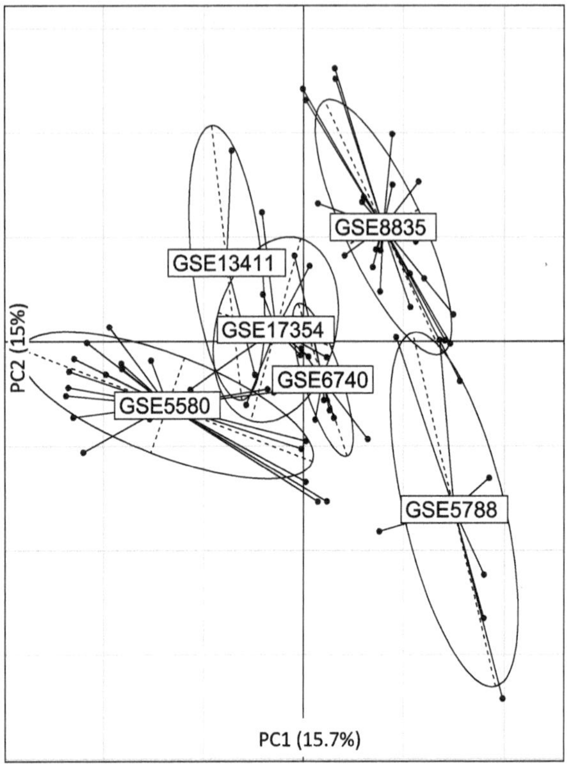


**Fig. 4** PCA on the normalized gene expression matrix for samples processed on the HG-U133A platform. Some datasets were well separated on the two first components. However, the dataset and cell type effect could be confounded, and the dataset effect was therefore not corrected.

### 2.3.1 Annotation

Expression matrices were annotated with manufacturer-specific IDs for which there was no direct match between platforms. For the purposes of combination, the official gene symbol was defined as the common identifier. The annotation slot of each single ExpressionSet was used for its annotation with this common identifier. Annotation slots were adjusted beforehand, with the Bioconductor-compliant annotation package where possible, and with annotations downloaded from GEO if not.

### 2.3.2 Collapse of redundant probe sets

Several probes or probe sets were designed to target the same gene in several different arrays. As a unique identifier was required to combine all the arrays into a single final dataset, these redundant rows had to be removed. The probe or probe set for which expression was the most variable was selected.

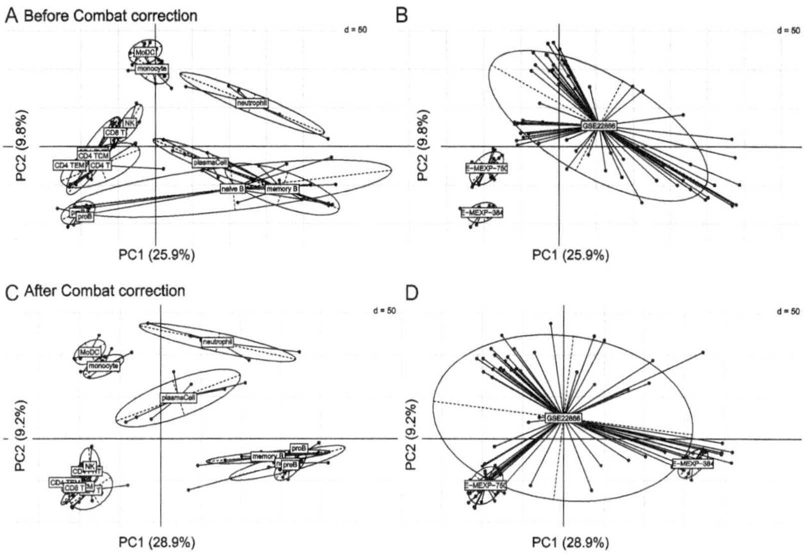


**Fig. 5** Before-After Combat correction. The 3 datasets from the U133 set platform are well separated on the two first components (top PCA plots). After combat correction, the dataset effect is no more visible and the cell types are now well separated on the 2 first components (bottom PCA plots).

### *2.3.3 Combination into a single virtual array*

This step involved the building of an expression matrix including only rows with unique identifiers available in all arrays. Rows with gene symbols present in only a subset of arrays were removed. This step greatly decreased the number of dimensions of the matrix, to 10,833 rows (i.e., genes).

### *2.3.4 Building a new ExpressionSet*

The expression matrix generated in this manner was integrated into an ExpressionSet, a fundamental data container in Bioconductor facilitating the coordination and manipulation of several kinds of connected data (e.g., expression matrix, features data, phenodata). The expression matrix was thus linked to a pData slot containing sample names, dataset names, array names, tissues and cell types.

### *2.3.5 Removal of batch effects*

The merged expression matrix was first quantile-normalized to adjust for the array effect (also known as the batch effect). PCA showed that this array effect persisted after quantile normalization. [Fig. 6 here].

The array effect was thus removed with the Combat function, using the model including cell type as a factor.

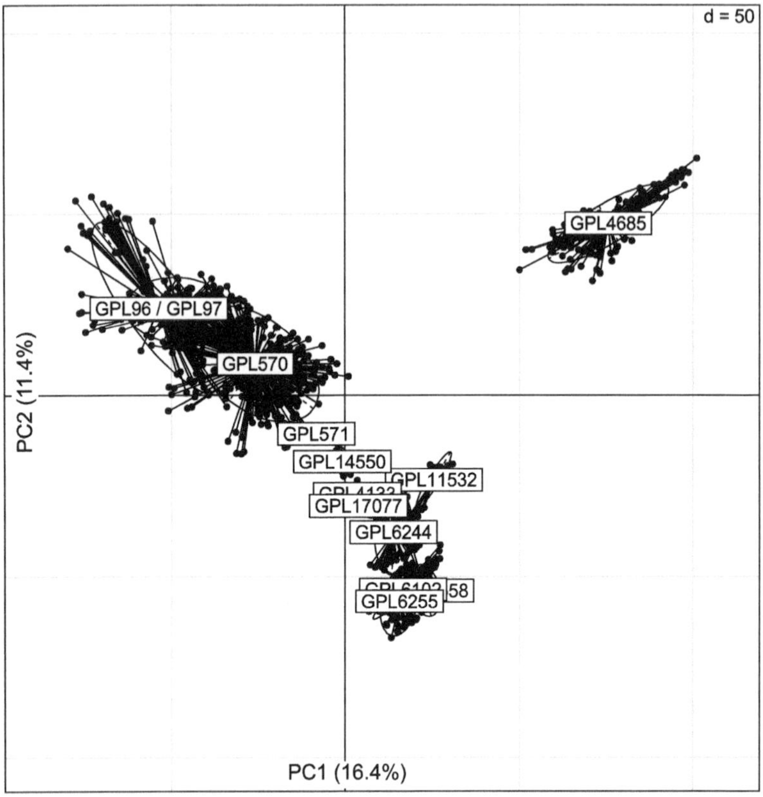


**Fig. 6** Platform effect highlighted by PCA. Different arrays are well separated on the first two components, with a trend toward clustering by manufacturer. This effect was estimated and corrected with Combat.

## 2.4 Step 4: Validation

Gene expression data were transformed during the generation of the compendium: several normalization and correction steps were applied. It was therefore essential to perform a final check to ensure that the expression profiles of known leukocyte marker were as expected. CD19, MS4A1 (also known as CD20), CD22 and PAX5 are known to be expressed by B cells. We used CD3E expression to validate data for T cells. The CD8B and CD4 markers were checked to validate the data for CD8 and CD4 T-cell subsets. In particular, we checked the expression data for FOXP3, a known marker of regulatory T cells, IFNG, a Th1-marker, and GATA3, a Th2-marker. NCAM1 (also called CD56) is a marker of NK cells. CD33 is a marker of myeloid cells, CD14 is expressed by the majority of monocytes and macrophages. CD1C (also called BDCA1), THBD (also called BDCA3), IL3RA (also called CD123) and CD207 (also called Langerin) were used to validate the expression profiles of dendritic cells. CSF3R (also known as CD114) was used to check the expression profiles of neutrophils, and CCR3 was used for eosinophils.

All expression profiles were consistent with the labels provided by dataset depositors (Fig. 7A–F).

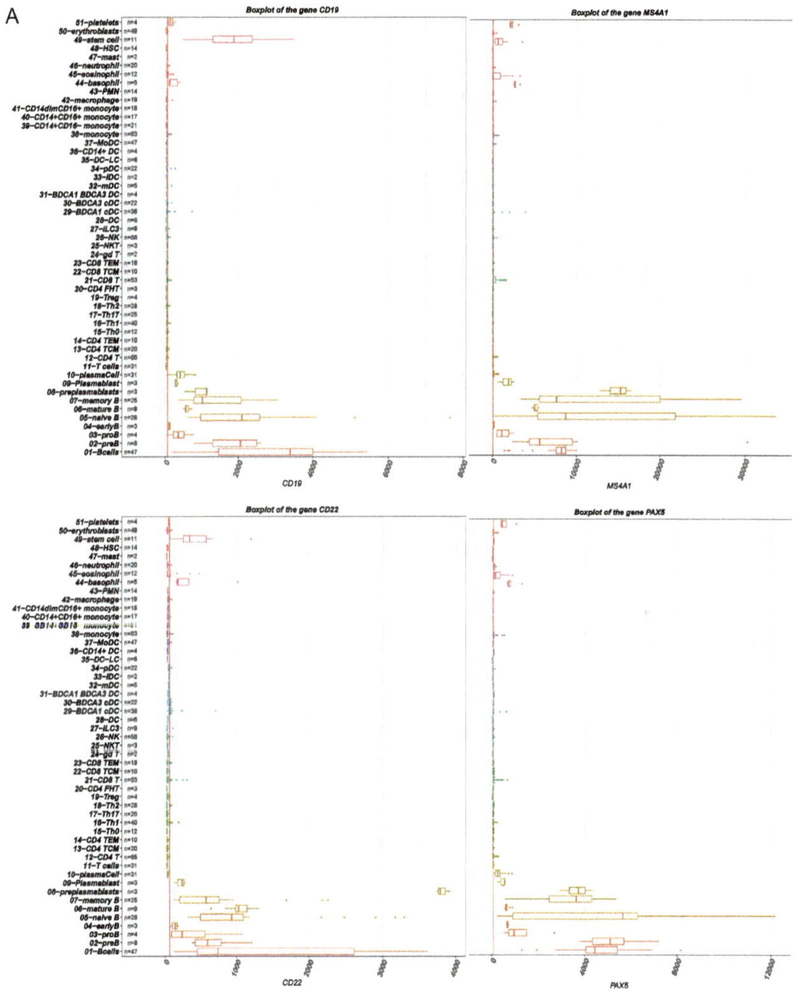


**Fig. 7** Validation by checking of cell type-specific marker expression profiles. (A) B-cell markers: expression profiles of CD19, MS4A1, CD22 and PAX5. (B) T-cell markers: expression profiles of CD3E, CD8B, CD4, FOXP3, IFNG and GATA3. (C) NK cell markers: expression profiles of NCAM1. (D) Monocyte and macrophage markers: expression profiles of CD33 and CD14. (E) Dendritic cell markers: expression profiles of CD1C, THBD, IL3RA and CD207. (F) Neutrophil and eosinophil markers: expression profiles of CSF3R and CCR3.

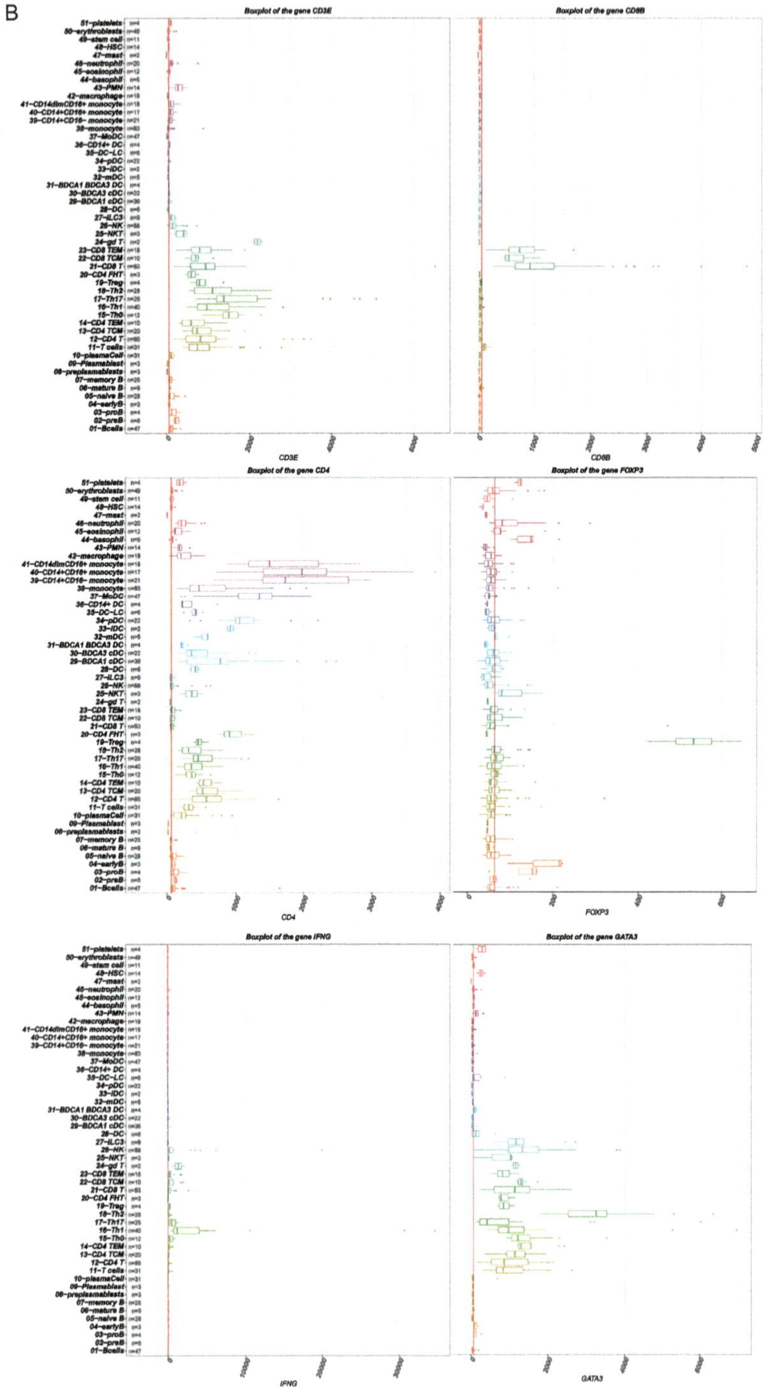
B
Boxplot of the gene CD3E
Boxplot of the gene CD8B
Boxplot of the gene CD4
Boxplot of the gene FOXP3
Boxplot of the gene IFNG
Boxplot of the gene GATA3
51-platelets
50-erythroblasts
49-stem cell
48-HSC
47-mast
46-neutrophil
45-eosinophil
44-basophil
43-PMN
42-macrophage
41-CD14dimCD16+ monocyte
40-CD14+CD16+ monocyte
39-CD14+CD16- monocyte
38-monocyte
37-MoDC
36-CD14+ DC
35-DC-LC
34-pDC
33-IDC
32-mDC
31-BDCA1 BDCA3 DC
30-BDCA3 cDC
29-BDCA1 cDC
28-DC
27-ILC3
26-NK
25-NKT
24-gd T
23-CD8 TEM
22-CD8 TCM
21-CD8 T
20-CD4 FHT
19-Treg
18-Th2
17-Th17
16-Th1
15-Th0
14-CD4 TEM
13-CD4 TCM
12-CD4 T
11-T cells
10-plasmaCell
09-Plasmablast
08-preplasmablasts
07-memory B
06-mature B
05-naive B
04-earlyB
03-proB
02-preB
01-Bcells
CD3E
CD8B
CD4
FOXP3
IFNG
GATA3

**Fig. 7—Cont'd**

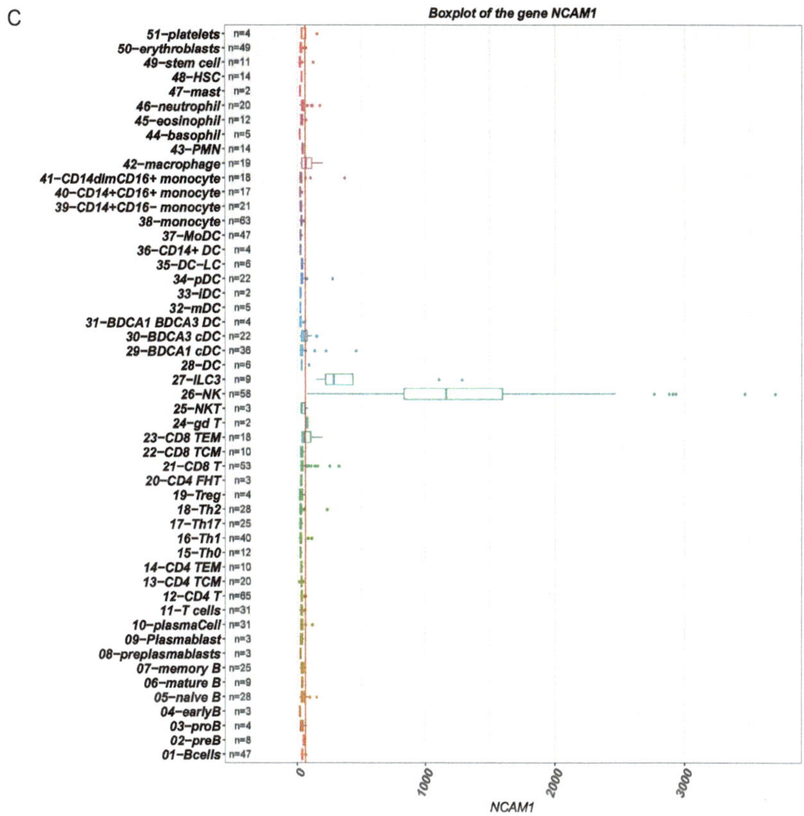


**Fig. 7—Cont'd**

The immune cell compendium was validated on the basis of the expression profiles of individual markers. Hierarchical clustering was performed on the full matrix, with distance calculated by Pearson's correlation analysis and Ward's method as linkage, to cluster and identify cell types on the basis of their whole transcriptomes (Fig. 8). The first cut-off in the dendrogram defined the lymphoid branch on one side and the myeloid branch on the other. Erythroblasts and stem cells clustered with lymphoid cells. In the lymphoid cell branch, B cells clustered together in two groups: plasma cells and other types of B cells. Innate lymphoid cells separated into two clusters, NK cells and ILC3, and were closer to steady-state T cells than to helper T cells. The immune cell compendium displayed the expected pattern of clustering and was therefore considered fully validated. It was then used to address several questions.

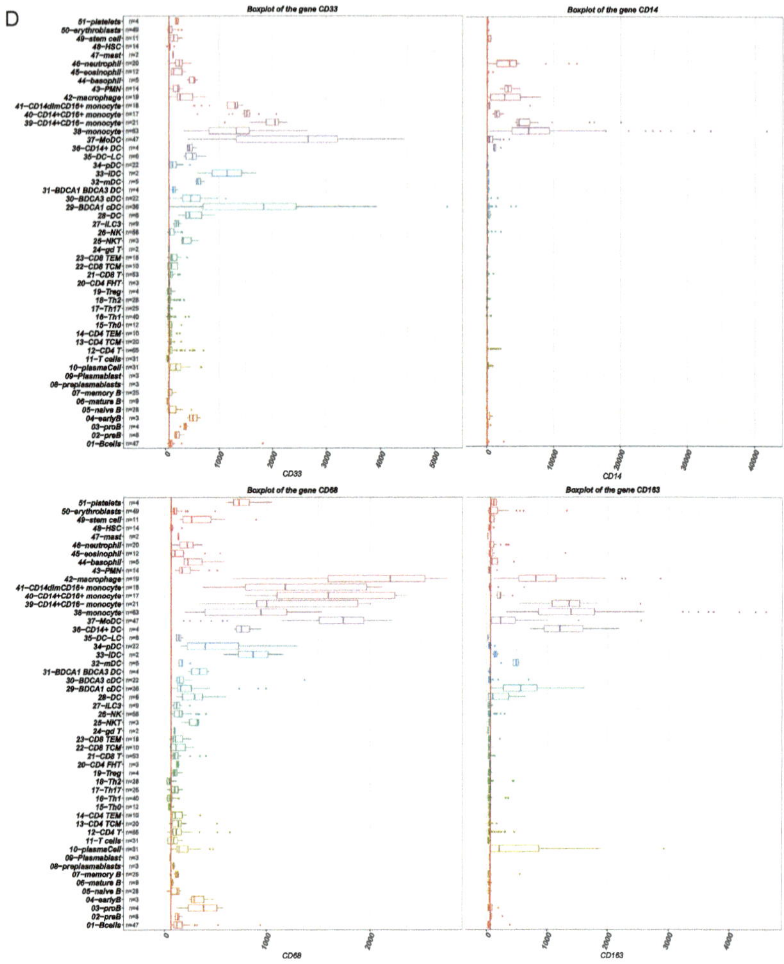


**Fig. 7—Cont'd**

## 3. Discussion

### 3.1 Selective depletion project

In inflammatory diseases, allergic disorders and cancers, the identification of markers expressed exclusively by the cells involved in the disease might make the selective depletion or the modulation of these cells possible, in the goal of inducing clinical benefits in patients. For example, eosinophils are known to play a critical role in asthma. The development of new

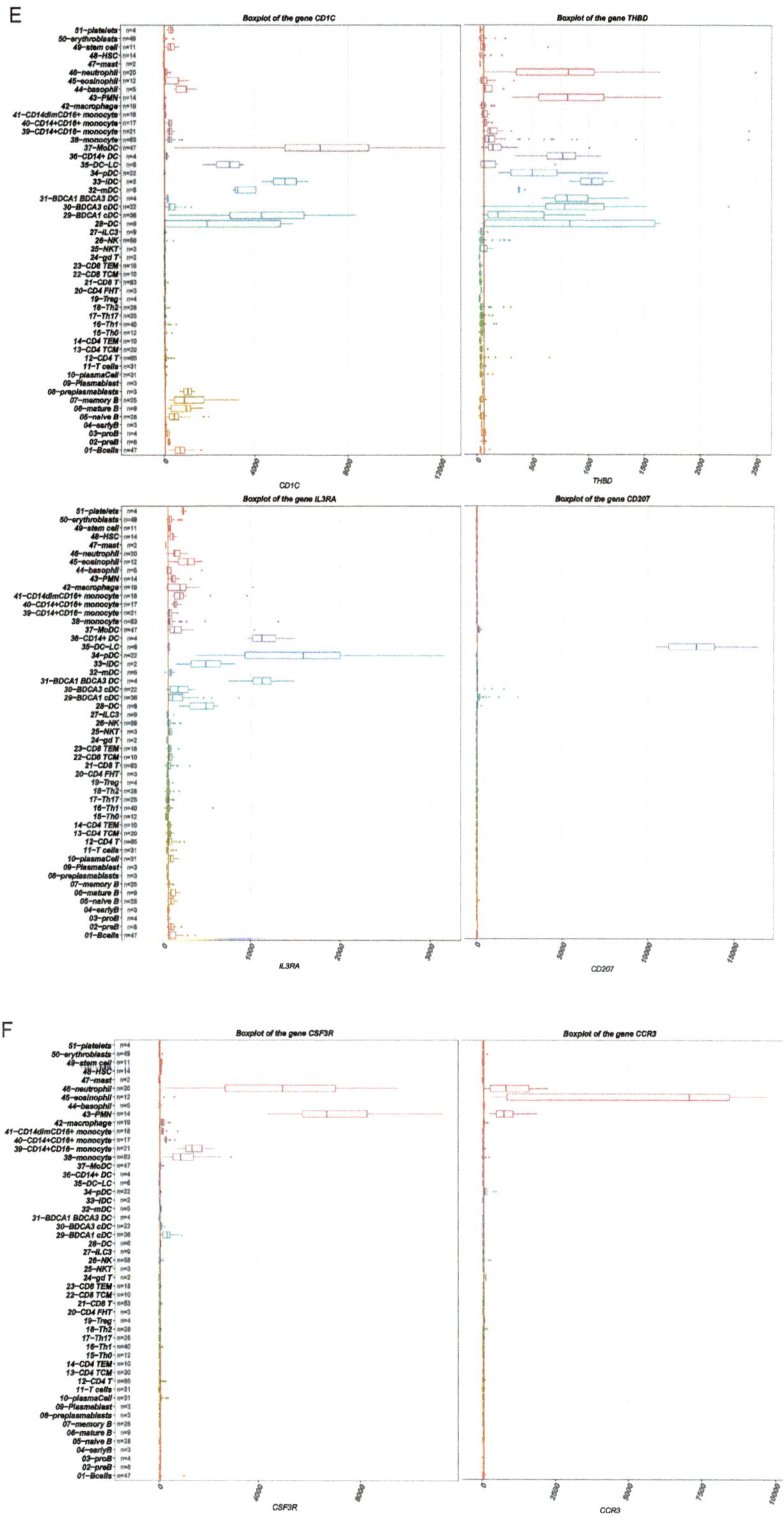
E
Boxplot of the gene CD1C
CD1C
Boxplot of the gene THBD
THBD
Boxplot of the gene IL3RA
IL3RA
Boxplot of the gene CD207
CD207
F
Boxplot of the gene CSF3R
CSF3R
Boxplot of the gene CCR3
CCR3

Fig. 7—Cont'd

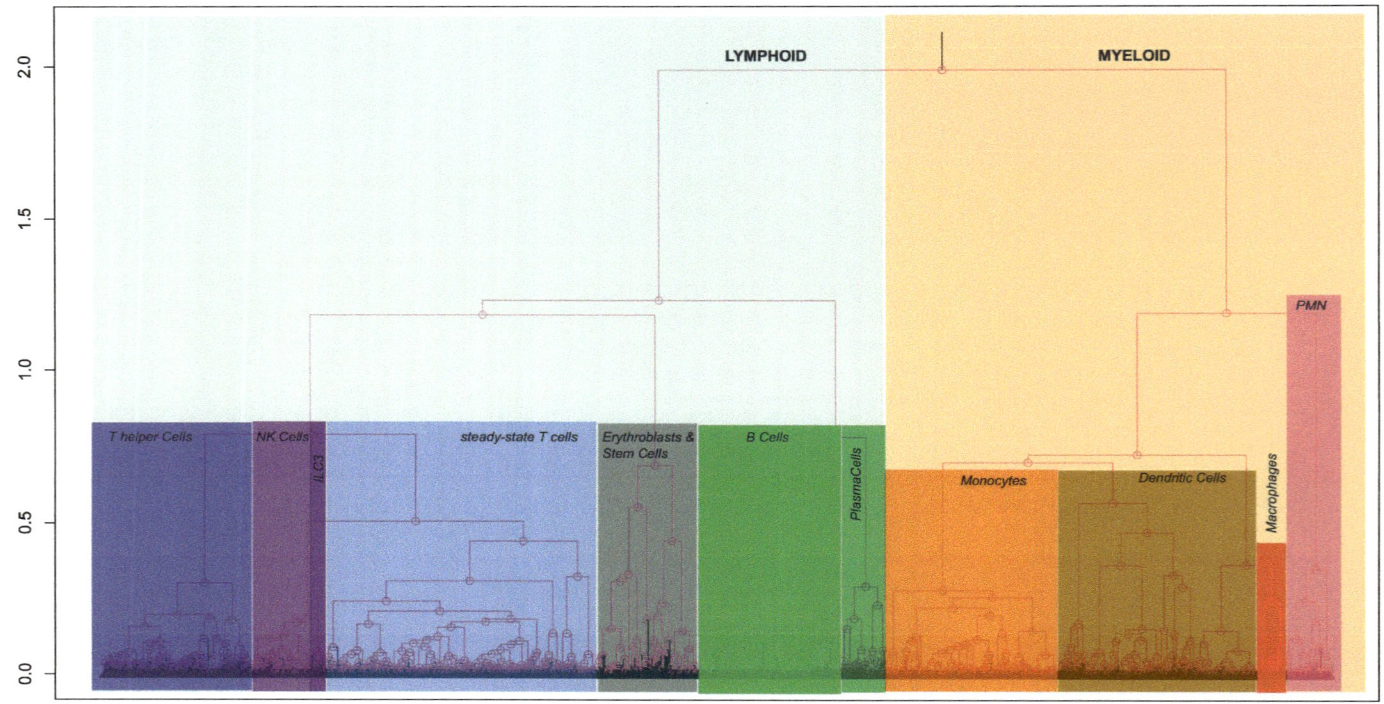


**Fig. 8** Hierarchical clustering (Pearson/Ward) of the compendium.

anti-eosinophilic drug that aims to selectively deplete eosinophils could offer the possibility to explore the role of this immune cells population in this disease condition and the consequences of their eradication. Deeper knowledge of the expression pattern of molecules can thus lead to the definition of different phenotypes and guide personalized medicine.

The immune cell compendium can also be used to discover new biomarkers expressed exclusively by a specific subtype. To ensure the specificity, the "min-max" method, can be applied to the full matrix. It involves selecting genes for which minimal expression in the immune cell type of interest is greater than maximal expression in all other cell types. The method does not allow any overlap between the profiles of the two populations to be compared and leads to a robust signature.

The algorithm has already highlighted obvious and already known markers, which we considered to be a good positive control of the approach.

### 3.2 Target evaluation

The research pipelines of diagnostic and pharmaceutical companies require both constant feeding and bioinformatics tools for evaluating new potential targets. The immune cell compendium can be uploaded onto an in-house graphical user interface (Fig. 9A and B). This interface displays gene expression profiles as box plots by cell type for a selected tissue (blood, skin, tonsil, spleen, lung, bone marrow). This makes it easy to determine whether the potential target is also expressed by other immune cells, which could be deleterious in some cases.

## 4. Future version

The immune cell compendium was based on datasets obtained with microarray technology. At the time, the choice of this technology was constrained by the availability of data. Many more datasets of sorted immune cells were available in microarray than in RNASeq. In addition to the technical issues inherent to microarrays probes such as cross-hybridization, non-specific hybridization and limited detection range, the signals obtained depended on the manufacturer's probe design. The arrays did not overlap fully in terms of gene availability, and it was not therefore possible to obtain the whole transcriptome. For these reasons, the next version of the immune cell compendium will make use of next-generation sequencing data. RNASeq avoids bias due to experimental design described above, is independent of probes design and demonstrated broader dynamic range than microarray.

A

B

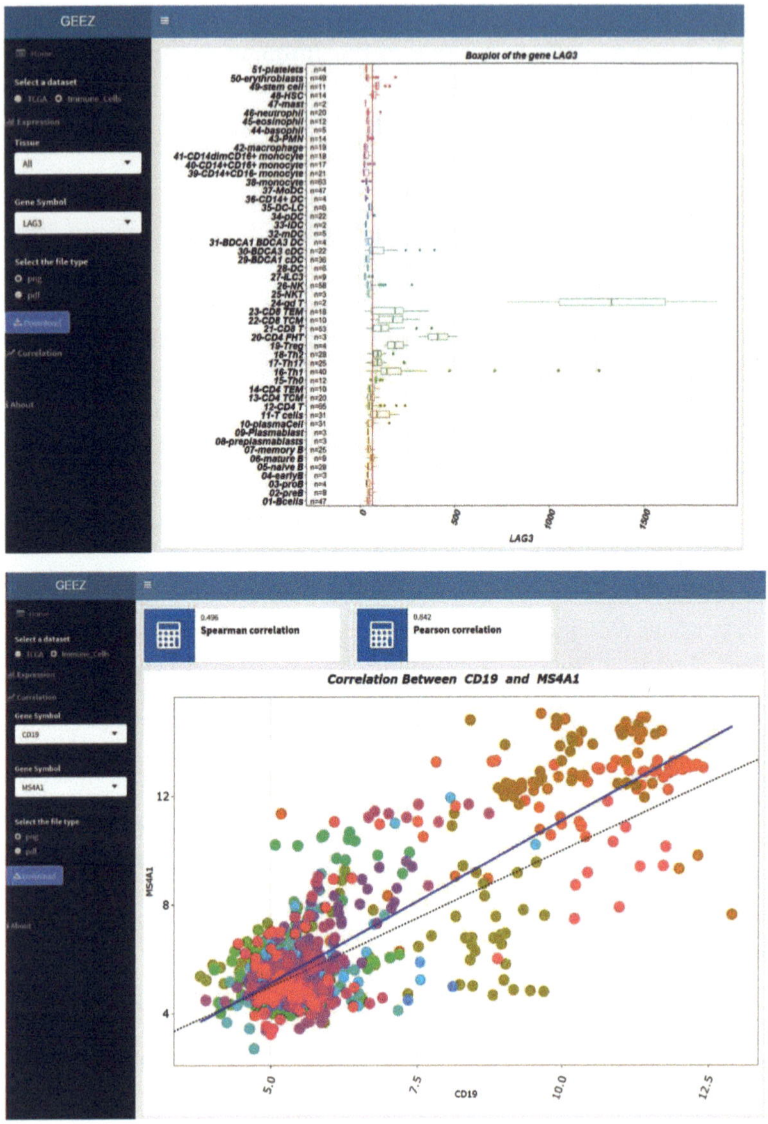


**Fig. 9** Graphical User Interface (GUI). (A) Expression profiles of markers specific to immune cell types. (B) Correlation between markers.

However, RNAseq as microarrays were processed for a bulk of sorted cells. The heterogeneity of expression between cells cannot be addressed with these two technologies because the gene expression is the average of all cells expression.

Single-cell RNASeq makes it possible to characterize single cells, determining their entire transcriptomes. The development of an immune cell compendium based on single-cell RNAseq will make it possible to meet the challenges described above with greater precision.

## Appendix A. Annexe: R scources

Main script for microarrays processing (Correction, normalization, quality controls)

## load the required packages

```
# Processing of datasets by array type
require(affy)
require(affyPLM)
require(yaqcaffy)
require(oligo)
require(limma)
require(lumi)
# Required packages for batch correction, PCA and clustering
require(sva)
require(ade4)
require(cluster)
require(Hmisc)
# List of all potential samples available to build the compendium
## some of them will not finally be included due to quality problems
gsm.Hs=read.csv2("ListeGSM-HsHealthy.csv")
```

## GPL96 Affymetrix

```
filesAff96=list.files("rawdata/Affymetrix/U133A/")
tab.Aff96=subset(gsm.Hs,Platform=="GPL96")
tab.Aff96=tab.Aff96[order(tab.Aff96$Sample.ID),]
rownames(tab.Aff96)=as.character(tab.Aff96$Sample.ID)
x.Aff96=ReadAffy(filenames=filesAff96,celfile.path="rawdata/
Affymetrix/U133A/")
eSet.Aff96=affy::rma(x.Aff96)
plotDensity(exprs(eSet.Aff96))
# some profiles look very different from the others
pset=fitPLM(x.Aff96)
NUSE(pset)
nuseStats.96=NUSE(pset,type="stats")
```

```
plotDensity(exprs(eSet.Aff96)[which(nuseStats.96["median",]
<1.05),])
# 5 samples are not OK (NUSE median > 1.05)
ya.Aff96=yaqc(x.Aff96)
affyDeg <- AffyRNAdeg(x.Aff96)
sumDeg <- summaryAffyRNAdeg(affyDeg)
tsumDeg <- t(sumDeg)
tsumDeg <- t(sumDeg)[,which(colnames(tsumDeg)=="slope")]
myDF <- cbind(nuseStats.96["median",],ya.Aff96@average.noise,ya.
Aff96@average.background,ya.Aff96@scale.factors,ya.
Aff96@percent.present,tsumDeg, t(getQCRatios(ya.Aff96)))
colnames(myDF) <- c("NUSEmedian","aveNoise","aveBG",
"scaleFactor","percPres","degSlope","bactin","GAPDH")
# the same 5 samples are not qualified : percent of present too low.
filesAff96=filesAff96[which(nuseStats.96["median",]<1.05)]
tab.Aff96=tab.Aff96[which(nuseStats.96["median",]<1.05),]
x.Aff96=ReadAffy(filenames=filesAff96,celfile.path="rawdata/
Affymetrix/U133A/")
eSet.Aff96=affy::rma(x.Aff96)
plotDensity(exprs(eSet.Aff96))
phenoData(eSet.Aff96)=new("AnnotatedDataFrame",data=tab.Aff96)
save(eSet.Aff96,file="esetAff96.RData")
pca.Aff96=dudi.pca(t(exprs(eSet.Aff96)),scannf=FALSE,nf=4)
s.class(pca.Aff96$li[,1:2],pData(eSet.Aff96)[,"cell.type"])
s.class(pca.Aff96$li[,1:2],pData(eSet.Aff96)[,"Dataset.ID"])
## GPL4685 Affymetrix
filesAff4685=list.files("rawdata/Affymetrix/U133AAofAv2/")
tab.Aff4685=subset(gsm.hs,Platform=="GPL4685")
tab.Aff4685=tab.Aff4685[order(tab.Aff4685$Sample.ID),]
rownames(tab.Aff4685)=as.character(tab.Aff4685$Sample.ID)
x.Aff4685=ReadAffy(filenames=filesAff4685,celfile.path="rawdata/
Affymetrix/U133AAofAv2/")
eSet.Aff4685=affy::rma(x.Aff4685)
plotDensity(exprs(eSet.Aff4685))
# some profiles are very different
pset=fitPLM(x.Aff4685)
NUSE(pset)
nuseStats.4685=NUSE(pset,type="stats")
```

```
plotDensity(exprs(eSet.Aff4685)[,which(nuseStats.4685["median",]
<1.05)])
# 14 not OK
ya.Aff4685=yaqc(x.Aff4685)
#Error : actin5: No probe found.
affyDeg <- AffyRNAdeg(x.Aff4685)
sumDeg <- summaryAffyRNAdeg(affyDeg)
tsumDeg <- t(sumDeg)
tsumDeg <- t(sumDeg)[,which(colnames(tsumDeg)=="slope")]
myDF.Aff4685 <- cbind(nuseStats.4685["median",],tsumDeg)
colnames(myDF.Aff4685) <- c("NUSEmedian","degSlope")
tab.Aff4685[which(nuseStats.4685["median",]>=1.05),]
filesAff4685=filesAff4685[which(nuseStats.4685["median",]<1.05)]
tab.Aff4685=tab.Aff4685[which(nuseStats.4685["median",]<1.05),]
x.Aff4685=ReadAffy(filenames=filesAff4685,celfile.path="rawdata/
Affymetrix/U133AAofAv2/")
eSet.Aff4685=affy::rma(x.Aff4685)
plotDensity(exprs(eSet.Aff4685))
phenoData(eSet.Aff4685)=new("AnnotatedDataFrame",data=tab.
Aff4685)
annotation(eSet.Aff4685)="hgu133a"
save(eSet.Aff4685,file="esetAff4685.RData")
## GPL570 Affymetrix
filesAff570=list.files("rawdata/Affymetrix/U133Plus2/")
tab.Aff570=subset(gsm.Hs,Platform=="GPL570")
tab.Aff570=tab.Aff570[order(tab.Aff570$Sample.ID),]
rownames(tab.Aff570)=as.character(tab.Aff570$Sample.ID)
x.Aff570=ReadAffy(filenames=filesAff570,celfile.path="rawdata/
Affymetrix/U133Plus2/")
eSet.Aff570=affy::rma(x.Aff570)
plotDensity(exprs(eSet.Aff570))
pset.570=fitPLM(x.Aff570)
NUSE(pset.570)
nuseStats.570=NUSE(pset.570,type="stats")
ya.Aff570=yaqc(x.Aff570)
affyDeg <- AffyRNAdeg(x.Aff570)
sumDeg <- summaryAffyRNAdeg(affyDeg)
tsumDeg <- t(sumDeg)
```

```
tsumDeg <- t(sumDeg)[,which(colnames(tsumDeg)=="slope")]
myDF.570   <-   cbind(nuseStats.570["median",],ya.Aff570@average.
noise,ya.Aff570@average.background,ya.Aff570@scale.factors,ya.
Aff570@percent.present,tsumDeg, t(getQCRatios(ya.Aff570)))
colnames(myDF) <- c("NUSEmedian","aveNoise",
"aveBG","scaleFactor","percPres","degSlope","bactin","GAPDH")
# comment: eSet.Aff570 was built after several updates. You can run
the command lines on the listed samples which were qualified.
phenoData(eSet.Aff570)=new("AnnotatedDataFrame",data=tab.Aff570)
save(eSet.Aff570,file="esetAff570v2.RData")
```

## GPL571 Affymetrix

```
filesAff571=list.files("rawdata/Affymetrix/U133A_2/")
tab.Aff571=subset(gsm.hs,Platform=="GPL571")
tab.Aff571=tab.Aff571[order(tab.Aff571$Sample.ID),]
rownames(tab.Aff571)=as.character(tab.Aff571$Sample.ID)
x.Aff571=ReadAffy(filenames=filesAff571,celfile.path="rawdata/
Affymetrix/U133A_2/")
eSet.Aff571=affy::rma(x.Aff571)
plotDensity(exprs(eSet.Aff571))
ya.Aff571=yaqc(x.Aff571)
#Erreur : actin5: No probe found.
affyDeg <- AffyRNAdeg(x.Aff571)
sumDeg <- summaryAffyRNAdeg(affyDeg)
tsumDeg <- t(sumDeg)
tsumDeg <- t(sumDeg)[,which(colnames(tsumDeg)=="slope")]
myDF   <-   cbind(ya.Aff571@average.noise,ya.Aff571@average.back-
ground,ya.Aff571@scale.factors,ya.Aff571@percent.present,
tsumDeg, t(getQCRatios(ya.Aff571)))
colnames(myDF) <- c("aveNoise","aveBG","scaleFactor","percPres",
"degSlope","bactin","GAPDH")
phenoData(eSet.Aff571)=new("AnnotatedDataFrame",data=tab.Aff571)
save(eSet.Aff571,file="esetAff571.RData")
```

## GPL6244 Affymetrix Exon

```
filesAff6244=list.celfiles("rawdata/Affymetrix/HuGene-1_0-
st/",listGzipped=TRUE,full.names=TRUE)
tab.Aff6244=subset(gsm.Hs,Platform=="GPL6244")
tab.Aff6244=tab.Aff6244[order(tab.Aff6244$Sample.ID),]
rownames(tab.Aff6244)=as.character(tab.Aff6244$Sample.ID)
x.Aff6244=oligo::read.celfiles(filenames=filesAff6244)
```

```
eSet.Aff6244=oligo::rma(x.Aff6244)
save(eSet.Aff6244,file="esetAff6244.RData")
plotDensity(exprs(eSet.Aff6244))
pset.6244=fitProbeLevelModel(x.Aff6244)
NUSE(pset.6244)
phenoData(eSet.Aff6244)=new("AnnotatedDataFrame",data=tab.Aff6244)
annotation(eSet.Aff6244)="hugene10sttranscriptcluster"
save(eSet.Aff6244,file="esetAff6244.RData")
pca.Aff6244=dudi.pca(t(exprs(eSet.Aff6244)),scannf=FALSE,nf=4)
s.class(pca.Aff6244$li[,1:2],pData(eSet.Aff6244)[,"cell.type"])
s.class(pca.Aff6244$li[,1:2],pData(eSet.Aff6244)[,"Dataset.ID"])
## GPL11532 Affymetrix Exon
filesAff11532=list.celfiles("rawdata/Affymetrix/HuGene-1_1-
st/",listGzipped=TRUE,full.names=TRUE)
tab.Aff11532=subset(gsm.Hs,Platform=="GPL11532")
tab.Aff11532=tab.Aff11532[order(tab.Aff11532$Sample.ID),]
rownames(tab.Aff11532)=as.character(tab.Aff11532$Sample.ID)
x.Aff11532=oligo::read.celfiles(filenames=filesAff11532)
eSet.Aff11532=oligo::rma(x.Aff11532)
save(eSet.Aff11532,file="esetAff11532.RData")
plotDensity(exprs(eSet.Aff11532))
pset.11532=fitProbeLevelModel(x.Aff11532)
NUSE(pset.11532)
phenoData(eSet.Aff11532)=new("AnnotatedDataFrame",data=tab.Aff11532)
annotation(eSet.Aff11532)="hugene11sttranscriptcluster"
save(eSet.Aff11532,file="esetAff11532.RData")
pca.Aff11532=dudi.pca(t(exprs(eSet.Aff11532)),scannf=FALSE,nf=4)
s.class(pca.Aff11532$li[,1:2],pData(eSet.Aff11532)[,"cell.type"])
s.class(pca.Aff11532$li[,1:2],pData(eSet.Aff11532)[,"Dataset.ID"])
## Merge GPL96 / GPL97
source("mergeABintoU133set.R")
pca.U133set=dudi.pca(t(exprs(normDataAcmbB)),scannf=FALSE,nf=4)
par(mfrow=c(1,2))
s.class(pca.U133set$li[,1:2],pData(normDataAcmbB)[,"cell.type"])
s.class(pca.U133set$li[,1:2],pData(normDataAcmbB)[,"Dataset.ID"])
tmp=ComBat(dat=exprs(normDataAcmbB),batch=factor(pData
(normDataAcmbB)[,"Dataset.ID"]),mod=model.matrix(~factor(pData
(normDataAcmbB)[,"cell.type"]) ,data=pData(normDataAcmbB)))
pca.tmp=dudi.pca(t(tmp),scannf=FALSE,nf=4)
```

```
par(mfrow=c(1,2))
s.class(pca.tmp$li[,1:2],pData(normDataAcmbB)[,"cell.type"])
s.class(pca.tmp$li[,1:2],pData(normDataAcmbB)[,"Dataset.ID"])
exprs(normDataAcmbB)=tmp
# Annotation
annotation(normDataAcmbB)="hgu133plus2" #U133set is included in
Hgu133plus2
save(normDataAcmbB,file="esetU133set.RData")
```

## ## GPL4133 Agilent-014850 Whole HG

```
filesAg4133=list.files("rawdata/Agilent/G4112F/")
x.Ag4133 <- read.maimages(filesAg4133, source="agilent",path=
"rawdata/Agilent/G4112F/", green.only=TRUE)
y.Ag4133 <- limma::backgroundCorrect(x.Ag4133,method="normexp")
y.Ag4133 <- limma::normalizeBetweenArrays(y.Ag4133,method=
"quantile")
exp.Ag4133= avereps(y.Ag4133, ID=y.Ag4133$genes$ProbeName) # here
we can average because the sequences of the multiple ProbeNames are
identical!
exp.Ag4133=exp.Ag4133$E[-which(exp.Ag4133$genes$ControlType!=0),]
tab.Ag4133=subset(gsm.hs,Platform=="GPL4133")
tab.Ag4133=tab.Ag4133[order(tab.Ag4133$Sample.ID),]
rownames(tab.Ag4133)=as.character(tab.Ag4133$Sample.ID)
pd.Ag4133=new("AnnotatedDataFrame",data=as.data.frame(tab.
Ag4133))
colnames(exp.Ag4133)=sampleNames(pd.Ag4133)
eSet.Ag4133 <- new("ExpressionSet", exprs = exp.Ag4133, phenoData
= pd.Ag4133)
save(eSet.Ag4133,file="eSetAg4133.RData")
plotDensity(exprs(eSet.Ag4133),addLegend=FALSE,logMode=FALSE)
pca.Ag4133=dudi.pca(t(exprs(eSet.Ag4133)),scannf=FALSE,nf=4)
s.class(pca.Ag4133$li[,1:2],pData(eSet.Ag4133)[,"cell.type"])
s.class(pca.Ag4133$li[,1:2],pData(eSet.Ag4133)[,"Dataset.ID"])
s.label(pca.Ag4133$li[,1:2],label=pData(eSet.Ag4133)[,"Sample.
ID"])
```

# # Annotation & probesets selection

```
annot.gpl4133=read.csv("../rawdata/Agilent/GPL4133-20424.csv")
z.Ag4133=eSet.Ag4133
exprs_z <- as.data.frame(exprs(z.Ag4133))
```

```
temp=c()
for(g in unique(annot.gpl4133$GENE_SYMBOL)){
  identifiers=annot.gpl4133[which(annot.gpl4133$
    GENE_SYMBOL==g),"NAME"]
  exp.g=exprs_z[as.character(identifiers),]
  temp=rbind(temp, exp.g[which.max(apply(exp.g,1,sd)),])
}
exprs_z=temp
rownames(exprs_z)=unique(annot.gpl4133$GENE_SYMBOL)
exprs(z.Ag4133) <- data.matrix(exprs_z, rownames.force = TRUE)
annotation(z.Ag4133)<- "Ag4133_org"
exprs(z.Ag4133) <- (exprs(z.Ag4133)/20*16) # transform in 16bit-
resolution
save(z.Ag4133,file="eset.Ag4133_annotORG.RData")
## GPL14550 Agilent-028004 SurePrint G3
filesAg14550=list.files("rawdata/Agilent/G4851A/")
x.Ag14550 <- read.maimages(filesAg14550, source="agilent",pat-
h="rawdata/Agilent/G4851A/", green.only=TRUE)
y.Ag14550 <- limma::backgroundCorrect(x.Ag14550,method="normexp")
y.Ag14550 <- limma::normalizeBetweenArrays(y.Ag14550,method=
"quantile")
exp.Ag14550= avereps(y.Ag14550, ID=y.Ag14550$genes$ProbeName) #
here we can average because the sequences of the multiple ProbeNames
are identical!
exp.Ag14550=exp.Ag14550$E[-which(exp.Ag14550$genes$ControlType!
=0),]
tab.Ag14550=subset(gsm.hs,Platform=="GPL14550")
tab.Ag14550=tab.Ag14550[order(tab.Ag14550$Sample.ID),]
rownames(tab.Ag14550)=as.character(tab.Ag14550$Sample.ID)
pd.Ag14550=new("AnnotatedDataFrame",data=as.data.frame(tab.
Ag14550))
colnames(exp.Ag14550)=sampleNames(pd.Ag14550)
eSet.Ag14550 <- new("ExpressionSet", exprs = exp.Ag14550,
phenoData = pd.Ag14550)
save(eSet.Ag14550,file="eSetAg14550.RData")
plotDensity(exprs(eSet.Ag14550),logMode=FALSE,cex=0.5)
pca.Ag14550=dudi.pca(t(exprs(eSet.Ag14550)),scannf=FALSE,nf=4)
s.class(pca.Ag14550$li[,1:2],pData(eSet.Ag14550)[,"cell.type"])
```

```
s.class(pca.Ag14550$li[,1:2],pData(eSet.Ag14550)[,"Dataset.ID"])
s.label(pca.Ag14550$li[,1:2],label=pData(eSet.Ag14550)
[,"comment"])
```

# Annotation

```
annot.gpl14550=read.delim("rawdata/Agilent/GPL14550-9757.txt",
skip=17,header=TRUE)
z.Ag14550=eSet.Ag14550
exprs_z <- as.data.frame(exprs(z.Ag14550))
temp=c()
for(g in unique(annot.gpl14550$GENE_SYMBOL)){
  identifiers=annot.gpl14550[which(annot.gpl14550
    $GENE_SYMBOL==g),"ID"]
  exp.g=exprs_z[as.character(identifiers),]
  temp=rbind(temp, exp.g[which.max(apply(exp.g,1,sd)),])
}
exprs_z=temp
rownames(exprs_z)=unique(annot.gpl14550$GENE_SYMBOL)
exprs(z.Ag14550) <- data.matrix(exprs_z, rownames.force = TRUE)
annotation(z.Ag14550)<- "Ag14550_org"
save(z.Ag14550,file="eset.Ag14550_annotORG.RData")
```

## GPL17077 Agilent-039494 SurePrint G3

```
filesAg17077=list.files("rawdata/Agilent/G4851B/")
x.Ag17077 <- read.maimages(filesAg17077, source="agilent",pat-
h="rawdata/Agilent/G4851B/", green.only=TRUE)
y.Ag17077 <- limma::backgroundCorrect(x.Ag17077,method="normexp")
y.Ag17077 <- limma::normalizeBetweenArrays(y.Ag17077,method=
"quantile")
exp.Ag17077= avereps(y.Ag17077, ID=y.Ag17077$genes$ProbeName) #
here we can average because the sequences of the multiple ProbeNames
are identical!
exp.Ag17077=exp.Ag17077$E[-which(exp.Ag17077$genes
$ControlType!=0),]
tab.Ag17077=subset(gsm.hs,Platform=="GPL17077")
tab.Ag17077=tab.Ag17077[order(tab.Ag17077$Sample.ID),]
rownames(tab.Ag17077)=as.character(tab.Ag17077$Sample.ID)
pd.Ag17077=new("AnnotatedDataFrame",data=as.data.frame(tab.
Ag17077))
colnames(exp.Ag17077)=sampleNames(pd.Ag17077)
```

```
eSet.Ag17077 <- new("ExpressionSet", exprs = exp.Ag17077,
phenoData = pd.Ag17077)
save(eSet.Ag17077,file="eSetAg17077.RData")
plotDensity(exprs(eSet.Ag17077))
pca.Ag17077=dudi.pca(t(exprs(eSet.Ag17077)),scannf=FALSE,nf=4,
scale=FALSE)
s.class(pca.Ag17077$li[,1:2],pData(eSet.Ag17077)[,"cell.type"])
s.class(pca.Ag17077$li[,1:2],pData(eSet.Ag17077)[,"Dataset.ID"])
s.label(pca.Ag17077$li[,1:2],label=pData(eSet.Ag17077)
[,"comment"])
# Annotation
annot.gpl17077=read.delim("rawdata/Agilent/GPL17077-17467.txt",
skip=16,header=TRUE)
z.Ag17077=eSet.Ag17077
exprs_z <- as.data.frame(exprs(z.Ag17077))
temp=c()
for(g in unique(annot.gpl17077$GENE_SYMBOL)){
  identifiers=annot.gpl17077[which(annot.gpl17077
    $GENE_SYMBOL==g),"ID"]
  exp.g=exprs_z[as.character(identifiers),]
  temp=rbind(temp, exp.g[which.max(apply(exp.g,1,sd)),])
}
exprs_z=temp
rownames(exprs_z)=unique(annot.gpl17077$GENE_SYMBOL)
exprs(z.Ag17077) <- data.matrix(exprs_z, rownames.force = TRUE)
annotation(z.Ag17077)<- "Ag17077_org"
save(z.Ag17077,file="eset.Ag17077_annotORG.RData")
## GPL6102 Illumina human-6 v2.0
# 1st dataset GSE25913
file.25913="rawdata/Illumina/Human-6v2/GSE25913_non-normalized_
data.txt"
x.gse25913=read.delim(file.25913,header=TRUE)
rownames(x.gse25913)=x.gse25913$ID_REF
x.gse25913=x.gse25913[,-1]
x.gse25913=x.gse25913[,-grep("Detection.Pval",colnames
(x.gse25913))]
tab.gse25913=subset(gsm.hs,Dataset.ID=="GSE25913")
tab.gse25913=tab.gse25913[order(tab.gse25913$Sample.ID),]
```

```
rownames(tab.gse25913)=as.character(tab.gse25913$Sample.ID)
# 2nd dataset GSE59295
file.59295="rawdata/Illumina/Human-6v2/GSE59295_non-normalized.
txt"
x.gse59295=read.delim(file.59295,header=TRUE)
samplenames=colnames(x.gse59295)
x.gse59295=read.delim(file.59295,header=TRUE,skip=1)
rownames(x.gse59295)=x.gse59295$ID_REF
x.gse59295=x.gse59295[,-1]
x.gse59295=x.gse59295[,-grep("Detection.Pval",colnames(x.
gse59295))]
tab.gse59295=subset(gsm.hs,Dataset.ID=="GSE59295")
tab.gse59295=tab.gse59295[order(tab.gse59295$Sample.ID),]
rownames(tab.gse59295)=as.character(tab.gse59295$Sample.ID)
# same columns order between x.gse59295(samplenames) and tab.
gse59295$comment
x.gse59295=x.gse59295[order(rownames(x.gse59295)),]
x.gse25913=x.gse25913[order(rownames(x.gse25913)),]
length(intersect(rownames(x.gse25913),rownames(x.gse59295)))
x.gse59295=x.gse59295[rownames(x.gse25913),]
mergedMat=cbind(x.gse25913,x.gse59295)
tab.Ill6102=rbind(tab.gse25913,tab.gse59295)
xN=lumiN(log2(as.matrix(mergedMat)))
pd.Ill6102=new("AnnotatedDataFrame",data=as.data.frame(tab.
Ill6102))
colnames(xN)=sampleNames(pd.Ill6102) # ordre a été préalablement
vérifié
eSet.Ill6102 <- new("ExpressionSet", exprs = xN, phenoData = pd.
Ill6102, annotation = "illuminaHumanv2")
save(eSet.Ill6102,file="esetIll6102v2.RData")
pca.Ill6102=dudi.pca(t(exprs(eSet.Ill6102)),scannf=FALSE,nf=4)
par(mfrow=c(1,2))
s.class(pca.Ill6102$li[,1:2],pData(eSet.Ill6102)[,"cell.type"])
s.class(pca.Ill6102$li[,1:2],pData(eSet.Ill6102)[,"Dataset.ID"])
## GPL10558 Illumina HumanHT-12 V4.0
# 3 datasets
file.35457="rawdata/Illumina/HumanHT-12v4/GSE35457_non-
normalized.txt"
```

```
file.36975="rawdata/Illumina/HumanHT-12v4/GSE36975_non-
normalized_data.txt"
file.43184="rawdata/Illumina/HumanHT-12v4/GSE43184_non-
normalized.txt"
x.gse35457=read.delim(file.35457,skip=1,header=TRUE)
rownames(x.gse35457)=x.gse35457$ID_REF
x.gse35457=x.gse35457[order(rownames(x.gse35457)),-1]
x.gse35457=x.gse35457[,-grep("Detection.Pval",colnames
(x.gse35457))]
linkwithGSM=read.delim("rawdata/Illumina/HumanHT-12v4/GSE35457-
seriesMatrix.csv")
x.gse35457=x.gse35457[,order(colnames(x.gse35457))]
colnames(x.gse35457)=linkwithGSM[order(linkwithGSM
$Sample_description),"Sample_geo_accession"]
x.gse36975=read.delim(file.36975,skip=2,header=TRUE)
rownames(x.gse36975)=x.gse36975$ID_REF
x.gse36975=x.gse36975[order(rownames(x.gse36975)),-1]
x.gse36975=x.gse36975[,-grep("Detection.Pval",colnames(x.
gse36975))]
x.gse36975=x.gse36975[,order(colnames(x.gse36975))]
# we only keep the unstimulated samples for each donor so those
that end with.1
x.gse36975=x.gse36975[,grep("1$",colnames(x.gse36975))]
x.gse43184=read.delim(file.43184,header=TRUE)
rownames(x.gse43184)=x.gse43184$ID_REF
x.gse43184=x.gse43184[order(rownames(x.gse43184)),-1]
x.gse43184=x.gse43184[,-grep("Detection.Pval",colnames
(x.gse43184))]
x.gse43184=x.gse43184[,9:18]
x.Il110558=cbind(x.gse35457[intersect(rownames(x.gse35457),
rownames(x.gse36975)),order(colnames(x.gse35457))],
      x.gse36975,
x.gse43184[intersect(rownames(x.gse43184),rownames(x.gse36975)),]
          )
xN.Il110558=lumiN(log2(as.matrix(x.Il110558)))
tab.GSE35457=subset(gsm.hs,Dataset.ID=="GSE35457")
tab.GSE36975=subset(gsm.hs,Dataset.ID=="GSE36975")
colnames(x.gse36975)=as.character(tab.GSE36975$Sample.ID)
```

```
tab.GSE43184=subset(gsm.hs,Dataset.ID=="GSE43184")
colnames(x.gse43184)=as.character(tab.GSE43184$Sample.ID)
tab.Il110558=rbind(tab.GSE35457,tab.GSE36975,tab.GSE43184)
rownames(tab.Il110558)=as.character(tab.Il110558$Sample.ID)
pd.Il110558=new("AnnotatedDataFrame",data=as.data.frame(tab.
Il110558))
eSet.Il110558 <- new("ExpressionSet", exprs = xN.Il110558,
phenoData = pd.Il110558, annotation = "illuminaHumanv4")
save(eSet.Il110558,file="eSetIl110558.RData")
pca.Il110558=dudi.pca(t(exprs(eSet.Il110558)),scannf=FALSE,nf=4)
par(mfrow=c(1,2))
s.class(pca.Il110558$li[,1:2],pData(eSet.Il110558)[,"cell.type"])
s.class(pca.Il110558$li[,1:2],pData(eSet.Il110558)[,"Dataset.ID"])
## GPL6255 Illumina humanRef-8 v2.0
geo.GPL6255=getGEO(GEO="GPL6255",filename="rawdata/Illumina//
HumanRef-8v2/GPL6255.annot.gz")
annotTable.6255=Table(dataTable(geo.GPL6255))
gse9820=getGEO("GSE9820",GSEMatrix=TRUE)
# samples were already quantiles normalized
#get subset of interest
tab.Il16255=subset(gsm.Hs,Platform=="GPL6255")
tab.Il16255=tab.Il16255[order(tab.Il16255$Sample.ID),]
rownames(tab.Il16255)=as.character(tab.Il16255$Sample.ID)
eSet.Il16255=gse9820[[1]][,which(colnames(exprs(gse9820[[1]]))
%in% tab.Il16255$Sample.ID)]
phenoData(eSet.Il16255)=new("AnnotatedDataFrame",data=tab.Il16255)
pData(eSet.Il16255)=pData(eSet.Il16255)[which(pData(eSet.Il16255)
$Sample.ID!="GSM247705"),]
save(eSet.Il16255,file="esetIl16255.RData")
pca.Il16255=dudi.pca(t(exprs(eSet.Il16255)),scannf=FALSE,nf=4)
s.class(pca.Il16255$li[,1:2],pData(eSet.Il16255)[,"cell.type"])
s.class(pca.Il16255$li[,1:2],pData(eSet.Il16255)[,"Dataset.ID"])
Source: mergeABintoU133set.R
####################
#
# merge U133A and U133B samples into U133set
# GSE22886, E-MEXP-384 and E-MEXP-750
#
######################
```

```
require(affy)
require(affyPLM)
linkfile=read.csv("MergeU133AB.csv")
files.A=paste(as.character(linkfile[which(linkfile
$Platform=="U133A"),1]),".CEL.gz",sep="")
files.B=paste(as.character(linkfile[which(linkfile
$Platform=="U133B"),1]),".CEL.gz",sep="")
x.A=ReadAffy(filenames=files.A,celfile.path="rawdata/Affymetrix/
U133set/")
x.B=ReadAffy(filenames=files.B,celfile.path="rawdata/Affymetrix/
U133set/")
eset.A=rma(x.A)
eset.B=rma(x.B)
nPairs = length(sampleNames(eset.A))
commonIndexA = which(rownames(exprs(eset.A)) %in% rownames
(exprs(eset.B)))
commonIndexB = which(rownames(exprs(eset.B)) %in% rownames
(exprs(eset.A)))
combined.chips = data.frame(matrix(nrow = length(union(rownames
(exprs(eset.A)),rownames(exprs(eset.B)))),
        ncol = nPairs))
row.names(combined.chips) = c(rownames(exprs(eset.A))
[-commonIndexA],
       rownames(exprs(eset.B)))
names(combined.chips) = sampleNames(eset.A)
pdf(file="plot-lm.pdf")
par(mfrow=c(2,2))
for (i in 1:nPairs) {
 print("log scale")
 x = (exprs(eset.B))[commonIndexB, i]
 y = (exprs(eset.A))[commonIndexA, i]
 m = lm(y ~ x)
 plot(m,main="log")
 print("unlog scale")
 x = (2^exprs(eset.B))[commonIndexB, i]
 y = (2^exprs(eset.A))[commonIndexA, i]
 m = lm(y ~ x)
 plot(m,main=" unlog ")
```

```
print("unlog scale and weighted linear model")
x = (2^exprs(eset.B))[commonIndexB, i]
y = (2^exprs(eset.A))[commonIndexA, i]
m = lm(y ~ x, weights = (x + y)^(-2))
plot(m,main="unlog+weight")
p = predict(m, data.frame(x = 2^exprs(eset.B)[, i]))
probestest=which(rownames(exprs(eset.A))%in%
rownames(exprs(eset.B)))
probestest2=which(rownames(as.data.frame(log2(p))) %in% rownames
(exprs(eset.A))[probestest])
plot(exprs(eset.A)[probestest,i],log2(p)[probestest2])
combined.chips[, i] = c(exprs(eset.A)[-commonIndexA,i], log2(p))
}
dev.off()
plotDensity(exprs(eset.A))
plotDensity(exprs(eset.A)[,-86])#index 86=TCM_U133A_3
# bad quality of TCM_U133B_3 sample
plotDensity(exprs(eset.B))
plotDensity(exprs(eset.B)[,-86])
# we remove the pair U133A and U133b for this sample
# the same script is run again
files.A=files.A[-86]
files.B=files.B[-86]
x.A=ReadAffy(filenames=files.A,celfile.path="rawdata/Affymetrix/
U133set/")
x.B=ReadAffy(filenames=files.B,celfile.path="rawdata/Affymetrix/
U133set/")
eset.A=rma(x.A)
eset.B=rma(x.B)
nPairs = length(sampleNames(eset.A))
commonIndexA = which(rownames(exprs(eset.A)) %in% rownames
(exprs(eset.B)))
commonIndexB = which(rownames(exprs(eset.B)) %in% rownames
(exprs(eset.A)))
combined.chips = data.frame(matrix(nrow = length(union(rownames
(exprs(eset.A)),rownames(exprs(eset.B)))),
        ncol = nPairs))
row.names(combined.chips) = c(rownames(exprs(eset.A))
[-commonIndexA],
       rownames(exprs(eset.B)))
```

```
names(combined.chips) = sampleNames(eset.A)
pdf(file="plot-lm.pdf")
par(mfrow=c(2,2))
for (i in 1:nPairs) {
 print("log scale")
 x = (exprs(eset.B))[commonIndexB, i]
 y = (exprs(eset.A))[commonIndexA, i]
 m = lm(y ~ x)
 plot(m,main="log")
 print("unlog scale")
 x = (2^exprs(eset.B))[commonIndexB, i]
 y = (2^exprs(eset.A))[commonIndexA, i]
 m = lm(y ~ x)
 plot(m,main="unlog")
 print("unlog sclae and weighted linear model")
 x = (2^exprs(eset.B))[commonIndexB, i]
 y = (2^exprs(eset.A))[commonIndexA, i]
 m = lm(y ~ x, weights = (x + y)^(-2))
 plot(m,main="unlog+weight")
 p = predict(m, data.frame(x = 2^exprs(eset.B)[, i]))
 probestest=which(rownames(exprs(eset.A))%in%
rownames(exprs(eset.B)))
 probestest2=which(rownames(as.data.frame(log2(p))) %in% rownames
(exprs(eset.A))[probestest])
 plot(exprs(eset.A)[probestest,i],log2(p)[probestest2])
 combined.chips[, i] = c(exprs(eset.A)[-commonIndexA,i], log2(p))
}
dev.off()
combinedChips= new("ExpressionSet", exprs = as.matrix(combined.
chips),
     phenoData = phenoData(eset.A))
normDataAcmbB <- normalize.ExpressionSet.quantiles(combinedChips)
save(normDataAcmbB,file="normDataAcmbB.RData")
# add phenoData : read GSM.csv
gsmfile=read.csv("../../ListeGSM.csv")
tab.AcmbB=gsmfile[which(gsmfile$Platform=="GPL96 / GPL97"),]
tab.AcmbB[60:71,]=tab.AcmbB[c(63,60,68,61,64,69,62,70,65,66,71,67),]
pData(normDataAcmbB)=tab.AcmbB
sampleNames(normDataAcmbB)=as.character(tab.AcmbB$Sample.ID)
```

```
save(normDataAcmbB,file="normDataAcmbB.RData")
```

### Combination of ExpressionSet objects

```
require(VirtualArray)
source("geneAnnotDataset.R")
# load each ExpressionSet object previously saved as RData
rdatafiles=list.files("DIR_RData",full.names=TRUE)
for( f in rdatafiles){
 load(f )
}
all_expression_sets=c("eSet.Aff96", "eSet.Aff4685", "eset.Aff570.
clean", "eSet.Aff571", "eSet.Aff6244", "eSet.Aff11532",
"normDataAcmbB", "z.Ag4133", "z.Ag14550", "z.Ag17077",
"eSet.Il16102", "eSet.Il110558" , "eSet.Il16255")
all_expression_sets <- sapply(X = all_expression_sets,
FUN = function(x) {
 as.symbol(x)}, USE.NAMES = TRUE)
expsts    <-    lapply(all_expression_sets,    geneAnnot.Dataset,
collapse_fun = "selectSD", identifier = "SYMBOL")
dtfrms <- lapply(expsts, virtualArrayBuildExprs)
#rm(expsts)
names_dtfrms <- names(dtfrms)
rownames_dtfrms <- lapply(dtfrms, FUN = function(X) {
 rownames(X)
})
names(rownames_dtfrms) <- names_dtfrms
rm(names_dtfrms)
message("Matching and merging all data sets. This could take some
time...",
  appendLF = TRUE)
merged <- virtualArrayMergeRecurse(dtfrms, by = "identifier",
          incomparables = NA)
rm(dtfrms)
rownames(merged) <- merged[, 1]
merged <- merged[2:dim(merged)[2]]
message("Size   of   expression   matrix   of   whole   dataset:   ",
dim(merged)[1],
  " rows and ", dim(merged)[2], " columns.", appendLF = TRUE)
rownames_dtfrms[["result"]] <- rownames(merged)
merged <- new("ExpressionSet", exprs = as.matrix(merged))
```

```
merged_norm <- normalize.ExpressionSet.quantiles(merged)
save(merged_norm,file="mergednorm.RData")
sample.info=c()
for(p in 1:length(expsts)){
 sample.info=rbind(sample.info,pData(expsts[[p]]))
   #print(dim(pData(expsts[[p]])))
}
pData(merged_norm)=as.data.frame(sample.info)
save(merged_norm,file="mergednorm.RData")
pca.mergednorm=dudi.pca(t(exprs(merged_norm)),scannf=FALSE,nf=4)
s.class(pca.mergednorm$li[,1:2],pData(merged_norm)[,"cell.type"])
s.class(pca.mergednorm$li[,1:2],pData(merged_norm)[,"Platform"])
# Platform effect to be corrected by Combat
# the platform GPL571 has the only 2 mast cells. To include the platform factor in the model avoiding a singular system: assign new platform to GPL571: all probesets of GPL96 are included in GPL571 which contains additional probesets.
merged_combat <- merged_norm
exprs(merged_combat)=na.omit(exprs(merged_combat))
Batch=pData(merged_combat)[,"Platform"]
Batch[which(Batch=="GPL571")]="GPL96"
Batch=factor(Batch)
pData(merged_combat)=cbind(pData(merged_combat),Batch)
exprs(merged_combat)=ComBat(dat=exprs(merged_combat),
batch=factor(pData(merged_combat)[,"Batch"]),mod=model.matrix
(~factor(pData(merged_combat)[,"cell.type"])))
```

Functions

```
geneAnnot.Dataset=function(x, identifier = "SYMBOL", collapse_fun
="selectSD"){
 print(x)
 message("Now preprocessing raw data of: Loading annotations...")
 identifier <- identifier
 z <- NULL
 z <- x
 annot <- annotation(z)
 annot_identifier <- paste(annot, identifier, sep = "")
 annot_package <- sub(" ", ".", paste(annot, "db"))
 if(length(grep("org",annot))>0){
```

```
 # => rownames are already Gene Symbols without duplication
 return(z)
} else {
 require(package = annot_package, character.only = TRUE)
 if (class(annot_identifier)[1] == "ProbeAnnDbBimap") {
  annot_identifier <- toggleProbes(value = "all", x = eval(as.
symbol(annot_identifier)))
 } else {
  annot_identifier <- eval(as.symbol(annot_identifier))
 }
 fData(z) <- toTable(annot_identifier)
 match_fData_y <- match(fData(z)[, 1], rownames(exprs(z)))
 exprs(z) <- as.matrix(exprs(z)[match_fData_y, ])
 exprs_z <- as.data.frame(exprs(z))
 exprs_z <- cbind(identifier = fData(z)[, 2], exprs_z, row.names =
NULL)
 if(collapse_fun=="selectSD"){
  message("Now preprocessing raw data of: Collapsing expression
values to their most variable: selection of the highest SD")
  exprs_z <- na.omit(exprs_z)
  temp=c()
  for(g in unique(exprs_z$identifier)){
   exp.g=exprs_z[which(exprs_z$identifier==g),]
   temp=rbind(temp, exp.g[which.max(apply(exp.g[,-1],1,sd)),])
  }
  exprs_z=temp
  rownames(exprs_z)=unique(exprs_z$identifier)
  exprs_z=exprs_z[,-1]
 }else{
  message("Now preprocessing raw data of: Collapsing expression
values to their ", as.character(as.list(collapse_fun)[[3]])[2],
    "...")
  exprs_z <- na.omit(exprs_z)
  exprs_z <- recast(data = exprs_z, formula = identifier ~
       variable, fun.aggregate = collapse_fun)
  message("Now preprocessing raw data of : Annotating expression
values with ", as.character(identifier),
    "...")
  rownames(exprs_z$data) <- exprs_z$labels[[1]][, 1]
```

```
      colnames(exprs_z$data) <- exprs_z$labels[[2]][, 1]
      exprs_z <- exprs_z$data
    }
    exprs(z) <- data.matrix(exprs_z, rownames.force = TRUE)
    match_fData_y2 <- match(rownames(exprs(z)), fData(z)[, 2])
    fData(z) <- fData(z)[match_fData_y2, ]
    rownames(fData(z)) <- fData(z)[, 2]
    rownames(exprs(z)) <- fData(z)[, 2]
    message("Size of expression matrix of: ", dim(exprs(z))[1],
      " rows and ", dim(exprs(z))[2],
      " columns.")
    x <- z
  }
  return(x)
}
```

# References

## Datasets

Abbas, A. R., Baldwin, D., Ma, Y., Ouyang, W., et al. (2005). Immune response in silico (IRIS): Immune-specific genes identified from a compendium of microarray expression data. *Genes and Immunity*, *6*(4), 319–331. 15789058.

Allantaz, F., Cheng, D. T., Bergauer, T., Ravindran, P., et al. (2012). Expression profiling of human immune cell subsets identifies miRNA-mRNA regulatory relationships correlated with cell type specific expression. *PLoS One*, 7(1), e29979. 22276136.

Ancuta, P., Liu, K. Y., Misra, V., Wacleche, V. S., et al. (2009). Transcriptional profiling reveals developmental relationship and distinct biological functions of CD16+ and CD16- monocyte subsets. *BMC Genomics*, *10*, 403. 19712453.

Rasheed, A., Rahn, H. P., Sallusto, F., Lipp, M., et al. (2006). Follicular B helper T cell activity is confined to CXCR5(hi)ICOS(hi) CD4 T cells and is independent of CD57 expression. *European Journal of Immunology*, *36*, 1892–1903. 200636136.

Balan, S., Ollion, V., Colletti, N., Chelbi, R., et al. (2014). Human XCR1+ dendritic cells derived in vitro from CD34+ progenitors closely resemble blood dendritic cells, including their adjuvant responsiveness, contrary to monocyte-derived dendritic cells. *Journal of Immunology*, *193*(4), 1622–1635. 25009205.

Baram, D., Dekel, O., Mekori, Y. A., & Sagi-Eisenberg, R. (2010). Activation of mast cells by trimeric G protein Gi3; coupling to the A3 adenosine receptor directly and upon T cell contact. *Journal of Immunology*, *184*(7), 3677–3688. 20190146.

Bernier, A., Cleret-Buhot, A., Zhang, Y., Goulet, J. P., et al. (2013). Transcriptional profiling reveals molecular signatures associated with HIV permissiveness in Th1Th17 cells and identifies peroxisome proliferator-activated receptor gamma as an intrinsic negative regulator of viral replication. *Retrovirology*, *10*, 160. 24359430.

Cassani, B., Montini, E., Maruggi, G., Ambrosi, A., et al. (2009). Integration of retroviral vectors induces minor changes in the transcriptional activity of T cells from ADA-SCID patients treated with gene therapy. *Blood*, *114*(17), 3546–3556. 19652199.

Chng, W. J., Kumar, S., Vanwier, S., Ahmann, G., et al. (2007). Molecular dissection of hyperdiploid multiple myeloma by gene expression profiling. *Cancer Research*, *67*(7), 2982–2989. 17409404.

Covens, K., Verbinnen, B., Geukens, N., Meyts, I., et al. (2013). Characterization of proposed human B-1 cells reveals pre-plasmablast phenotype. *Blood*, *121*(26), 5176–5183. 23613519.

Dürig, J., Bug, S., Klein-Hitpass, L., Boes, T., et al. (2007). Combined single nucleotide polymorphism-based genomic mapping and global gene expression profiling identifies novel chromosomal imbalances, mechanisms and candidate genes important in the pathogenesis of T-cell prolymphocytic leukemia with inv.(14)(q11q32). *Leukemia*, *21*(10), 2153–2163. PMID: 17713554.

Dybkaer, K., Iqbal, J., Zhou, G., Geng, H., et al. (2007). Genome wide transcriptional analysis of resting and IL2 activated human natural killer cells: Gene expression signatures indicative of novel molecular signaling pathways. *BMC Genomics*, *8*, 230. 17623099.

Emonts M., Maat M., de Ridder D., Schonewille T., Hazelzet J., Laman J., et al., Gene expression profiling in pediatric meningococcal sepsis reveals dynamic changes in NK-cell and cytotoxic molecules, 2008, Erasmus MC: Rotterdam, The Netherlands.

Fujisaki, H., Kakuda, H., Shimasaki, N., Imai, C., et al. (2009). Expansion of highly cytotoxic human natural killer cells for cancer cell therapy. *Cancer Research*, *69*(9), 4010–4017. 19383914.

Garg, T. K., Szmania, S. M., Khan, J. A., Hoering, A., et al. (2012). Highly activated and expanded natural killer cells for multiple myeloma immunotherapy. *Haematologica*, *97*(9), 1348–1356. 22419581.

Glatzer, T., Killig, M., Meisig, J., Ommert, I., et al. (2013). RORγt+ innate lymphoid cells acquire a proinflammatory program upon engagement of the activating receptor NKp44. *38*(6), 1223–1235. 23791642.

Good, K. L., Avery, D. T., & Tangye, S. G. (2009). Resting human memory B cells are intrinsically programmed for enhanced survival and responsiveness to diverse stimuli compared to naive B cells. *Journal of Immunology*, *182*(2), 890–901. 19124732.

Görgün, G., Holderried, T. A., Zahrieh, D., Neuberg, D., et al. (2005). Chronic lymphocytic leukemia cells induce changes in gene expression of CD4 and CD8 T cells. *The Journal of Clinical Investigation*, *115*(7), 1797–1805. 15965501.

Gutiérrez, N. C., Ocio, E. M., de Las Rivas, J., Maiso, P., et al. (2007). Gene expression profiling of B lymphocytes and plasma cells from Waldenström's macroglobulinemia: Comparison with expression patterns of the same cell counterparts from chronic lymphocytic leukemia, multiple myeloma and normal individuals. *Leukemia*, *21*(3), 541–549. 17252022.

Haniffa, M., Shin, A., Bigley, V., McGovern, N., et al. (2012). Human tissues contain CD141hi cross-presenting dendritic cells with functional homology to mouse CD103 + nonlymphoid dendritic cells. *Immunity*, *37*(1), 60–73. 22795876.

Hutter, C., Kauer, M., Simonitsch-Klupp, I., Jug, G., et al. (2012). Notch is active in Langerhans cell histiocytosis and confers pathognomonic features on dendritic cells. *Blood*, *120*(26), 5199–5208. 23074278.

Hyrcza, M. D., Kovacs, C., Loutfy, M., Halpenny, R., et al. (2007). Distinct transcriptional profiles in ex vivo CD4+ and CD8+ T cells are established early in human immunodeficiency virus type 1 infection and are characterized by a chronic interferon response as well as extensive transcriptional changes in CD8+ T cells. *Journal of Virology*, *81*(7), 3477–3486. 17251300.

Jourdan, M., Caraux, A., De Vos, J., Fiol, G., Larroque, M., et al. (2009). An in vitro model of differentiation of memory B cells into plasmablasts and plasma cells including detailed phenotypic and molecular characterization. *Blood*, *114*(25), 5173–5181. PMID: PMID:19846886.

Juelke, K., Killig, M., Luetke-Eversloh, M., Parente, E., et al. (2010). CD62L expression identifies a unique subset of polyfunctional CD56dim NK cells. *Blood*, *116*(8), 1299–1307. 20505160.

Kassambara, A., Rème, T., Jourdan, M., Fest, T., et al. (2015). GenomicScape: An easy-to-use web tool for gene expression data analysis. Application to investigate the molecular events in the differentiation of B cells into plasma cells. *PLoS Computational Biology*, *11*, e1004077. 25633866.

Laudanski, K., Miller-Graziano, C., Xiao, W., Mindrinos, M. N., et al. (2006). Cell-specific expression and pathway analyses reveal alterations in trauma-related human T cell and monocyte pathways. *Proceedings of the National Academy of Sciences of the United States of America*, *103*(42), 15564–15569. 17032758.

Le Gallou, S., Caron, G., Delaloy, C., Rossille, D., et al. (2012). IL-2 requirement for human plasma cell generation: Coupling differentiation and proliferation by enhancing MAPK-ERK signaling. *Journal of Immunology*, *189*(1), 161–173. 22634617.

Lee, M. N., Ye, C., Villani, A. C., Raj, T., et al. (2014). Common genetic variants modulate pathogen-sensing responses in human dendritic cells. *Science*, *343*(6175) 1246980. 24604203.

Lindstedt, M., Lundberg, K., & Borrebaeck, C. (2005). Gene family clustering identifies functionally associated subsets of human in vivo blood and tonsillar dendritic cells. *The Journal of Immunology*, *175*(8), 4839. 16210585.

López-Corral, L., Corchete, L. A., Sarasquete, M. E., Mateos, M. V., et al. (2014). Transcriptome analysis reveals molecular profiles associated with evolving steps of monoclonal gammopathies. *Haematologica*, *99*(8), 1365–1372. 24816239.

Mabbott, N. A., Baillie, J. K., Brown, H., Freeman, T. C., et al. (2013). An expression atlas of human primary cells: Inference of gene function from coexpression networks. *BMC Genomics*, *14*, 632. 24053356.

Merryweather-Clarke, A. T., Atzberger, A., Soneji, S., Gray, N., et al. (2011). Global gene expression analysis of human erythroid progenitors. *Blood*, *117*(13), e96–108. 21270440.

Mezger, M., Wozniok, I., Blockhaus, C., Kurzai, O., et al. (2008). Impact of mycophenolic acid on the functionality of human polymorphonuclear neutrophils and dendritic cells during interaction with Aspergillus fumigatus. *Antimicrobial Agents and Chemotherapy*, *52*(7), 2644–2646. 18426895.

Miyara, M., Yoshioka, Y., Kitoh, A., Shima, T., et al. (2009). Functional delineation and differentiation dynamics of human CD4+ T cells expressing the FoxP3 transcription factor. *Immunity*, *30*(6), 899–911. 19464196.

Montaldo, E., Teixeira-Alves, L. G., Glatzer, T., Durek, P., et al. (2014). Human RORγt(+)CD34(+) cells are lineage-specified progenitors of group 3 RORγt(+) innate lymphoid cells. *Immunity*, *41*(6), 988–1000. 25500367.

Nodland, S. E., Berkowska, M. A., Bajer, A. A., Shah, N., et al. (2011). IL-7R expression and IL-7 signaling confer a distinct phenotype on developing human B-lineage cells. *Blood*, *118*(8), 2116. 21680796.

Novershtern, N., Subramanian, A., Lawton, L. N., Mak, R. H., et al. (2011). Densely interconnected transcriptional circuits control cell states in human hematopoiesis. *Cell*, *144*(2), 296–309. 21241896.

Ono, C., Yu, Z., Kasahara, Y., Kikuchi, Y., et al. (2014). Fluorescently activated cell sorting followed by microarray profiling of helper T cell subtypes from human peripheral blood. *PLoS One*, *9*(11), e111405, 25379667.

Raghavachari, N., Xu, X., Harris, A., Villagra, J., et al. (2007). Amplified expression profiling of platelet transcriptome reveals changes in arginine metabolic pathways in patients with sickle cell disease. *Circulation*, *115*(12), 1551–1562. 17353439.

Santarlasci, V., Maggi, L., Capone, M., Querci, V., et al. (2012). Rarity of human T helper 17 cells is due to retinoic acid orphan receptor-dependent mechanisms that limit their expansion. *Immunity*, *36*(2), 201–214. 22326581.

Schirmer, S. H., Fledderus, J. O., & van der Laan, A. M. (2009). van der Pouw-Kraan TC et al. Suppression of inflammatory signaling in monocytes from patients with coronary artery disease. *Journal of Molecular and Cellular Cardiology*, *46*(2), 177–185. 19059264.

Segura, E., Touzot, M., Bohineust, A., Cappuccio, A., et al. (2013). Human inflammatory dendritic cells induce Th17 cell differentiation. *Immunity*, *38*(2), 336–348. 23352235.

Sirois, M., Robitaille, L., Allary, R., Shah, M., et al. (2011). TRAF6 and IRF7 control HIV replication in macrophages. *PLoS One*, *6*(11), e28125, 22140520.

Streicher, K., Morehouse, C. A., Groves, C. J., Rajan, B., et al. (2014). The plasma cell signature in autoimmune disease. *Arthritis & Rhematology*, *66*(1), 173–184. 24431284.

Széles, L., Póliska, S., Nagy, G., Szatmari, I., et al. (2010). Research resource: Transcriptome profiling of genes regulated by RXR and its permissive and nonpermissive partners in differentiating monocyte-derived dendritic cells. *Molecular Endocrinology*, *24*(11), 2218–2231, PMID: 20861222.

Touzot, M., Grandclaudon, M., Cappuccio, A., Satoh, T., et al. (2014). Combinatorial flexibility of cytokine function during human T helper cell differentiation. *Nature Communications*, *5*, 3987. 24865484.

Vinh, D. C., Patel, S. Y., Uzel, G., Anderson, V. L., et al. (2010). Autosomal dominant and sporadic monocytopenia with susceptibility to mycobacteria, fungi, papillomaviruses, and myelodysplasia. *Blood*, *115*(8), 1519–1529. 20040766.

Wang, L., Shalek, A. K., Lawrence, M., Ding, R., et al. (2014). Somatic mutation as a mechanism of Wnt/β-catenin pathway activation in CLL. *Blood*, *124*(7), 1089–1098. 24778153.

Wong, K. L., Tai, J. J., Wong, W. C., Han, H., et al. (2011). Gene expression profiling reveals the defining features of the classical, intermediate, and nonclassical human monocyte subsets. *Blood*, *118*(5), e16–e31. 21653326.

Yu, C. I., Becker, C., Wang, Y., Marches, F., et al. (2013). Human CD1c + dendritic cells drive the differentiation of CD103+ CD8+ mucosal effector T cells via the cytokine TGF-β. *Immunity*, *38*(4), 818–830. 23562160.

Zhang, H., Nestor, C. E., Zhao, S., Lentini, A., et al. (2013). Profiling of human CD4+ T-cell subsets identifies the TH2-specific noncoding RNA GATA3-AS1. *The Journal of Allergy and Clinical Immunology*, *132*(4), 1005–1008. 23870669.

Zhang, W., Ferguson, J., Ng, S. M., Hui, K., et al. (2012). Effector CD4+ T cell expression signatures and immune-mediated disease associated genes. *PLoS One*, 7(6), e38510, 22715389.

Zhang, Y., Ohyashiki, J. H., Shimizu, N., & Ohyashiki, K. (2010). Aberrant expression of NK cell receptors in Epstein-Barr virus-positive gammadelta T-cell lymphoproliferative disorders. *Hematology*, *15*(1), 43–47. 20132662.

## Methods

Bolstad, B. M., Irizarry, R. A., Astrand, M., & Speed, T. P. (2003). A comparison of normalization methods for high density oligonucleotide array data based on variance and bias. *Bioinformatics*, *19*(2), 185–193. 12538238.

Bolstad, B. M. (2004). *Low Level Analysis of High-density Oligonucleotide Array Data: Background, Normalization and Summarization.* PhD Dissertation Berkeley: University of California.

Dray, S., & Dufour, A. (2007). The ade4 Package: Implementing the duality diagram for ecologists. *Journal of Statistical Software*, *22*(4), 1–20.

Du, P., Kibbe, W. A., & Lin, S. M. (2008). lumi: A pipeline for processing Illumina microarray. *Bioinformatics*, *24*(13). 18467348.

Edgar, R., Domrachev, M., & Lash, A. E. (2002). Gene expression omnibus: NCBI gene expression and hybridization array data repository. *Nucleic Acids Research*, *30*(1), 207–210. 11752295.

Gatto, L. (2018). *yaqcaffy: Affymetrix expression data quality control and reproducibility analysis.* R package version 1.42.0.

Gauthier, L., Cope, L., Bolstad, B. M., & Irizarry, R. A. (2004). Affy—analysis of Affymetrix GeneChip data at the probe level. *Bioinformatics*, *20*(3), 307–315. 14960456.

Heider, A., & Alt, R. (2013). virtualArray: A R/bioconductor package to merge raw data from different microarray platforms. *BMC Bioinformatics*, *14*, 75. 23452776.

Huber, W., Carey, V. J., Gentleman, R., Anders, S., et al. (2015). Orchestrating high-throuput genomic analysis with bioconductor. *Nature Methods*, *12*, 115–121. 25633503.

Irizarry, R. A., Hobbs, B., Collin, F., Beazer-Barclay, Y. D., Antonellis, K. J., Scherf, U., et al. (2003). Exploration, normalization, and summaries of high density oligonucleotide array probe level data. *Biostatistics*, *4*(2), 249–264. 12925520.

Johnson, W. E., Li, C., & Rabinovic, A. (2007). Adjusting batch effects in microarray expression data using empirical Bayes methods. *Biostatistics*, *8*(1), 118–127. 16632515.

Leek, J. T., Johnson, W. E., Parker, H. S., Jaffe, A. E., & Storey, J. D. (2012). The sva package for removing batch effects and other unwanted variation in high-throughput experiments. *Bioinformatics*, *28*(6), 882–883.

Parkinson, H., Sarkans, U., Kolesnikov, N., Abeygunawardena, N., et al. (2011). ArrayExpress update–An archive of microarray and high-throughput sequencing-based functional genomics experiments. *Nucleic Acids Research*, *39*, D1002–D1004. 21071405.

Ritchie, M. E., Silver, J., Oshlack, A., Holmes, M., Diyagama, D., Holloway, A., et al. (2007). A comparison of background correction methods for two-color microarrays. *Bioinformatics*, *23*(20), 2700–2707. 17720982.

Ritchie, M. E., Phipson, B., Wu, D., Hu, Y., Law, C. W., Shi, W., et al. (2015). limma powers differential expression analyses for RNA-sequencing and microarray studies. *Nucleic Acids Research*, *43*(7), e47. 25605792.

## Further reading

Menssen, A., Edinger, G., Grün, J. R., Haase, U., et al. (2009). SiPaGene: A new repository for instant online retrieval, sharing and meta-analyses of GeneChip expression data. *BMC Genomics*, *10*, 98. 19265543.

CHAPTER TWO

# DIY: Visualizing the immune landscape of tumors using transcriptome and methylome data

**Steven F. Gameiro[a], Farhad Ghasemi[b], John W. Barrett[b], James Koropatnick[a,c,d,e,f], Anthony C. Nichols[b,c,f], Joe S. Mymryk[a,b,c,f,*], Saman Maleki Vareki[c,f,*]**

[a]Department of Microbiology and Immunology, The University of Western Ontario, London, ON, Canada
[b]Department of Otolaryngology, Head & Neck Surgery, The University of Western Ontario, London, ON, Canada
[c]Department of Oncology, The University of Western Ontario, London, ON, Canada
[d]Department of Pathology, The University of Western Ontario, London, ON, Canada
[e]Department of Physiology and Pharmacology, The University of Western Ontario, London, ON, Canada
[f]London Regional Cancer Program, Lawson Health Research Institute, London, ON, Canada
*Corresponding authors: e-mail address: jmymryk@uwo.ca; saman.malekivareki@lhsc.on.ca

## Contents



*Methods in Enzymology*, Volume 636
ISSN 0076-6879
https://doi.org/10.1016/bs.mie.2019.05.040

## Abstract

The importance of the immune system in combating many different types of malignancies has been firmly established. Indeed, the success of immune checkpoint-blocking antibodies in cancer treatment has focused the attention of many researchers on the tumor immune microenvironment. As one consequence, the analysis of the number, type, and activation status of tumor-infiltrating leukocytes (TILs) has become an area of great interest. Detailed knowledge of the immunological profile of a given tumor has the potential to predict patient response to therapy and overall survival. This chapter is intended to summarize the steps necessary for basic comparative analysis of the infiltration and activation status of immune cells in various human solid tumors, using freely available data obtained from The Cancer Genome Atlas (TCGA).

## 1. Introduction

Tumors are not homogeneous aggregates of malignant cells: rather, they are heterogeneous collections of cancer cells, supporting stromal cells, and a diverse array of tumor-infiltrating leukocytes (TILs) (Hanahan & Weinberg, 2011). Although it has been long recognized that the immune system plays key roles in cancer, recent data has illustrated that understanding the number, type, and activation status of TILs has the potential to predict patient response to therapy and overall survival (Emens, 2018; Fridman, Zitvogel, Sautès-Fridman, & Kroemer, 2017; Gooden, de Bock, Leffers, Daemen, & Nijman, 2011; Uryvaev, Passhak, Hershkovits, Sabo, & Bar-Sela, 2018). In addition, characterization of the immune landscape of tumors is helping guide the rational design of therapies that mobilize the immune system to induce long-term remission in cancers previously considered to be terminal (Byrne & Fisher, 2017; Costa, Das, Kini Bailur, Dhodapkar, & Dhodapkar, 2018; George et al., 2017; Hölzel, Bovier, & Tüting, 2013).

Historically, TIL detection within human solid tumors has been primarily performed using immunohistochemistry (IHC) or flow cytometry (Llosa et al., 2015). Both of these techniques have taught us a great deal about the tumor immune profile, but they require antibodies to specific phenotypic markers of cell type or activation status, can be challenging to standardize, difficult to implement with small tissue samples, and expensive to perform for the dozens of markers necessary for a comprehensive view of the immune landscape of a given tumor. More recent alternatives capitalize on the plunging costs of "-omics" technologies, including next-generation sequencing (NGS) for genomics and transcriptomics, as well as genome wide methylome analysis using high density arrays. Indeed, massive collaborative efforts such as The Cancer Genome Atlas (TCGA) have generated genomic, transcriptomic, and methylomic data for over 11,000 tumor samples from 33 cancer types and subtypes (Liu et al., 2018). Capitalizing on this freely available data, it is possible to estimate the total fraction of tumor composed of leukocytes, based on differences in DNA methylation between leukocytes and other cell types (Hoadley et al., 2018). More specifically, it is also possible to estimate the detailed composition and activation status of tumor-infiltrating immune cells in solid tumors from bulk tumor RNA-sequencing (RNA-seq) data using computational approaches based on a set of immune-specific marker genes or expression signatures that correlate with distinct leukocyte types (Bindea et al., 2013; Danaher et al., 2017; Gameiro, Zhang, et al., 2017; Gameiro et al., 2018; Newman et al., 2015; Rooney, Shukla, Wu, Getz, & Hacohen, 2015; Wolf & Samuels, 2018).

The goal of this chapter is to summarize the steps for performing comparative analyses of the infiltration and activation status of immune cells in various human solid tumors using freely available TCGA data. Although analyses of immune gene expression in bulk tumors may not be definitive, they can often identify underlying differences in the tumor immune landscape, leading to the generation of new hypotheses to be investigated using more traditional immunological methods.

## 2. A dummy's guide to the TCGA

The TCGA was a large-scale collaboration between the National Cancer Institute (NCI) and the National Human Genome Research Institute (NHGRI) to comprehensively characterize the molecular events in primary cancers. This effort, which ended in 2017, generated publicly available, comprehensive, multidimensional maps of genomic changes in >11,000 tumor samples. The TCGA dataset encompasses 33 types of human tumors, all processed via a uniform genomic data analysis pipeline that collected, selected, and analyzed these cancer samples in detail. These analyses included whole exome and/or whole genome sequencing, mRNA sequencing, microRNA sequencing, DNA copy number profiling, DNA methylation profiling, and reverse-phase protein array expression profiling. Although comprehensive clinical annotation was not possible for every TCGA sample, substantial clinical data is available, allowing comparison of many clinical variables and, in most cases, reasonably robust outcomes assessment.

There are 33 tumor types included in the TCGA and a breakdown of the sample numbers and available data for each type of analysis is provided in Table 1. Unfortunately, limitations in available normal control tissues seriously limits the power for some types of comparisons.

Although each TCGA sample is identified by a 22-digit barcode, most available files use a shortened version, which is in the format **TCGA-01-0001-01A**, as illustrated in Fig. 1. This identifier indicates: Tissue Source Site, Study Participant Number, Sample Type, and Vial. A frequent point of confusion is related to the Sample Type identifier. For example, the same study participant may have provided multiple tissue samples, such as primary solid tumor (code 01), metastatic lesion (code 06), and/or normal control (code 11) tissues. Inclusion of data from the metastatic lesions may or may not be appropriate for any given analysis, and inadvertent swapping of normal control values for primary tumor values could invalidate any conclusions.

## 3. Where to get the TCGA data and what information is available?

As the TCGA represents the largest/richest cancer data set so far assembled, this massive data resource can be applied to study many aspects of human cancer. However, laboratories that want to use these data are simply not prepared to invest the time and other resources into working with low level/raw data. Importantly, other options exist that do not require advanced bioinformatics training, yet still allow detailed customized comparisons to be performed beyond the scope of generalized web-based tools such as cBioPortal (http://www.cbioportal.org/) (Cerami et al., 2012). One of these useful options is the online tool OncoLnc (http://www.oncolnc.org) (Anaya, 2016) which allows researchers to generate Kaplan-Meier plots from 21 different cancer studies performed by the TCGA, and then

**Table 1** TCGA cohorts and their composition. The composition of each of the 33 TCGA cohorts, including total samples, number of normal tissue control samples, number of metastatic samples and a breakdown of the available samples processed for each analysis type.

| TCGA cohort | Disease type | Total cases | Normal tissue | Metastatic | mRNASeq tumor | mRNASeq normal tissue | Methylation tumor | Methylation normal tissue |
|---|---|---|---|---|---|---|---|---|
| ACC | Adrenocortical carcinoma | 92 | 5 | 0 | 79 | 0 | 80 | 0 |
| BLCA | Bladder urothelial carcinoma | 412 | 36 | 1 | 408 | 19 | 412 | 21 |
| BRCA | Breast invasive carcinoma | 1098 | 162 | 7 | 1093 | 112 | 1097 | 124 |
| CESC | Cervical squamous cell carcinoma and endocervical adenocarcinoma | 307 | 8 | 2 | 304 | 3 | 307 | 3 |
| CHOL | Cholangiocarcinoma | 51 | 15 | 0 | 36 | 9 | 36 | 9 |
| COAD | Colon adenocarcinoma | 460 | 93 | 2 | 457 | 41 | 457 | 75 |
| DLBC | Lymphoid neoplasm diffuse large B-cell lymphoma | 58 | 0 | 0 | 48 | 0 | 48 | 0 |
| ESCA | Esophageal carcinoma | 185 | 65 | 1 | 184 | 13 | 185 | 16 |
| GBM | Glioblastoma multiforme | 613 | 49 | 0 | 160 | 5 | 420 | 2 |
| HNSC | Head and neck squamous cell carcinoma | 528 | 82 | 2 | 520 | 44 | 528 | 50 |
| KICH | Kidney chromophobe | 113 | 57 | 0 | 66 | 25 | 66 | 0 |
| KIRC | Kidney renal clear cell carcinoma | 537 | 440 | 0 | 533 | 72 | 535 | 359 |
| KIRP | Kidney renal papillary cell carcinoma | 323 | 110 | 0 | 290 | 32 | 291 | 50 |
| LAML | Acute myeloid leukemia | 200 | 200 | 0 | 179 | 0 | 194 | 0 |
| LGG | Brain lower grade glioma | 516 | 2 | 0 | 516 | 0 | 516 | 0 |
| LIHC | Liver hepatocellular carcinoma | 377 | 89 | 0 | 371 | 50 | 377 | 50 |
| LUAD | Lung adenocarcinoma | 585 | 184 | 0 | 515 | 59 | 578 | 56 |

*Continued*

**Table 1** TCGA cohorts and their composition. The composition of each of the 33 TCGA cohorts, including total samples, number of normal tissue control samples, number of metastatic samples and a breakdown of the available samples processed for each analysis type.—cont'd

| TCGA cohort | Disease type | Total cases | Normal tissue | Metastatic | mRNASeq tumor | mRNASeq normal tissue | Methylation tumor | Methylation normal tissue |
|---|---|---|---|---|---|---|---|---|
| LUSC | Lung squamous cell carcinoma | 504 | 249 | 0 | 501 | 51 | 503 | 69 |
| MESO | Mesothelioma | 87 | 1 | 0 | 87 | 0 | 87 | 0 |
| OV | Ovarian serous cystadenocarcinoma | 602 | 133 | 0 | 304 | 0 | 594 | 12 |
| PAAD | Pancreatic adenocarcinoma | 185 | 37 | 1 | 178 | 4 | 184 | 10 |
| PCPG | Pheochromocytoma and paraganglioma | 179 | 5 | 2 | 179 | 3 | 179 | 3 |
| PRAD | Prostate adenocarcinoma | 499 | 117 | 1 | 497 | 52 | 498 | 50 |
| READ | Rectum adenocarcinoma | 171 | 17 | 0 | 166 | 10 | 165 | 12 |
| SARC | Sarcoma | 261 | 23 | 1 | 259 | 2 | 261 | 4 |
| SKCM | Skin cutaneous melanoma | 470 | 3 | 368 | 469 | 1 | 470 | 2 |
| STAD | Stomach adenocarcinoma | 443 | 101 | 0 | 415 | 37 | 443 | 27 |
| TGCT | Testicular germ cell tumors | 150 | 0 | 0 | 150 | 0 | 150 | 0 |
| THCA | Thyroid carcinoma | 503 | 100 | 8 | 501 | 59 | 503 | 56 |
| THYM | Thymoma | 124 | 13 | 0 | 120 | 2 | 124 | 2 |
| UCEC | Uterine corpus endometrial carcinoma | 560 | 51 | 0 | 545 | 35 | 547 | 47 |
| UCS | Uterine carcinosarcoma | 57 | 6 | 0 | 57 | 0 | 57 | 0 |
| UVM | Uveal melanoma | 80 | 0 | 0 | 80 | 0 | 80 | 0 |
| Totals | | 11368 | 2453 | 396 | 10267 | 740 | 10972 | 1109 |

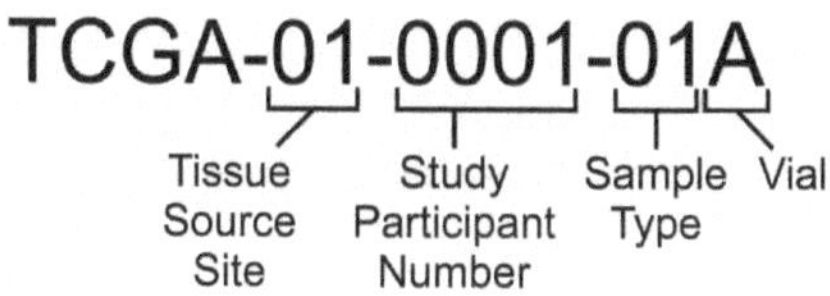


| Sample Type | Definition | Sample Type | Definition |
|---|---|---|---|
| 1 | Primary Solid Tumor | 8 | Human Tumor Original Cells |
| 2 | Recurrent Solid Tumor | 9 | Primary Blood Derived Cancer - Bone Marrow |
| 3 | Primary Blood Derived Cancer - Peripheral Blood | 10 | Blood Derived Normal |
| 4 | Recurrent Blood Derived Cancer - Bone Marrow | 11 | Solid Tissue Normal |
| 5 | Additional - New Primary | 12 | Buccal Cell Normal |
| 6 | Metastatic | 14 | Bone Marrow Normal |
| 7 | Additional Metastatic | 40 | Recurrent Blood Derived Cancer - Peripheral Blood |

**Fig. 1** Interpreting the basic TCGA barcode. The TCGA barcode contains key information that not only uniquely identifies the sample used for each analysis, but also provides key information regarding sample workflow for quality control. For many applications, the basic information required can be reduced to the first 22 digits of the code, which comprises the Tissue Source Site (TSS), Study Participant Number, Sample Type, and Vial. The TSS represents the TCGA affiliated institution that provided the sample and a table interpreting these codes can be found at: https://gdc.cancer.gov/resources-tcga-users/tcga-code-tables/tissue-source-site-codes. The Study Participant Number is simply a number representing the individual from which the sample was collected at the TSS. This number is not unique across all TCGA samples but is unique for each TSS. Sample Type reflects the type of sample and is defined using the abbreviated table shown. As an example, a given study might use primary solid tumor (code 01) results in comparison with solid tissue normal (code 11) data. Vial represents which fraction of the original sample was used for the analysis, as samples were routinely subdivided into multiple aliquots.

to dichotomize patients based on expression data for mRNAs, miRNAs, or long noncoding RNAs (lncRNAs). Another useful web-based tool is MEXPRESS (https://mexpress.be) (Koch, De Meyer, Jeschke, & Van Criekinge, 2015): it allows users to readily visualize and interpret gene expression and DNA methylation data coupled to various clinical variables for different TCGA datasets. However, even though they provide an easy-to-use, user-friendly interface, the number of varied analyses that can be performed are relatively limited.

Alternatively, many different investigations can be performed using higher level data that has been analyzed using consistent and widely accepted methods. One useful source of higher-level data is the Firehose pipeline implemented by the Broad Institute Genome Data Analysis Center.

Firehose provides access to standardized higher-level data for each individual TCGA cohort as of the final build of 2016_01_28 on their server (https://gdac.broadinstitute.org/). For the purposes of this chapter, key files include mRNA-seq and methylation data that has been analyzed and normalized using a standardized pipeline that allows comparison within and between TCGA cohorts. This data can be freely downloaded by simply clicking on the "Browse" link located in rightmost column of a specific TCGA cohort. A new window labeled "Archives" will appear with a list of GZ archive files that can be downloaded and uncompressed using any tool that uses the standard GNU zip (gzip) compression algorithm. The TAR file is subsequently uncompressed to yield two text files: the data file and a manifest file. The data is in the form of a tab-delimited text file and can typically be opened and manipulated in a spreadsheet application such as Microsoft Excel. Thus, high level data can be downloaded and ready to use immediately by nearly any group, and with minimal user learning time. Alternatively, for some specific analyses, the Broad Institute's Firebrowse website (http://firebrowse.org/) can be used as a powerful, yet simple, tool to compare genomic alterations or gene expression across the various TCGA cohorts.

For gene expression comparisons using the TCGA Illumina mRNA-seq data, multiple types of high level files are available from Firehose. Typical comparisons would likely utilize the gene level data, which contains RNA-seq results for ~20,500 genes for each sample in each specific TCGA cohort. In some cases, a second set of gene expression data that may be important is the gene isoforms data. This archive contains expression data for ~73,600 gene isoforms for every sample in each TCGA cohort. This second dataset allows comparison of expression levels of each isoform product of a given gene, which may be required in some instances (Gameiro, Kolendowski, et al., 2017). For either case, it is important to download the archives that are designated as RSEM-normalized. RSEM is an acronym for RNA-Seq by Expectation Maximization (Li & Dewey, 2011). The use of mRNA expression data that has undergone RSEM normalization via a uniform pipeline allows comparison of samples within and between TCGA cohorts. Effectively, these files represent tables listing a quantitative RNA-seq read value for each gene or isoform for each TCGA sample. Note that individual cohort members may have contributed multiple samples in addition to primary tumor, such as matched normal tissue and/or metastatic lesions. This can be determined via the barcoding described above.

For the TCGA genome methylation data, Illumina Infinium DNA methylation Beadchip array data for either the 27K, 450K (Bibikova et al., 2011), or both platforms is available via Firehose for each TCGA cohort. These data archives contain a beta value ranging from 0 to 1 for each of the ~27,000 or ~450,000 CpG probes in the array for each TCGA sample in that cohort. Beta values represent an estimate of methylation level of a given CpG, using the ratio of intensities between methylated and unmethylated alleles. For example, a beta value of 1 represents a fully methylated CpG site in the sample. Conversely a beta value of 0 represents a fully unmethylated CpG in the sample. Another key set of files available from Firehose is the Clinical Merge archive. These contain files listing all clinical data for each member of a given TCGA cohort, including basic patient data like age, gender and race, information on clinical outcomes and specialized data related to treatment modality, tumor pathology characteristics, smoking status, etc.

## 4. Using specific genes for estimating relative tumor-infiltrating lymphocyte (TIL) numbers and immune status

The intratumoral abundance of various immune cell populations and, potentially, their activation status, can be inferred using expression of various genes that are not normally expressed by carcinoma cells, stroma, or vascular tissue present in bulk tumor. Numerous individual marker genes or gene expression signatures have been successfully utilized for this purpose (Bindea et al., 2013; Danaher et al., 2017; Gameiro, Zhang, et al., 2017; Gameiro et al., 2018; Newman et al., 2015; Rooney et al., 2015; Wolf & Samuels, 2018) and a list of specific candidates that might be useful in profiling tumor immune status, grouped by leukocyte subtype or functional significance are listed in detail below and summarized in Table 2. Note that not all genes listed will be relevant or useful in all analyses. Caution must be exercised, as a number of these candidates are expressed by different immune cell types. Typically, multiple markers should be used for each class of lymphocyte and correlations between expression of two or more markers can help establish the utility of a given immune marker in a specific class of solid tumors. Predeveloped lymphocyte gene signatures can also be exploited for this via a number of software packages (see Section 5), but these solutions function as a "black box" providing little understanding to the user regarding which genes in these signatures are expressed/not expressed, and their relative levels.

### 4.1 Exploring the relative status of helper T-cells in the tumor microenvironment

$CD4^+$ T Helper cells ($T_H$) are perhaps the most crucial component of adaptive immunity, shaping both the humoral and cytotoxic responses toward

**Table 2** Potentially useful markers for the characterization of tumor-infiltrating leukocytes. Summary of individual marker genes or gene expression signatures that might be useful in profiling tumor immune status, grouped by leukocyte subtype or functional significance.

| Immune cell subtype | Gene expression signature |
|---|---|
| $CD4^+$ Helper T-cells | *CD4* |
| $T_H1$ | *TBX21, IL12RB2, GZMK, IFNG, PRF1, CCL5* |
| $T_H2$ | *GATA3, STAT6, IL4, IL10, IL13* |
| $T_H9$ | *BATF, IRF4, IL9* |
| $T_H17$ | *RORA, RORC, KLRB1, CTSH, PTPN13* |
| $T_H22$ | *IL22, FGF1, FGF12, FGF13, CCL7* |
| $CD8^+$ T-cells (presence) | *CD8a, CD8b* |
| $CD8^+$ T-cells (activated) | *IFNG, TNF, GZMA, GZMB, PRF1, TNFRSF9, CD69, CD44, HLA- DRB1, HLA-DRA, CD38, ITGAE, ENTPD1* |
| $CD8^+$ T-cells (exhausted) | *PDCD1, HAVCR2, LAG3, TIGIT, ENTPD1, CTLA4, BTLA, CD244, CD160, VSIR* |
| Anti-tumor suppressors | *INDO, CD274, TGFB1, VEGF, HIF1A, ARG1, CTNNB1, STAT3, PTGES2* |
| Regulatory T-cells | *FOXP3, IL2RA* |
| Dendritic cells (presence) | *ITGAX, ITGAE, MRC1, NOD1, CD1A* |
| Dendritic cells (activated) | *CD80, CD83, CD86, IL12A, STAT4, IL23A, EBI3* |
| Natural killer cells (presence) | *NKAM1, SH2D1A, XCL2, KIR2DL3, KIR3DL1, KIR3DL2* |
| Natural killer cells (activated) | *KLRK1, NCR1, NCR2, NCR3* |
| Macrophages (general) | *CD68* |
| M1 Macrophages | *CD40, HLA-DRB1* |
| M2 Macrophages | *CD163, CD206, CD204* |

external (microbes) and internal (cancerous cells) threats in the body. Although the strength of the T-cell receptor (TCR) signaling (signal 1) and engagement of costimulatory molecules (signal 2) influence naïve $CD4^+$ T-cell polarization, it is the presence of key cytokines in the microenvironment (signal 3) that mainly dictate in which specific direction naive

$T_H$ cells differentiate (Kaplan, Hufford, & Olson, 2015; Luckheeram, Zhou, Verma, & Xia, 2012). Cytokine activation of signal transducer and activator of transcription (STAT) family members are what mainly decides the fate of a naïve T-cell. Some cytokines such as IL-2 are important for the development of various $T_H$ cell families, while other cytokines, particularly IL-12, are only important for one $T_H$ lineage ($T_H1$).

$T_H1$ cells direct the cytotoxic T-cell responses that are protective against cancer. A patient's tumor with a dominant $T_H1$ signature often has more $CD8^+$ cytotoxic T-cells (CTLs) and is associated with better patient outcome (Dai, Hellstrom, Yip, Sjögren, & Hellstrom, 2018; Johansson, Denardo, & Coussens, 2008). On the other hand, a $T_H2$ signature in the tumor microenvironment is associated with poor patient outcome (Zhao, Wu, Wu, Wang, & Huang, 2015). $T_H17$ cells play a more complex role in tumor immunity either suppressing or promoting tumor growth depending on the type of cancer and therapy (Bailey et al., 2014). Moreover, depending on the type of malignancy, analyzing $T_H9$ or $T_H22$ signatures could be informative. For example, $T_H9$ cells are characterized by their ability to produce IL-9 and IL-21. These cells have remarkable antitumor activity against melanoma; however, they exhibit tumor-promoting effects in leukemia and lymphomas (Chen, Lv, Li, Lu, & Wang, 2014; Végran, Apetoh, & Ghiringhelli, 2015). $T_H22$ cells are recognized by their ability to produce IL-22 and these cells are linked to the development of colorectal cancer via the activation of STAT3 (Huang, Cao, Jiang, Zhang, & Gao, 2015). The frequency of $T_H22$ cells and IL-22 levels are also found to be significantly increased in the peripheral blood of patients with late stage ovarian cancer (stages III-IV) and this is positively correlated with tumor progression (Wang et al., 2017).

While the level of *CD4* expression can be used as a general marker of $T_H$ cell presence in a tumor sample, expression levels of the following genes have potential utility for determining the relative weighting of each of the $T_H$ signatures in the TCGA dataset:

$T_H1$: *TBX21*, *IL12RB2*, *GZMK*, *IFNG*, *PRF1*, and *CCL5* (Duhen, Ni, & Campbell, 2014; Gameiro et al., 2018; Hu et al., 2017; Oestreich & Weinmann, 2012)

$T_H2$: *GATA3*, *STAT6*, *IL4*, *IL10*, and *IL13* (Dai et al., 2018; Gameiro et al., 2018)

$T_H9$: *BATF*, *IRF4*, and *IL9* (Jabeen et al., 2013)

$T_H17$: *RORA*, *RORC*, *KLRB1*, *CTSH*, and *PTPN13* (Castro et al., 2017; Duhen et al., 2014; Gameiro et al., 2018)

$T_H22$: *IL22*, *FGF1*, *FGF12*, *FGF13*, and *CCL7* (Eyerich et al., 2009)

## 4.2 Exploring the relative infiltration and activation status of cytotoxic T lymphocytes in the tumor microenvironment

One important class of T-cells that are critical for antitumor immunity and are the main target for cancer immunotherapeutics today are the CTLs. These cells can identify and destroy malignant cells based on the recognition of specific peptides presented on the surface of tumor cells by major histocompatibility complex class I (MHC I). Exclusion of CTLs from the tumor microenvironment is one mechanism that allows tumors to escape protective immunity (Jiang et al., 2018). As such, the presence of $CD8^+$ CTLs in the tumor is generally a positive prognostic biomarker in many cancers (Fukunaga et al., 2004; Kärjä et al., 2005; Mahmoud, Lee, Ellis, & Green, 2012; Mlecnik et al., 2011; Sato et al., 2005; Sharma et al., 2007; van Oijen et al., 2004). Total levels of infiltrating $CD8^+$ CTLs can be estimated by expression of *CD8a* and *CD8b*. The primary function of CTLs is to identify and destroy tumor cells, which is carried out primarily via secretion of effector molecules such as IFN-γ, TNF, granzymes A and B, and perforin. Assessing the mRNA expression levels of these effector genes also provides vital information on the status of CTLs in the tumor microenvironment. The following genes can be examined to determine the status of the aforementioned molecules: *IFNG*, *TNF*, *GZMA*, *GZMB*, and *PRF1*.

The effector function of $CD8^+$ T-cells is directly linked to their activation status. We expect T-cells with effector functions to be fully activated. Therefore, the activation status of T-cells can be also informative when examining the immune landscape of a tumor. A member of the tumor necrosis factor receptor (TNFR)-family, CD137 (often called 4-1BB) has T-cell costimulatory functions that promote T-cell expansion, survival, and enhances their cytolytic functions. *CD137* expression is unique to activated T-cells and is absent on naïve T-cells. CD137 is often considered a superior marker for TILs and has been used to selectively isolate tumor-specific T-cells from TILs from ovarian tumors and ascites (Ye et al., 2014).

In addition to CD137, other markers of TIL activation can be informative. Indeed, examining the coordinate expression of two activation markers of TILs (explained in more detail below) can identify specific tumor-reactive subsets with important ramifications on clinical outcome. For example, CD103, integrin αE, is expressed on a subset of $CD8^+$ TILs and plays a role in recruiting these cells into the tumor upon activation (Djenidi et al., 2015). CD39 is an ectonucleotidase that is expressed on various immune subsets including activated as well as exhausted $CD8^+$ T-cells (Gupta et al., 2015; Simoni et al., 2018). CD39 has also been shown to be uniquely expressed on tumor-specific $CD8^+$ T-cells within the tumor microenvironment and not by bystander $CD8^+$ T-cells that have been recruited to the tumor but recognize antigens unrelated to the tumor (Simoni et al., 2018). The coexpression of CD103 and CD39 identifies a unique population of $CD8^+$ TILs with a tissue-resident memory phenotype, tumor reactivity, exhausted phenotype, distinct TCR repertoire. Notably, head and neck cancer patients with greater frequency of these cells at the time of surgery had higher overall survival (Duhen et al., 2018). Therefore, examining the correlation of expression of CD39 and CD103 can provide valuable insight on activated T-cells in the tumor microenvironment. The following genes can be examined to determine the status of activated T-cells in the tumor microenvironment: *TNFRSF9* (CD137), *CD69*, *CD44*, *HLA-DRB1* (HLA-DR), *HLA-DRA* (HLA-DR), *CD38*, *ITGAE* (CD103), and *ENTPD1* (CD39).

The activity of $CD8^+$ CTLs can be suppressed by various mechanisms including suppressive molecules expressed by tumor cells or cells of the tumor microenvironment, including indoleamine 2,3-dioxygenase (IDO), transforming growth factor beta (TGF-β), adenosine, galectin, vascular endothelial growth factor (VEGF), and hypoxia-inducible factor 1-alpha (HIF-1α). These molecules can either directly suppress tumor-reactive T-cells or they can activate certain immune regulatory pathways that can either induce anergy on antitumor immune cells (mainly T-cells) or exclude them from the tumor microenvironment and create the immune excluded phenotype. For example, IDO expressed by tumor cells or by mycloid-derived suppressor cells (MDSCs) and macrophages depletes tryptophan, the most essential amino acid, from the tumor microenvironment. CTLs depend on tryptophan for their function and survival (Lee et al., 2002). Additionally, tryptophan catabolism results in the production of kynurenine metabolites that can either directly suppress CTLs or induce regulatory T cells (Tregs) that can further induce apoptosis in CTLs (Frumento et al., 2002; Mezrich et al., 2010). Marker genes upregulated on tumor cells to suppress anti-tumor responses include: *INDO* (IDO1), *CD274* (PD-L1), *TGFB1*, *VEGF*, *HIF1A*, *ARG1*, *CTNNB1*, *STAT3*, and *PTGES2*.

Tregs play a crucial role in maintaining immune homeostasis and self-tolerance. In cancer, however, they help to establish an immunosuppressive tumor microenvironment that promotes tumor immune evasion. Tregs suppress antitumor immunity via various contact-dependent or contact-independent mechanisms and their accumulation in tumors (increased Treg/$CD8^+$ ratio) is associated with poor patient outcome in many cancers

(Chaudhary & Elkord, 2016). Various mechanisms that Tregs use to suppress tumor-reactive CTLs include: (1) directly consuming IL-2 and reducing its bioavailability, which induces metabolic disruption of CTLs' activity; (2) producing immunosuppressive cytokines such as IL-10, TGFβ, and IL-35; (3) direct expression of checkpoint molecules (CTLA-4, PD-1, LAG-3, TIGIT, TIM-3, IDO); (4) induction of tolerogenic DCs; and (5) induction of apoptosis directly on CTLs (Chaudhary & Elkord, 2016; McNally, Hill, Sparwasser, Thomas, & Steptoe, 2011). Expression of genes such as *FOXP3* and *IL2RA* can be used to determine the presence of Tregs in the tumor microenvironment.

Chronic antigen exposure along with the harsh metabolic conditions present in the tumor microenvironment can induce T-cell exhaustion that is accompanied by surface expression of various checkpoint molecules on tumor-reactive T-cells (Gameiro et al., 2018; Zhang et al., 2017). These exhaustion molecules are currently the main target of novel cancer immunotherapeutics such as anti-CTLA-4 and anti-PD-1/PD-L1 drugs. Furthermore, several other novel checkpoint inhibitors are being developed that block exhaustion molecules other than PD-1 and PD-L1, such as LAG-3, TIM-3, and TIGIT. These checkpoint molecules are often coregulated, hence they can be coexpressed on exhausted T-cells, which leads to resistance to monotherapy with only one checkpoint inhibitor (Borrie & Maleki Vareki, 2018). The expression of these exhaustion molecules can also be induced by specific tumor conditions. For example, the hypoxic tumor microenvironment can induce LAG-3 expression on TILs via HIF-1α, independent of PD-1 expression. LAG-3 expression on TILs impairs their function and can be rescued via downregulation of HIF-1α in T-cells (Zhang et al., 2017). T-cell exhaustion genes that may be useful for characterizing the immune microenvironment include: *PDCD1* (PD1), *HAVCR2* (TIM3), *LAG3*, *TIGIT*, *ENTPD1* (CD39), *CTLA4*, *BTLA*, *CD244* (2B4), *CD160*, and *VSIR* (VISTA) (Borrie & Maleki Vareki, 2018).

Note that immunosuppressive molecules can also be found on stromal and tumor-supportive immune cells such as M2 macrophages and MDSCs.

Importantly, the presence or absence of tumor-reactive T-cells in the tumor can determine whether a patient would respond to immunotherapy. As one example, melanoma patients with a T-cell-inflamed tumor, defined by a higher percentage of $CD4^+$ and $CD8^+$ T-cells and an IFN signature, have a much higher objective response rate (ORR) and improved overall survival (OS) compared to patients with a non-T-cell-inflamed phenotype (Gajewski, Schreiber, & Fu, 2013; Maleki Vareki, 2018). This is because the mechanism of action of immune checkpoint inhibitors mainly involves reinvigorating the already existing anti-tumor immunity, rather than generating new tumor-reactive T-cells (Maleki Vareki, 2018).

## 4.3 Characterizing the number and status of dendritic cells in the tumor microenvironment

Initiation of antitumor immunity mainly depends on a subset of conventional dendritic cells (cDC1;CD141$^{hi}$, CD13$^{hi}$) that are able to pick up tumor antigens, travel to draining lymph nodes, and present those antigens to naïve T-cells in a process called "cross-priming" (Gardner & Ruffell, 2016; Granot et al., 2017). cDC1s also provide those T-cells with costimulatory molecules (CD80/CD86) and inflammatory cytokines such as IL-12, which drive T-cell proliferation and differentiation to induce an optimal antitumor response. The activated T-cells then travel to the site of the tumor and other parts of the body and destroy the tumor cells that express the same antigen.

Other DC subsets including cDC2 (CD11c, SIRPα, and CD1c), plasmacytoid DCs (CD123, CD303/BDCA2, and CD304/BDCA4), and monocyte-derived DCs (CD14$^{hi}$) do not have the same capacity as cDC1 to activate antitumor immunity. Given their transient presence in the tumor it is difficult to reliably determine the presence of cDC1 cells in tumors via gene expression profiling. However, it is possible to couple the expression of their lineage marker CD141 gene (*THBD*) with the transcription factor IRF8 (*IRF8*) to get a better picture of their presence in the tumor microenvironment. We should note that IRF8 is also important in the development of plasmacytoid DCs (Collin & Bigley, 2018; Gardner & Ruffell, 2016) and CD141 is also expressed on macrophages and other cells. Expression of several other DC markers can be used to determine the general presence of DCs in the tumor including: *ITGAX* (CD11c), *ITGAE* (CD103), *MRC1* (CD206), *NOD1*, and *CD1A*. The following genes may be useful in determining the activation status of DCs in the tumor: *CD80*, *CD83*, *CD86*, *IL12A*, *STAT4*, *IL23A*, and *EBI3* (Schinnerling, García-González, & Aguillón, 2015).

## 4.4 Characterizing the number and status of natural killer cells in the tumor microenvironment

Natural Killer (NK) cells are a subset of Innate Lymphoid Cells (ILCs) that are distinguished from adaptive immune cells by lack of antigen specificity and lineage specific markers. Similar to CD8$^{+}$ T-cells, NK cells are part of the immunosurveillance system against viral infections and tumors.

NK cells express the transcription factor T-bet (*TBX21*) and secrete effector cytokines including IFN-γ and TNF (Nicholson, Keating, & Belz, 2017). NK cells lyse virally-infected cells or tumor cells mainly via granzyme B and perforin, but also with Fas/FasL and TRAIL/TRAIL receptor triggering. NK cells cytolytic functions are regulated via the activatory molecules (e.g., NKp30, NKG2D, NKp44, DNAM-1, 2B4, and CD16) (Martinet & Smyth, 2015; Yokoyama & Plougastel, 2003) and MHC class I-mediated inhibitory receptors (e.g., KLRG1, Ly49 receptors, and leukocyte immunoglobulin-like receptors). NK cells depend on IL-15 for their proliferation, survival, and effector function (Nicholson et al., 2017). The following genes could be examined to determine the presence and activation of NK cells in the tumor microenvironment:

NK cell markers: *NKAM1*(CD56), *SH2D1A*, *XCL2*, *KIR2DL3*, *KIR3DL1*, and *KIR3DL2* (Gameiro et al., 2018).

NK cell activation genes: *KLRK1*, *NCR1* (NKp46), *NCR2* (NKp44), and *NCR3* (NKp30) (Campbell, Yusa, Kikuchi-Maki, & Catina, 2004; Freud et al., 2013).

## 4.5 Characterizing the number and status of macrophage in the tumor microenvironment

Macrophages are a subset of myeloid cells that are often among the most abundant leukocytes in any solid tumor. Undifferentiated macrophages are called M0 whereas differentiated macrophages are divided into M1 and M2 subtypes based on pro- and anti-inflammatory properties, respectively. Tumor-associated macrophages (TAMs) are typically immunosuppressive (mainly an M2 phenotype) and play a prominent role in tumor progression and resistance to therapy (Mantovani, Marchesi, Malesci, Laghi, & Allavena, 2017). These cells can directly or indirectly suppress TILs via the expression of (1) immune checkpoint molecules such as PD-L1 and VISTA (Gao et al., 2017); (2) secretion of immunoregulatory cytokines (IL-10 and TGF-β); (3) production of suppressive molecules (IDO, arginase, and VEGF); (4) induction of Tregs; and (5) inhibition of DC maturation. M1 macrophages are tumor inhibitory with antigen presenting capacities and are commonly induced in the presence of IFN-γ. Given the important role of TAMs in oncology and their prognostic value (Aras & Zaidi, 2017), it is often crucial to determine the presence and status of these cells in the tumor microenvironment. Expression of the *CD68* gene can be used to broadly identify all macrophages in a tumor.

The following genes could be examined to determine the presence and type of macrophages in the tumor microenvironment (Newman et al., 2015).

M1 macrophages: *CD40* and *HLA-DRB1*.

M2 macrophages: *CD163*, *CD206*, and *CD204*.

## 5. Mining existing TCGA-wide analyses of RNA expression as a tool for estimating various immune parameters

In addition to using single specific genes "a la carte" to estimate the intratumoral abundance of various immune cell populations, multiple groups have proposed multigene signatures as more powerful approaches (Finotello & Trajanoski, 2018). This has led to the evolution of complex tools, such as Cell-type Identification By Estimating Relative Subsets Of RNA Transcripts (CIBERSORT), which uses a 547 gene panel to deconvolute complex mixtures of cell types and distinguish 22 human hematopoietic cell phenotypes. This analysis estimates the relative levels of seven T-cell types, naïve and memory B cells, plasma cells, NK cells, and myeloid subsets (Newman et al., 2015). In addition, this tool has been used on every sample for the 20 solid cancers in the TCGA (Charoentong et al., 2017) and the fully-analyzed data is available online from the Pan-Cancer Atlas. The CIBERSORT immune fraction data file (TCGA.Kallisto.fullIDs.cibersort.relative.tsv) can be downloaded from the National Cancer Institute Cancer Genome Commons (https://gdc.cancer.gov/about-data/publications/panimmune). More importantly, the CIBERSORT algorithm generates a false discovery rate (FDR) value, providing a statistical measure of confidence for the estimation of each of the 22 immune cell types in each TCGA sample. Selection of a relatively stringent value (e.g., $P<0.01$) may help restrict any subsequent comparisons to the subset of data with the highest statistical confidence. Numerous other algorithms that deconvolute the immune infiltration of cancer samples have been reported, and can similarly be applied to analyze TCGA data (Finotello & Trajanoski, 2018).

## 6. Immunomethylomics: Mining existing TCGA-wide analyses of DNA methylation as a tool for estimating various immune parameters

As an alternative to using the levels of expression of leukocyte specific genes as a means to estimate the relative level of immune infiltration of a given tumor sample, differential genomic methylation can also be used

(Chakravarthy et al., 2018; Thorsson et al., 2018). In its simplest form, the total leukocyte fraction has been estimated by the Pan-Cancer Atlas for all TCGA samples with available methylation data and this file (TCGA_all_leuk_estimate.masked.20170107.tsv) can be downloaded from the National Cancer Institute Cancer Genome Commons (https://gdc.cancer.gov/about-data/publications/panimmune). This file contains the estimated leukocyte content expressed as a fraction for nearly 10,000 of the TCGA samples. The availability of this "ready-to-go" data means that a variety of comparisons can be performed to determine if correlations exist between various gene expression patterns, clinical features, and overall TIL levels.

A more specific tool using deconvolution of methylation signatures has been recently developed along the lines of CIBERSORT, and is referred to as methylCIBERSORT (Chakravarthy et al., 2018). This approach estimates not only leukocyte fraction, but also immune subtype populations, allowing the classification of tumors into "immune hot" vs "immune cold." A more thorough description of the use of immunomethylomics can be found in the accompanying chapter by De Carvalho et al.

## 7. Taking advantage of the clinical data

Given the relatively short 10-year time frame of the TCGA program, the acquisition of comprehensive clinical annotation was not possible for every TCGA sample. Nevertheless, substantial clinical data is available, allowing comparison of many clinical variables and, in most cases, reasonably robust outcomes assessment. Like the RNA-seq archives, the clinical merge files for each TCGA cohort are also freely available via Firehose. These clinical merge files contain standardized patient information, such as age, gender, race, smoking status, as well as cohort specific information related to anatomical subsite, treatment information, viral infection status, etc. This data can be extracted and used in correlative studies with the RNA-seq data to discover correlations between gene expression and clinical variables.

One commonly desired set of analyses based on clinical data are related to patient survival. In many clinical studies, 2, 5, or even 10-year survival rates are calculated to provide prognostic information or to compare the effectiveness of alternative treatments. Survival rates are typically based on mortality or disease progression events. As such, sufficient time posttreatment must elapse to capture these events. However, the minimum follow-up time required varies depending on both the aggressiveness of each type of cancer being investigated and which clinical endpoint is utilized. Thus, data collected by the TCGA might have been collected sufficiently long enough for enough patients in one cancer cohort to allow statistically significant outcome differences to be detected, but not long enough for another cohort based on the vagaries of patient accrual and disease specific outcome characteristics.

A recent study across all the TCGA cohorts has made studies using clinical outcome substantially easier (Liu et al., 2018). This systematic analysis of the TCGA clinical data across all 33 cohorts assessed the adequacy of the clinical follow-up interval for four survival measures, including overall survival and its surrogate survival measures including disease-specific survival, disease-free interval, and progression-free interval. Overall, this study recommends the use of all four endpoints without reservation for 12 of the TCGA solid tumor cancer types and recommends various subsets for the others. This paper also provides integrated, curated, and filtered clinical and survival outcome data for all TCGA patients, which serves as a useful resource for any studies comparing survival. This file is available at: https://www.cell.com/cms/10.1016/j.cell.2018.02.052/attachment/f4eb6b31-8957-4817-a41f-e46fd2a1d9c3/mmc1.xlsx.

## 8. Practical examples of gene expression data used to characterize the tumor microenvironment and correlate results with clinical outcomes

### 8.1 An example analysis using differential gene expression to infer altered immune infiltration in a TCGA cohort

As a practical example of one application of the TCGA data, we could estimate infiltration of kidney cancers by $CD8^+$ T-cells. There are three types of kidney cancers in the TCGA: Kidney Chromophobe (KICH), Kidney Renal Clear Cell Carcinoma (KIRP), and Kidney Renal Papillary Cell Carcinoma (KIRC). Each cohort has a significant number of tumor and normal tissue samples as indicated in Table 1. Using RSEM normalized mRNA-seq data downloaded from Firehose for each cohort for *CD8a* expression (a marker of $CD8^+$ T-cells as described above) we can calculate a median expression for tumor and normal control samples and plot these values as a box plot, as shown in Fig. 2. These data suggest that the relative

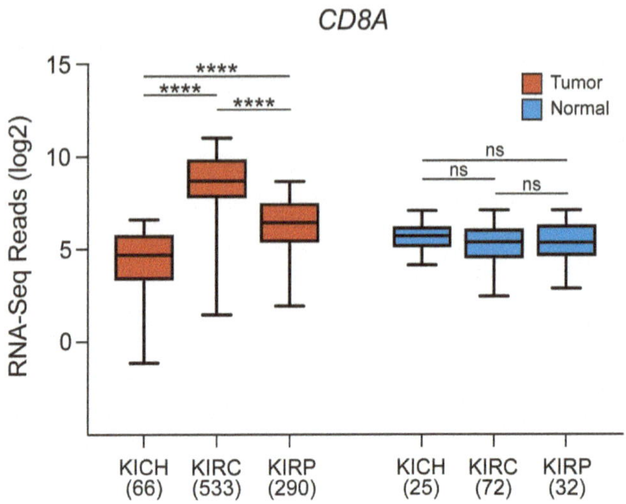


**Fig. 2** Example estimation of $CD8^+$ tumor cell infiltration in three TCGA kidney cancer cohorts. RSEM normalized mRNA-seq data for the KICH, KIRP, and KIRC cohorts was downloaded from the Firehose server at the Broad GDAC. Values for *CD8A* expression were extracted for tumor and normal samples and analyzed as described in the text. Results are presented using boxplot diagrams and suggest that there are significant differences in $CD8^+$ TIL infiltration across the cohorts, such that KIRC > KIRP > KICH. In contrast, no significant differences exist between the normal tissue control samples across the three cohorts.

infiltration of $CD8^+$ CTLs is quite different between these three types of kidney cancers, but not in the normal control tissues. For the box plots, center lines show the medians, box limits indicate the 25th and 75th percentiles as determined by standard statistical software, and whiskers extend 1.5 times the interquartile range from the 25th and 75th percentiles. As the data from these samples do not generally exhibit a normal distribution, statistical significance is typically calculated using a nonparametric test. This is often done using the Mann-Whitney *U* test. If no significant difference is observed, a *post hoc* power calculation should be performed to determine if the analysis is sufficiently powered. This can be done with G*Power software (Faul, Erdfelder, Lang, & Buchner, 2007), using *post hoc* *t*-test family calculations, with effect size selected as 0.8 and $\alpha = 0.05$. Importantly, if a given comparison does not achieve a power value >0.8, the analysis is likely underpowered and the lack of significance is inconclusive as it may simply be related to insufficient sample numbers. This is primarily an issue in comparisons between tumor and normal tissue in cohorts like CESC, GBM, and PAAD, for which the TCGA was unable to collect many normal tissue samples.

## 8.2 An example analysis for determining if expression of two immune marker genes are coexpressed in tumor samples from the TCGA cohort

As a practical example of another application of the TCGA data, we could determine if the expression of two immunologically related genes are expressed coordinately. The T-cell co-receptor expressed by CD8$^+$ T-cells is comprised of an alpha and beta chain, encoded by the *CD8a* and *CD8b* genes, respectively. Tumor-infiltrating CD8$^+$ CTLs should express both genes and we expect that patient samples with high levels of one would also express high levels of the other. When the expression values for *CD8a* are plotted against *CD8b* for each of the three TCGA kidney cancer cohorts (Fig. 3), a clear correlation between expression of these two markers is observed. Statistical significance can be determined using Spearman's Rho analysis, with $P<0.05$ as a cutoff.

## 8.3 Using the clinical data to correlate immune activity based on gene expression markers with clinical outcome

As a practical example of one application of the TCGA outcomes data, we can stratify patients from the cervical squamous cell carcinoma and endocervical adenocarcinoma (CESC) cohort into two groups based on median expression of the T-cell exhaustion marker *TIGIT*, with their associated clinical outcomes identified from the outcomes file from reference (Liu et al., 2018). Plotting the overall survival, disease-specific survival, disease-free interval, and progression-free interval for patients expressing

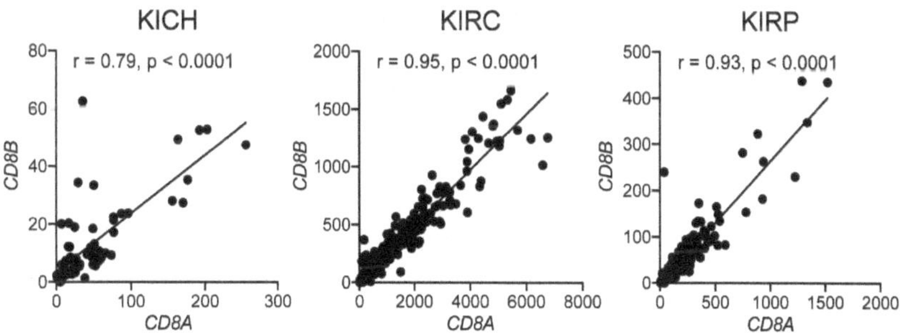


**Fig. 3** Example analysis to determine if two markers of CD8$^+$ CTLs are coordinately expressed in the TCGA kidney cancer cohorts. RSEM normalized mRNA-seq data for the KICH, KIRC, and KIRP cohorts was downloaded from the Firehose server at the Broad GDAC. Expression values for *CD8A* and *CD8B*, two markers of CD8$^+$ CTLs, were plotted as a scatterplot and subjected to a Spearman's Rho analysis. Expression of both genes is significantly correlated across all three cohorts.

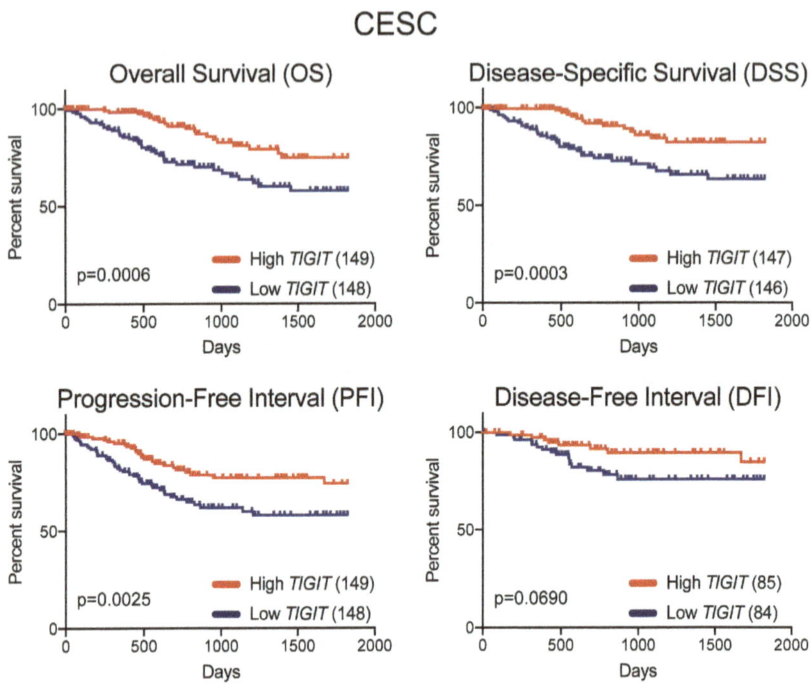


**Fig. 4** Example comparison of clinical outcomes with expression of the immune exhaustion marker *TIGIT* in the TCGA cervical cancer cohort. RSEM normalized mRNA-seq data for the CESC cohort was downloaded from the Firehose server at the Broad GDAC. Samples were dichotomized based on median values for *TIGIT* expression and plotted as Kaplan-Meier curves based on clinical outcome data reported in Liu et al. (2018). Improved overall survival, disease-specific survival, and progression-free interval, but not disease-free interval, are correlated with high levels of *TIGIT* expression.

higher than median levels of *TIGIT* vs those expressing lower than median levels in a Kaplan-Meier curve, a clear survival advantage is seen for the group with high *TIGIT* expression by all measures except disease-free interval (Fig. 4).

## 9. Summary and conclusions

Although completed in 2017, the TCGA project provides an abundant source of free data available for mining by cancer researchers in many different ways. This chapter focuses on how to use this information to visualize the immune landscape of various solid tumors and to attempt to correlate changes in TIL infiltration with clinical parameters such as overall survival. Importantly, many other questions unrelated to tumor immunity can be similarly addressed, such as correlations between gene expression and chemotherapy response, alterations induced by oncogenic pathogens, or expression of markers of differentiation, proliferation or tumor aggressiveness. Additionally, batch-wise analysis of this valuable data set for various correlations can also generate novel hypotheses that can be pursued experimentally. Literally, the seeds of thousands of important analyses await discovery in this collection of "big" data.

## Acknowledgments

This work was supported from grants provided by the Canadian Institutes of Health Research (MOP#142491) to J.S.M. and A.C.N. and (MOP#389137) to J.K. and S.M.V. Steven F. Gameiro was supported in part from a Cancer Research and Technology Transfer (CaRTT) studentship.

## References

Anaya, J. (2016). OncoLnc: Linking TCGA survival data to mRNAs, miRNAs, and lncRNAs. *PeerJ Computer Science*, *2*, e67. https://doi.org/10.7717/peerj-cs.67.

Aras, S., & Zaidi, M. R. (2017). TAMeless traitors: Macrophages in cancer progression and metastasis. *British Journal of Cancer*, *117*(11), 1583–1591. https://doi.org/10.1038/bjc.2017.356.

Bailey, S. R., Nelson, M. H., Himes, R. A., Li, Z., Mehrotra, S., & Paulos, C. M. (2014). Th17 cells in cancer: The ultimate identity crisis. *Frontiers in Immunology*, *5*(2), 276. https://doi.org/10.3389/fimmu.2014.00276.

Bibikova, M., Barnes, B., Tsan, C., Ho, V., Klotzle, B., Le, J. M., et al. (2011). High density DNA methylation array with single CpG site resolution. *Genomics*, *98*(4), 288–295. https://doi.org/10.1016/j.ygeno.2011.07.007.

Bindea, G., Mlecnik, B., Tosolini, M., Kirilovsky, A., Waldner, M., Obenauf, A. C., et al. (2013). Spatiotemporal dynamics of intratumoral immune cells reveal the immune landscape in human cancer. *Immunity*, *39*(4), 782–795. https://doi.org/10.1016/j.immuni.2013.10.003.

Borrie, A. E., & Maleki Vareki, S. (2018). T lymphocyte-based cancer immunotherapeutics. *International Review of Cell and Molecular Biology*, *341*, 201–276. https://doi.org/10.1016/bs.ircmb.2018.05.010.

Byrne, E. H., & Fisher, D. E. (2017). Immune and molecular correlates in melanoma treated with immune checkpoint blockade. *Cancer*, *123*(S11), 2143–2153. https://doi.org/10.1002/cncr.30444.

Campbell, K. S., Yusa, S.-I., Kikuchi-Maki, A., & Catina, T. L. (2004). NKp44 triggers NK cell activation through DAP12 association that is not influenced by a putative cytoplasmic inhibitory sequence. *Journal of Immunology*, *172*(2), 899–906.

Castro, G., Liu, X., Ngo, K., De Leon-Tabaldo, A., Zhao, S., Luna-Roman, R., et al. (2017). RORγt and RORα signature genes in human Th17 cells. *PLoS One*, *12*(8). e0181868. https://doi.org/10.1371/journal.pone.0181868.

Cerami, E., Gao, J., Dogrusoz, U., Gross, B. E., Sumer, S. O., Aksoy, B. A., et al. (2012). The cBio cancer genomics portal: An open platform for exploring multidimensional cancer genomics data. *Cancer Discovery*, *2*(5), 401–404. https://doi.org/10.1158/2159-8290.CD-12-0095.

Chakravarthy, A., Furness, A., Joshi, K., Ghorani, E., Ford, K., Ward, M. J., et al. (2018). Pan-cancer deconvolution of tumour composition using DNA methylation. *Nature Communications*, *9*(1), 3220. https://doi.org/10.1038/s41467-018-05570-1.

Charoentong, P., Finotello, F., Angelova, M., Mayer, C., Efremova, M., Rieder, D., et al. (2017). Pan-cancer immunogenomic analyses reveal genotype-immunophenotype relationships and predictors of response to checkpoint blockade. *Cell Reports*, *18*(1), 248–262. https://doi.org/10.1016/j.celrep.2016.12.019.

Chaudhary, B., & Elkord, E. (2016). Regulatory T cells in the tumor microenvironment and cancer progression: Role and therapeutic targeting. *Vaccine*, *4*(3), 28. https://doi.org/10.3390/vaccines4030028.

Chen, N., Lv, X., Li, P., Lu, K., & Wang, X. (2014). Role of high expression of IL-9 in prognosis of CLL. *International Journal of Clinical and Experimental Pathology*, 7(2), 716–721.

Collin, M., & Bigley, V. (2018). Human dendritic cell subsets: An update. *Immunology*, *154*(1), 3–20. https://doi.org/10.1111/imm.12888.

Costa, F., Das, R., Kini Bailur, J., Dhodapkar, K., & Dhodapkar, M. V. (2018). Checkpoint inhibition in myeloma: Opportunities and challenges. *Frontiers in Immunology*, *9*, 2204. https://doi.org/10.3389/fimmu.2018.02204.

Dai, M., Hellstrom, I., Yip, Y. Y., Sjögren, H. O., & Hellstrom, K. E. (2018). Tumor regression and cure depends on sustained Th1 responses. *Journal of Immunotherapy (Hagerstown, MD):1997*, *41*(8), 369–378. https://doi.org/10.1097/CJI.0000000000000231.

Danaher, P., Warren, S., Dennis, L., D'Amico, L., White, A., Disis, M. L., et al. (2017). Gene expression markers of tumor infiltrating leukocytes. *Journal for Immunotherapy of Cancer*, *5*(1), 18. https://doi.org/10.1186/s40425-017-0215-8.

Djenidi, F., Adam, J., Goubar, A., Durgeau, A., Meurice, G., de Montpréville, V., et al. (2015). CD8+CD103+ tumor-infiltrating lymphocytes are tumor-specific tissue-resident memory T cells and a prognostic factor for survival in lung cancer patients. *Journal of Immunology*, *194*(7), 3475–3486. https://doi.org/10.4049/jimmunol.1402711.

Duhen, T., Duhen, R., Montler, R., Moses, J., Moudgil, T., de Miranda, N. F., et al. (2018). Co-expression of CD39 and CD103 identifies tumor-reactive CD8 T cells in human solid tumors. *Nature Communications*, *9*(1), 2724. https://doi.org/10.1038/s41467-018-05072-0.

Duhen, T., Ni, C., & Campbell, D. (2014). Identification of a specific gene signature in human Th1/17 cells (BA13P.126). *Journal of Immunology*, 192(1 Suppl), 177.12.

Emens, L. A. (2018). Breast cancer immunotherapy: Facts and hopes. *Clinical Cancer Research*, *24*(3), 511–520. https://doi.org/10.1158/1078-0432.CCR-16-3001.

Eyerich, S., Eyerich, K., Pennino, D., Carbone, T., Nasorri, F., Pallotta, S., et al. (2009). Th22 cells represent a distinct human T cell subset involved in epidermal immunity and remodeling. *The Journal of Clinical Investigation*, *119*(12), 3573–3585. https://doi.org/10.1172/JCI40202.

Faul, F., Erdfelder, E., Lang, A.-G., & Buchner, A. (2007). G*Power 3: A flexible statistical power analysis program for the social, behavioral, and biomedical sciences. *Behavior Research Methods*, *39*(2), 175–191.

Finotello, F., & Trajanoski, Z. (2018). Quantifying tumor-infiltrating immune cells from transcriptomics data. *Cancer Immunology, Immunotherapy*, *67*(7), 1031–1040. https://doi.org/10.1007/s00262-018-2150-z.

Freud, A. G., Zhao, S., Wei, S., Gitana, G. M., Molina-Kirsch, H. F., Atwater, S. K., et al. (2013). Expression of the activating receptor, NKp46 (CD335), in human natural killer and T-cell neoplasia. *American Journal of Clinical Pathology*, *140*(6), 853–866. https://doi.org/10.1309/AJCPWGG69MCZOWMM.

Fridman, W. H., Zitvogel, L., Sautès-Fridman, C., & Kroemer, G. (2017). The immune contexture in cancer prognosis and treatment. *Nature Reviews. Clinical Oncology*, *14*(12), 717–734. https://doi.org/10.1038/nrclinonc.2017.101.

Frumento, G., Rotondo, R., Tonetti, M., Damonte, G., Benatti, U., & Ferrara, G. B. (2002). Tryptophan-derived catabolites are responsible for inhibition of T and natural killer cell proliferation induced by indoleamine 2,3-dioxygenase. *The Journal of Experimental Medicine, 196*(4), 459–468. https://doi.org/10.1084/jem.20020121.

Fukunaga, A., Miyamoto, M., Cho, Y., Murakami, S., Kawarada, Y., Oshikiri, T., et al. (2004). CD8+ tumor-infiltrating lymphocytes together with CD4+ tumor-infiltrating lymphocytes and dendritic cells improve the prognosis of patients with pancreatic adenocarcinoma. *Pancreas, 28*(1), e26–e31.

Gajewski, T. F., Schreiber, H., & Fu, Y.-X. (2013). Innate and adaptive immune cells in the tumor microenvironment. *Nature Immunology, 14*(10), 1014–1022. https://doi.org/10.1038/ni.2703.

Gameiro, S. F., Ghasemi, F., Barrett, J. W., Koropatnick, J., Nichols, A. C., Mymryk, J. S., et al. (2018). Treatment-naïve HPV+ head and neck cancers display a T-cell-inflamed phenotype distinct from their HPV− counterparts that has implications for immunotherapy. *Oncoimmunology,* 7(10), e1498439. https://doi.org/10.1080/2162402X.2018.1498439.

Gameiro, S. F., Kolendowski, B., Zhang, A., Barrett, J. W., Nichols, A. C., Torchia, J., et al. (2017). Human papillomavirus dysregulates the cellular apparatus controlling the methylation status of H3K27 in different human cancers to consistently alter gene expression regardless of tissue of origin. *Oncotarget, 8*(42), 72564–72576. https://doi.org/10.18632/oncotarget.19885.

Gameiro, S. F., Zhang, A., Ghasemi, F., Barrett, J. W., Nichols, A. C., & Mymryk, J. S. (2017). Analysis of class I major histocompatibility complex gene transcription in human tumors caused by human papillomavirus infection. *Viruses, 9*(9), 252. https://doi.org/10.3390/v9090252.

Gao, J., Ward, J. F., Pettaway, C. A., Shi, L. Z., Subudhi, S. K., Vence, L. M., et al. (2017). VISTA is an inhibitory immune checkpoint that is increased after ipilimumab therapy in patients with prostate cancer. *Nature Medicine, 23*(5), 551–555. https://doi.org/10.1038/nm.4308.

Gardner, A., & Ruffell, B. (2016). Dendritic cells and cancer immunity. *Trends in Immunology, 37*(12), 855–865. https://doi.org/10.1016/j.it.2016.09.006.

George, S., Miao, D., Demetri, G. D., Adeegbe, D., Rodig, S. J., Shukla, S., et al. (2017). Loss of PTEN is associated with resistance to anti-PD-1 checkpoint blockade therapy in metastatic uterine leiomyosarcoma. *Immunity, 46*(2), 197–204. https://doi.org/10.1016/j.immuni.2017.02.001.

Gooden, M. J. M., de Bock, G. H., Leffers, N., Daemen, T., & Nijman, H. W. (2011). The prognostic influence of tumour-infiltrating lymphocytes in cancer: A systematic review with meta-analysis. *British Journal of Cancer, 105*(1), 93–103. https://doi.org/10.1038/bjc.2011.189.

Granot, T., Senda, T., Carpenter, D. J., Matsuoka, N., Weiner, J., Gordon, C. L., et al. (2017). Dendritic cells display subset and tissue-specific maturation dynamics over human life. *Immunity, 46*(3), 504–515. https://doi.org/10.1016/j.immuni.2017.02.019.

Gupta, P. K., Godec, J., Wolski, D., Adland, E., Yates, K., Pauken, K. E., et al. (2015). CD39 expression identifies terminally exhausted CD8+ T cells. *PLoS Pathogens, 11*(10), e1005177. https://doi.org/10.1371/journal.ppat.1005177.

Hanahan, D., & Weinberg, R. A. (2011). Hallmarks of cancer: The next generation. *Cell, 144*(5), 646–674. https://doi.org/10.1016/j.cell.2011.02.013.

Hoadley, K. A., Yau, C., Hinoue, T., Wolf, D. M., Lazar, A. J., Drill, E., et al. (2018). Cell-of-origin patterns dominate the molecular classification of 10,000 tumors from 33 types of cancer. *Cell, 173*(2). https://doi.org/10.1016/j.cell.2018.03.022. 291-304.e6.

Hölzel, M., Bovier, A., & Tüting, T. (2013). Plasticity of tumour and immune cells: A source of heterogeneity and a cause for therapy resistance? *Nature Reviews. Cancer, 13*(5), 365–376. https://doi.org/10.1038/nrc3498.

Hu, D., Notarbartolo, S., Croonenborghs, T., Patel, B., Cialic, R., Yang, T.-H., et al. (2017). Transcriptional signature of human pro-inflammatory TH17 cells identifies reduced IL10 gene expression in multiple sclerosis. *Nature Communications*, *8*(1), 1600. https://doi.org/10.1038/s41467-017-01571-8.

Huang, Y.-H., Cao, Y.-F., Jiang, Z.-Y., Zhang, S., & Gao, F. (2015). Th22 cell accumulation is associated with colorectal cancer development. *World Journal of Gastroenterology*, *21*(14), 4216–4224. https://doi.org/10.3748/wjg.v21.i14.4216.

Jabeen, R., Goswami, R., Awe, O., Kulkarni, A., Nguyen, E. T., Attenasio, A., et al. (2013). Th9 cell development requires a BATF-regulated transcriptional network. *The Journal of Clinical Investigation*, *123*(11), 4641–4653. https://doi.org/10.1172/JCI69489.

Jiang, P., Gu, S., Pan, D., Fu, J., Sahu, A., Hu, X., et al. (2018). Signatures of T cell dysfunction and exclusion predict cancer immunotherapy response. *Nature Medicine*, *24*(10), 1550–1558. https://doi.org/10.1038/s41591-018-0136-1.

Johansson, M., Denardo, D. G., & Coussens, L. M. (2008). Polarized immune responses differentially regulate cancer development. *Immunological Reviews*, *222*(1), 145–154. https://doi.org/10.1111/j.1600-065X.2008.00600.x.

Kaplan, M. H., Hufford, M. M., & Olson, M. R. (2015). The development and in vivo function of T helper 9 cells. *Nature Reviews. Immunology*, *15*(5), 295–307. https://doi.org/10.1038/nri3824.

Kärjä, V., Aaltomaa, S., Lipponen, P., Isotalo, T., Talja, M., & Mokka, R. (2005). Tumour-infiltrating lymphocytes: A prognostic factor of PSA-free survival in patients with local prostate carcinoma treated by radical prostatectomy. *Anticancer Research*, *25*(6C), 4435–4438.

Koch, A., De Meyer, T., Jeschke, J., & Van Criekinge, W. (2015). MEXPRESS: Visualizing expression, DNA methylation and clinical TCGA data. *BMC Genomics*, *16*(1), 636. https://doi.org/10.1186/s12864-015-1847-z.

Lee, G. K., Park, H. J., Macleod, M., Chandler, P., Munn, D. H., & Mellor, A. L. (2002). Tryptophan deprivation sensitizes activated T cells to apoptosis prior to cell division. *Immunology*, *107*(4), 452–460. https://doi.org/10.1046/j.1365-2567.2002.01526.x.

Li, B., & Dewey, C. N. (2011). RSEM: Accurate transcript quantification from RNA-Seq data with or without a reference genome. *BMC Bioinformatics*, *12*(1), 323. https://doi.org/10.1186/1471-2105-12-323.

Liu, J., Lichtenberg, T., Hoadley, K. A., Poisson, L. M., Lazar, A. J., Cherniack, A. D., et al. (2018). An integrated TCGA Pan-cancer clinical data resource to drive high-quality survival outcome analytics. *Cell*, *173*(2). https://doi.org/10.1016/j.cell.2018.02.052. 400-416.e11.

Llosa, N. J., Cruise, M., Tam, A., Wicks, E. C., Hechenbleikner, E. M., Taube, J. M., et al. (2015). The vigorous immune microenvironment of microsatellite instable colon cancer is balanced by multiple counter-inhibitory checkpoints. *Cancer Discovery*, *5*(1), 43–51. https://doi.org/10.1158/2159-8290.CD-14-0863.

Luckheeram, R. V., Zhou, R., Verma, A. D., & Xia, B. (2012). $CD4^+$ T cells: Differentiation and functions. *Clinical & Developmental Immunology*, *2012*, 925135. https://doi.org/10.1155/2012/925135.

Mahmoud, S., Lee, A., Ellis, I., & Green, A. (2012). CD8(+) T lymphocytes infiltrating breast cancer: A promising new prognostic marker? *Oncoimmunology*, *1*(3), 364–365. https://doi.org/10.4161/onci.18614.

Maleki Vareki, S. (2018). High and low mutational burden tumors versus immunologically hot and cold tumors and response to immune checkpoint inhibitors. *Journal for Immunotherapy of Cancer*, *6*(1), 157. https://doi.org/10.1186/s40425-018-0479-7.

Mantovani, A., Marchesi, F., Malesci, A., Laghi, L., & Allavena, P. (2017). Tumour-associated macrophages as treatment targets in oncology. *Nature Reviews. Clinical Oncology*, *14*(7), 399–416. https://doi.org/10.1038/nrclinonc.2016.217.

Martinet, L., & Smyth, M. J. (2015). Balancing natural killer cell activation through paired receptors. *Nature Reviews Immunology*, *15*(4), 243–254. https://doi.org/10.1038/nri3799.

McNally, A., Hill, G. R., Sparwasser, T., Thomas, R., & Steptoe, R. J. (2011). CD4+CD25+ regulatory T cells control CD8+ T-cell effector differentiation by modulating IL-2 homeostasis. *Proceedings of the National Academy of Sciences of the United States of America*, *108*(18), 7529–7534. https://doi.org/10.1073/pnas.1103782108.

Mezrich, J. D., Fechner, J. H., Zhang, X., Johnson, B. P., Burlingham, W. J., & Bradfield, C. A. (2010). An interaction between kynurenine and the aryl hydrocarbon receptor can generate regulatory T cells. *Journal of Immunology*, *185*(6), 3190–3198. https://doi.org/10.4049/jimmunol.0903670.

Mlecnik, B., Tosolini, M., Kirilovsky, A., Berger, A., Bindea, G., Meatchi, T., et al. (2011). Histopathologic-based prognostic factors of colorectal cancers are associated with the state of the local immune reaction. *Journal of Clinical Oncology: Official Journal of the American Society of Clinical Oncology*, *29*(6), 610–618. https://doi.org/10.1200/JCO.2010.30.5425.

Newman, A. M., Liu, C. L., Green, M. R., Gentles, A. J., Feng, W., Xu, Y., et al. (2015). Robust enumeration of cell subsets from tissue expression profiles. *Nature Methods*, *12*(5), 453–457. https://doi.org/10.1038/nmeth.3337.

Nicholson, S. E., Keating, N., & Belz, G. T. (2017). Natural killer cells and anti-tumor immunity. *Molecular Immunology*, *110*, 40–47. https://doi.org/10.1016/j.molimm.2017.12.002.

Oestreich, K. J., & Weinmann, A. S. (2012). Transcriptional mechanisms that regulate T helper 1 cell differentiation. *Current Opinion in Immunology*, *24*(2), 191–195. https://doi.org/10.1016/j.coi.2011.12.004.

Rooney, M. S., Shukla, S. A., Wu, C. J., Getz, G., & Hacohen, N. (2015). Molecular and genetic properties of tumors associated with local immune cytolytic activity. *Cell*, *160*(1–2), 48–61. https://doi.org/10.1016/j.cell.2014.12.033.

Sato, E., Olson, S. H., Ahn, J., Bundy, B., Nishikawa, H., Qian, F., et al. (2005). Intraepithelial CD8+ tumor-infiltrating lymphocytes and a high CD8+/regulatory T cell ratio are associated with favorable prognosis in ovarian cancer. *Proceedings of the National Academy of Sciences of the United States of America*, *102*(51), 18538–18543. https://doi.org/10.1073/pnas.0509182102.

Schinnerling, K., García-González, P., & Aguillón, J. C. (2015). Gene expression profiling of human monocyte-derived dendritic cells—Searching for molecular regulators of tolerogenicity. *Frontiers in Immunology*, *6*(10), 528. https://doi.org/10.3389/fimmu.2015.00528.

Sharma, P., Shen, Y., Wen, S., Yamada, S., Jungbluth, A. A., Gnjatic, S., et al. (2007). CD8 tumor-infiltrating lymphocytes are predictive of survival in muscle-invasive urothelial carcinoma. *Proceedings of the National Academy of Sciences of the United States of America*, *104*(10), 3967–3972. https://doi.org/10.1073/pnas.0611618104.

Simoni, Y., Becht, E., Fehlings, M., Loh, C. Y., Koo, S.-L., Teng, K. W. W., et al. (2018). Bystander CD8+ T cells are abundant and phenotypically distinct in human tumour infiltrates. *Nature*, *557*(7706), 575–579. https://doi.org/10.1038/s41586-018-0130-2.

Thorsson, V., Gibbs, D. L., Brown, S. D., Wolf, D., Bortone, D. S., Ou Yang, T.-H., et al. (2018). The immune landscape of cancer. *Immunity*, *48*(4). https://doi.org/10.1016/j.immuni.2018.03.023. 812-830.e14.

Uryvaev, A., Passhak, M., Hershkovits, D., Sabo, E., & Bar-Sela, G. (2018). The role of tumor-infiltrating lymphocytes (TILs) as a predictive biomarker of response to anti-PD1 therapy in patients with metastatic non-small cell lung cancer or metastatic melanoma. *Medical Oncology (Northwood, London, England)*, *35*(3), 25. https://doi.org/10.1007/s12032-018-1080-0.

van Oijen, M., Bins, A., Elias, S., Sein, J., Weder, P., de Gast, G., et al. (2004). On the role of melanoma-specific CD8+ T-cell immunity in disease progression of advanced-stage melanoma patients. *Clinical Cancer Research*, *10*(14), 4754–4760. https://doi.org/10.1158/1078-0432.CCR-04-0260.

Végran, F., Apetoh, L., & Ghiringhelli, F. (2015). Th9 cells: A novel CD4 T-cell subset in the immune war against cancer. *Cancer Research*, *75*(3), 475–479. https://doi.org/10.1158/0008-5472.CAN-14-2748.

Wang, T., Zhang, Z., Xing, H., Wang, L., Zhang, G., Yu, N., et al. (2017). Elevated Th22 cells and related cytokines in patients with epithelial ovarian cancer. *Medicine*, *96*(43), e8359. https://doi.org/10.1097/MD.0000000000008359.

Wolf, Y., & Samuels, Y. (2018). Cancer research in the era of immunogenomics. *ESMO Open*, *3*(7), e000475. https://doi.org/10.1136/esmoopen-2018-000475.

Ye, Q., Song, D.-G., Poussin, M., Yamamoto, T., Best, A., Li, C., et al. (2014). CD137 accurately identifies and enriches for naturally occurring tumor-reactive T cells in tumor. *Clinical Cancer Research*, *20*(1), 44–55. https://doi.org/10.1158/1078-0432.CCR-13-0945.

Yokoyama, W. M., & Plougastel, B. F. M. (2003). Immune functions encoded by the natural killer gene complex. *Nature Reviews Immunology*, *3*(4), 304–316. https://doi.org/10.1038/nri1055.

Zhang, Y., Kurupati, R., Liu, L., Zhou, X. Y., Zhang, G., Hudaihed, A., et al. (2017). Enhancing CD8+ T cell fatty acid catabolism within a metabolically challenging tumor microenvironment increases the efficacy of melanoma immunotherapy. *Cancer Cell*, *32*(3). https://doi.org/10.1016/j.ccell.2017.08.004. 377-391.e9.

Zhao, S., Wu, D., Wu, P., Wang, Z., & Huang, J. (2015). Serum IL-10 predicts worse outcome in cancer patients: A meta-analysis. *PLoS One*, *10*(10), e0139598. https://doi.org/10.1371/journal.pone.0139598.

CHAPTER THREE

# Using epigenetic data to estimate immune composition in admixed samples

**Ankur Chakravarthy**[a,*], **Daniel D. De Carvalho**[a,b,*]
[a]UHN Princess Margaret Cancer Centre, Toronto, ON, Canada
[b]Department of Medical Biophysics, University of Toronto, Toronto, ON, Canada
*Corresponding authors: e-mail address: ankur.chakravarthy.10@ucl.ac.uk; daniel.decarvalho@uhnresearch.ca

## Contents



## Abstract

Variation in cell-type decomposition can be of interest to changes in biology by themselves, or a confounder in comparing admixed samples with different phenotypes. Here, I present and review various algorithms that may be used for purposes of deconvolving epigenomic data from admixed samples to yield estimates that may be factored into downstream analyses. More specifically, I present a protocol based on an adaptation of the CIBERSORT algorithm, previously developed for gene-expression, to methylation data.

*Methods in Enzymology*, Volume 636
ISSN 0076-6879
https://doi.org/10.1016/bs.mie.2019.07.004

## 1. Introduction

Immunological studies may be conducted on a broad range of tissue samples, ranging from blood to cell-free DNA. So-called omics based assays that depend on mixed samples often raise questions about whether the observed changes are due to shifts in cellular composition or are due to shifts within specific cell types. While single cell analyses have permitted these possibilities to be distinguished from each other, the application of these techniques to large cohorts can be prohibitively expensive. On the other hand, in some circumstances, it may be possible to borrow information from purified populations and admixed samples in order to separate the two, or to examine the impact of either factor independently, through the use of deconvolution approaches.

Here, I describe methods that can be used to evaluate shifts in cell type composition and factor this into subsequent downstream analysis in studies based on -omics data. While I focus on DNA methylation data, similar concepts and methods are applicable to other kinds of epigenetic data or to transcriptomic data. Indeed, some of the methods described in this chapter were originally developed for the analysis of transcriptomic datasets and may readily be repurposed to epigenetic data. These methods promise to expand the utility of already existing epigenomic profiles of tissue in order to permit deeper characterization of changes in the cellular and epigenetic profiles as drivers of immunological phenomena.

## 2. Epigenetic profiles as a marker of cellular lineage and identity

All cells in an individual largely share the same genetic sequence, and yet, in the course of development from a zygote to a complex, multicellular organism, a remarkable range of transcriptional plasticity is seen. Cells in specific lineages have well defined patterns of transcription that are often inherited by descendants of a stem or progenitor cell. Further, as cellular differentiation proceeds, specific populations may emerge, displaying transcriptional profiles that involve reprogramming from the perspective of the progenitor's expression programs.

This transcriptional plasticity is mediated in large part by epigenetic processes. These can involve either localized changes such as histone modifications or DNA methylation, and large scale interactions that lead to chromatin compaction in large blocks as well as the physical separation of DNA into different compartments that can govern whether interactions between distally located enhancers and gene promoters happen. It is of no surprise that these changes have been shown to include a tissue specific component at the level of large scale chromatin organization, histone modifications, DNAse hyperaccessibility, transposase accessibility, enhancer expression and DNA methylation (Andersson et al., 2014; Kundaje et al., 2015). Moreover, epigenetic profiles have also been correlated with stemness potential across a wide range of stem cells and in turn relates to aggressiveness and dedifferentiation across a range of malignancies (Malta et al., 2018). In essence, epigenetic profiles encode the cellular identity of a given cell type, which is of utility in the application and development of deconvolution algorithms.

## 3. DNA methylation

DNA methylation, in humans, involves the modification of the 5′ carbon atom in cytosine to carry a methyl group and is orchestrated by DNA methyltransferases, which may either carry this reaction out *de novo* or do so on the basis of a methylated template strand when DNA is replicated, depending on the enzyme involved (Gujar, Weisenberger, & Liang, 2019). In humans, cytosine DNA methylation is mostly restricted to CpG sequences, with the occasional modification of cytosines in a CHH context in neurons, mediated by a different class of enzymes.

DNA methylation notably performs several key regulatory roles. In enhancers, its presence is strongly anticorrelated with that of the activating histone mark H3k27ac (Kundaje et al., 2015). Further, promoter methylation has generally been associated with the repression of gene expression, and indeed, targeted methylation can silence gene expression (Amabile et al., 2016). DNA methylation has also emerged as a candidate driver of transcription factor binding and the subsequent activation of specific transcriptional programs with the discovery that many transcription factors can exhibit altered affinities toward cognate DNA motifs based on whether the CpGs within the motif are methylated or not (Yin et al., 2017).

A wide range of approaches based on molecular profiling, including microarray based approaches as well as sequencing based approaches, have been developed for profiling DNA methylation at scale (Yong, Hsu, & Chen, 2016). The utility of these approaches can be markedly extended in the study of immunological processes by leveraging the association between cellular identity and methylation to estimate, and account for, cell type. This can be used to either remove the influence of cell-type changes on methylation profiles for the purposes of Epigenome-wide association studies, or it can be used to estimate cell proportions to infer shifts in populations as a function of pathology.

## 4. There are two broad categories of deconvolution algorithms

Deconvolution approaches are based on the idea that signal intensity in epigenomic data, be it accessibility, enrichment of peak signals in chromatin, or methylation as measured in beta-values, from a bulk sample with an admixture of cell types, may be modeled as the sum of contributions from different cell types with a typical epigenomic profile of their own. In the following sections, I review many of the methodological advances that have applied this principle to infer cell-type composition from epigenomic data. There are fundamentally two distinct classes of deconvolution algorithms that can perform this task, namely, "reference-based" methods, wherein pure cell populations and their epigenetic profiles are used as a reference in order to estimate proportions, and "reference-free" algorithms, where populations and the features that characterize them are both estimated from the admixed data.

The two classes of algorithms have distinct advantages and disadvantages, which are summarized in the table below.

| Feature | Reference based | Reference free |
|---|---|---|
| Dependence on purified cell populations | Yes—reference-based methods typically decompose the signal in a mixed sample (B) as a product of the sums of average methylation per cell type and their proportions | No—this makes these algorithms particularly useful when only admixed data are available and purified cell populations to serve as a reference are absent |
| Interpretability | Yes—where most cell types that are present in the admixed sample are also present in the reference, the estimated proportions returned can be highly correlated with their absolute abundance in the sample | No as a general principle, however, there are specific methods for reference free algorithms that do produce interpretable output |

| | | |
|---|---|---|
| Algorithmic flexibility | Yes—a wide range of well-characterized statistical approaches, such as linear modeling, constrained projection, quadratic programming and support vector machine regression are available | These algorithms are less well-characterized, however, some outstanding methods exist nonetheless |

## 5. Reference-based methods for the deconvolution of methylation data

A wide range of algorithms have been developed for reference-based deconvolution of methylation data. In part, this was motivated by the finding that the composition of blood cell types was a major confounder in Epigenome Wide Association Studies, where blood cells are often used as a surrogate tissue to discover epigenetic changes that track with pathology. A short review of a few of these approaches follows.

The first instance of an approach developed for deconvolution of methylation array profiles was proposed for the Illumina 27k microarray platform. This approach selected, for a given panel of purified cell types, up to 300 features each and then used a quadratic programming approach to estimate the fractions of cell types that would, in combination, yield the observed methylation beta-values (Houseman et al., 2012). This was subsequently followed by an extension of the algorithm to the Illumina HumanMethylation 450k array, which surveys an order of magnitude more CpG sites. Here, the top 600 cell type specific CpGs (MVPs or methylation variable positions) were used to estimate the cell type composition and then identify probes where cell type variability confounded the identification and interpretation of methylation changes otherwise correlated with age (Jaffe & Irizarry, 2014). This approach is implemented in the estimateCellCounts function in the widely used *minfi* R package (Aryee et al., 2014) for the analysis of Illumina BeadChip methylation arrays. For technical clarity, DMRs (differentially methylated regions) are genomic loci with one or more MVPs.

Particularly in the case of cancer, several illuminating examples of methylation-based deconvolution approaches exist. These range from approaches that aim to estimate few cell fractions to those that estimate the proportions of large numbers of cell fractions. The Cancer Genome Atlas used a simple kernel-based method to estimate global leukocyte infiltration into tumors (Carter et al., 2012) while other prominent examples of low resolution deconvolution include approaches such as EDec (Onuchic et al., 2016). On the other end of the spectrum, an adaptation of the CIBERSORT algorithm (Newman et al., 2015) the author developed, called MethylCIBERSORT, was used to estimate cell type proportions for 12 different cell types across 23 tumor types in TCGA, which was pretty close to estimating absolute fractions given the high correlations between the estimated cancer cell fractions and the purity estimated orthogonally using whole exome sequencing and copy number data (Chakravarthy et al., 2018).

When the original implementation of CIBERSORT was developed for expression data, its ability to deconvolute up to 22 cell types simultaneously was demonstrated, suggesting that the methods that exist will be able to estimate markedly greater numbers of cell types as long as they are present at fractions that exceed the noise threshold of methylation microarrays or sequencing data.

It is anticipated that methylation will produce more robust estimates because it is intrinsically less noisy than transcription. This is because the former may be present in just three states within any given cell (unmethylated, hemimethylated or fully methylated) whereas expression can take a broader range of values. Indeed, when we developed MethylCIBERSORT, we performed comparisons between expression-based and methylation-based deconvolution on cell mixtures of similar complexity, and found that error rates were significantly lower for the latter (Chakravarthy et al., 2018).

## 6. MethylCIBERSORT

MethylCIBERSORT is an algorithm the author developed in order to adapt the state-of-the-art CIBERSORT algorithm to the deconvolution of methylation data. However, the package may be used for feature selection that creates reference matrices which are compatible with a wide range of deconvolution algorithms downstream. The reason CIBERSORT, which implements Support Vector Regression is preferable, however, is that it generally outperforms a range of other deconvolution algorithm both in terms of transcriptional deconvolution and methylation deconvolution, as demonstrated previously using simulated data and gold-standard mixtures with flow-cytometry estimates (Chakravarthy et al., 2018; Newman et al., 2015; Teschendorff, Breeze, Zheng, & Beck, 2017).

For feature selection, in brief, MethylCIBERSORT takes in a matrix of cell types, both stromal and tumor, and then uses pairwise *limma* fits (Smyth, 2004) on the beta-values to select Methylation Variable Positions (differentially methylated CpG sites, in other words) that meet given thresholds for False Discovery Rate and median beta-value difference. Further, the number of CpG sites selected for the reference matrix is determined by the user, but is automatically capped based on the cell-type with the fewest MVPs associated with it in order to prevent imbalanced representation of cell types in the reference matrix.The package currently includes a collection of nine cell types, but the reference data should be easily extensible given the collection of cell types profiled in other recent studies. Once a reference matrix is created, it can be used not only with CIBERSORT but a rage of other deconvolution algorithms to estimate proportions. We have also tested the protocol on the updated version of CIBERSORT, called CIBERSORTx (Newman et al., 2019) and confirm forward-compatibility of the protocol.

## 6.1 Protocol for running MethylCIBERSORT

This protocol requires basic competence in the R programming language, and some familiarity with the minfi Bioconductor package. There is a wet-lab component that may need to be carried out if one is working either with novel cell types or with new samples from which methylation data are not yet available. In short, the process includes three distinct stages—preparation of the bulk data to be deconvolved, including array preprocessing, preparation of a cell-type specific matrix using purified cell type methylation data, and finally deconvolution and downstream analysis.

**(a)** Generate data, if needed, on either the Illumina 450k or Illumina EPIC methylation microarrays. The use of IDAT files rather than GenomeStudio normalization is recommended.

**(b)** Using minfi's read.metharray function, parse the IDATs into an R session. Then, quality control may be carried out using the shinyMethylQC package or the 'qcReport function in *minfi*, which exports a PDF of the QC plots. Interpretation of these plots is beyond the scope of this chapter, but the documentation of the *minfi* package should be illuminating.

**(c)** Normalize the microarrays. Single-sample Noob normalization (Fortin, Triche, & Hansen, 2017), or Functional Normalization (Fortin et al., 2014) are recommended instead of quantile normalization. This is because quantile normalization shifts global distributions. This is undesirable when, in phenotypes like cancer, global shifts that define these samples may artificially bias other samples if quantile normalization is applied.

**(d)** Obtain beta-value matrices from the normalized dataset using the getBeta function in the minfi R package. This will be the matrix of bulk samples that will be deconvolved in the subsequent steps.

**(e)** Repeat the preprocessing analyses for any novel purified populations you also wish to estimate if they are not already present in the collection of cell types included in the package. The MethylCIBERSORT R package includes uniformly preprocessed samples from neutrophils, eosinophils, CD4 effector cells, Tregs, fibroblasts, CD19+ B-cells, CD14+ Monocytes, dermal fibroblasts, HUVEC vascular endothelial cell lines, and CD8+ lymphocytes.

**(f)** Install the MethylCIBERSORT package from the Zenodo record at the doi (digital object identifier): https://doi.org/10.5281/zenodo.1284582. Load the package into R with the command "`library (MethylCIBERSORT)`"

**(g)** Load the reference dataset of cell-types built into MethylCIBERSORT using the R command "`data("V2_Signatures")`". This loads a vector of phenotypes and a matrix of beta-values with samples in columns and probe identifiers in rows. Append any novel cell types not already in the conclusion and a character vector designating the cell type of these samples to the reference matrix.

**(h)** Use the FeatureSelect.V4 function in order to create a reference matrix. An example syntax call is "`Signature <- FeatureSelect.V4(CellLines.matrix = NULL, Heatmap = FALSE, export = TRUE, sigName = "ExampleType", Stroma.matrix = Stromal_v2, deltaBeta = 0.2, FDR = 0.01, MaxDMRs = 100, Phenotype.stroma = Stromal_v2.pheno)`"

In this syntax call, CellLines.matrix is a matrix of beta values/ expression measurements from cell lines, if any. This needs to be set to NULL and the cell types need to be combined into a single "stromal" matrix if the number of novel cell types in addition to that already present in the reference collection is more than one. The Heatmap variable, if set to true, plots a heatmap of the reference matrix if it has been calculated.

deltaBeta specifies the threshold for median delta-beta cutoff for MVPs for each pairwise comparison between cell types and FDR is the false discovery rate cutoff for the same comparisons. If export is set to true, a text file is written out with the name specified by "sigName" and will be required for deconvolution using the CIBERSORT server. In previous publications, we have used 0.3 as the median delta-Beta cutoff and an FDR of 1% or less. Depending on how distinct the cell types are from each other, exploration with various thresholds is recommended.

Stroma.matrix is a matrix of beta values consisting of beta-values from the pure cell types the bulk data is meant to be decomposed into. Rows represent features (or CpG probes) and columns represent samples. Phenotype.stroma is a character vector consisting of the cell type allocations for each of the samples in the corresponding matrix. Finally, MaxDMRs specifies the number of CpG sites/probes picked for each cell type

**(i)** Read the exported signature file into R with the read.delim function in R. For purposes of our example, we will assign this to an object called Reference, with the signature file being called "SigFile". The example syntax would be `"Reference <- read.delim("SigFile.txt")"`

**(j)** Prepare and export the bulk data for deconvolution using CIBERSORT. If Mat is the name of the beta-value matrix of the admixed samples for deconvolution, and Reference is the read-in reference matrix prepared in the preceding steps, the bulk data can be prepared for deconvolution using the "Prep.CancerType" function in the MethylCIBERSORT R package with `"Prep.CancerType(Mat, probes = Reference$NAME, fname = "Mixture")`, which writes out a text file that can be used with CIBERSORT.

**(k)** Acquire an account for the use of CIBERSORT at cibersort.stanford.edu. Select the Menu dropdown and select "Upload files." Upload the signature (SigFile) and mixture files (Mixture). Then select "run CIBERSORT" from the Menu drop-down list. Here, select the files you uploaded, set the number of permutations to the desired number (the higher the number, the better) and then run CIBERSORT. Once the job is complete, a visual summary heatmap of the deconvolution estimates is displayed, and files may be downloaded in various formats, including tab-delimited text files and Microsoft Excel spreadsheets for further analysis. These results are also accessible later using the "Job Results" option from the "Menu" drop-down.

## 7. Downstream applications for deconvolution estimates

The deconvolution estimates, once derived, may be used in multiple different ways. Initially, however, some questions include whether deconvolution was reliable, and whether the estimates returned encompass all of the cell types in a mixture or not. The advantage of the online implementation of CIBERSORT is that it performs an estimation procedure using Monte-Carlo resampling to estimate whether the observed correlation between the signature and mixture files conditional on the inferred estimates is significantly different from random data or not. This yields a global Pearson's correlation for deconvolution.

The closer this is to 1, the greater the inference that these estimates are likely to represent absolute, rather than relative cell proportions. This is because a high correlation implies that the cell proportion estimates and the features in the original matrix were sufficient to recapitulate methylation profiles of the whole mixture. Second, the deconvolution process also returns a *P*-value, indicating whether the deconvolution accuracy is greater than expected given randomly permuted signature matrices. It may be useful to set cutoffs for both of these quality controls.

To illustrate how these estimates may then be used, I illustrate a couple of key examples.

In the MethylCIBERSORT publication (Chakravarthy et al., 2018), the cell type estimates that were derived in HPV-negative Head and Neck Cancers were used for clustering to produce subgroups of "immune hot" and "immune cold" tumors and a classifier was then developed to classify the rest of the TCGA dataset into these classes. This is an example wherein the estimates themselves may be used to discover novel patterns on the basis of cell composition, which may then be correlated with other molecular features or pathology of interest.

In another case study (Chakravarthy et al., 2018), we combined deconvolution based estimates from MethylCIBERSORT with transcriptional data to infer that fibroblasts were the likely source of a novel program of extracellular matrix genes which in turn was associated with adverse outcomes and immunotherapy failure. In a third case, it is possible to define tumor or other subgroups using orthogonal methods, such as consensus clustering, and then examine whether particular patterns of cellular composition are associated with it, as is demonstrated with transcriptional data in TCGA analyses of sarcomas (Cancer Genome Atlas Research Network, 2017).

In summary, reference-based deconvolution methods using DNA methylation data may be applied in a user-friendly manner and contribute to unbiased discovery of disease subgroups, or they may be used to add biological context to orthogonally derived measurements of diseases and disease related processes.

## 8. Simultaneous estimation of proportions and underlying methylation state using supervised algorithms

The naïve application of approaches such as linear modeling to bulk data may result in methylation changes specific to certain cell-types being obscured by shifts in the cell composition *per se*. However, there are statistical approaches that leverage deconvolution to resolve this issue, and permit inference of expression or methylation changes specifically within cell populations. Two key examples are csSAM (cell-specific Significance Analysis of Microarrays) (Shen-Orr et al., 2010) and cellDMC (Zheng, Breeze, Beck, & Teschendorff, 2018), implemented in the EpiDISH R package. csSAM first uses deconvolution, and then uses between-sample permutation based on ratios of gene expression to estimate significance of association. For purposes of adaptation to methylation data, the values may need to be transformed to unlogged M values instead of beta-values (see Du et al., 2010, for a review of these measures and the relationship between them). It is recommended that unlogged transformations of M values be used because deconvolution in log-linear space, rather than linear space, may degrade performance (Zhong & Liu, 2011).

CellDMC, on the other hand, has been specifically designed for the deconvolution and estimation of DNA methylation data. This approach couples the ability of deconvolution to estimate cell type specific proportions and the subsequent use of interaction terms in linear modeling to estimate cell-type specific DNA methylation changes. Further, this package implements a variant of EpiDISH designed to account for the hierarchical nature of relationships between cell types insofar their development is concerned in order to estimate cell type fractions, which can contribute to improved deconvolution accuracy. This package is therefore suitable for the simultaneous deconvolution and cell-type specific differential methylation analyses and is valuable when examining data consisting of cellular mixtures.

## 9. Reference-free methods for methylation data

A simple limitation of reference-based methods for deconvolution is that accurate inference relies both on the existence of methylation patterns in purified cell types, as well as correspondence between the pure populations and those found in admixed samples. In the case of transcriptional profiles and deconvolution, it has been demonstrated that the choice of reference profile can have a marked influence on deconvolution accuracy. Further, as discussed above, the failure to have a full account of the cell-types present in admixed samples can yield relative estimates in some samples and absolute ones in others, which raises questions of comparability for downstream statistical analysis. In this setting, reference-based deconvolution methods may be useful, since they operate with no *a priori* presumption of what cellular populations are present or what their characteristic methylation profiles are. I discuss a few distinct approaches to illustrate the statistical methodologies that are often used in this setting, but this should not be taken to be an exhaustive list of such methods.

Surrogate Variable Analysis, using Singular Value Decomposition, produces various singular variable scores that may be used to automatically correct for cell-type heterogeneity by partitioning the residuals of a linear model fit that are then used to correct the fit in an iterative process. RefFreeEWAS (Houseman, Molitor, & Marsit, 2014), on the other hand, uses the unadjusted linear model coefficients from a regression fit for decomposition. The number of latent variables to decompose the data into is estimated using Random Matrix Theory and standard errors are computed using bootstrapping. The limitation of these approaches is that they do not explicitly account for confounding between technical variation and cell-type composition variation, and further enable supervised analysis between phenotypes, but do not estimate cell composition itself to be used for further analysis or interpretation.

On the other hand, approaches have been developed to also estimate the number and proportion of distinct cell types present in a dataset consisting of admixed samples. RefFreeEWAS2.0 (Houseman et al., 2016) implements a version of Non-negative matrix factorization that is constrained to the unit interval in order to decompose a matrix of methylation beta-values into a given number of cell types and their contributions to the dataset. A bootstrapping procedure and multiple iterations fitting the data to different numbers of clusters is used to derive these parameters. This method also identifies CpG sites associated with the populations, which may be, if reference-based sets of features are available, may be linked to known-cell types. Where analyses of cell-type composition and their shifts are desired, these estimates can be used in the absence of reference datasets, or as a secondary layer of validation for whether the estimates derived from reference-based deconvolution are absolute or relative.

## 10. Adaptation to other sequencing data/assays

Many of these methods, including CIBERSORT, SVA and csSAM, were designed originally with transcriptomic data in mind and may be used for data such as ChIP or ATAC-seq peak intensity signals with ease as long as one is not operating in logarithmic space. However, other methods, such as RefFreeEWAS 2.0, enforce a sum to one constraint, which is violated for other epigenomic or transcriptomic data. In general, as long as other methylation profiling methods produce methylation measurements in terms of beta-values, methods developed around methylation microarrays should be reliably applicable.

## 11. Summary and resources

In essence, a wide range of computational methods is applicable to the problem of deconvolving epigenomic (and transcriptomic) data that is likely to be confounded both by phenotypic variation and cell-type composition in the samples. The following table summarizes a range of packages, mostly written in the R programming language, that may be useful to people contemplating the use of deconvolution in their own analyses of epigenomic data.

| Method | Notes | Reference |
|---|---|---|
| EstimateCellCounts in the *minfi* R package | Reference-based deconvolution of blood | Aryee et al. (2014), Fortin et al. (2017) Package available on Bioconductor |
| MethylCIBERSORT | Framework to prepare signature files for deconvolution from pure populations, can be used with CIBERSORT or other methods | Chakravarthy et al. (2018) Package available on Zenodo at doi: https://doi.org/10.5281/zenodo.1284582 |
| RefFreeEWAS (v1.0) | SVD-based method for performing EWAS after accounting for cell-type composition, reference free | Houseman et al. (2014, 2016) The R package is available on CRAN as "RefFreeEWAS" |

| | | |
|---|---|---|
| RefFreeEWAS (v2.0) | NMF based reference-free deconvolution | See above |
| EDec | User friendly framework for deconvolving and integrating estimates into gene-expression analyses | Onuchic et al. (2016) Available on GitHub and may be installed using devtools using "BRL-BCM/EDec" as the path |
| EpiDISH and CellDMC | Implemented in the EpiDISH package, these methods implement Robust Partial Correlations and CIBERSORT for deconvolution and estimation of cell-type specific methylation changes in the same framework | Teschendorff et al. (2017), Zheng et al. (2018) Available on Bioconductor as "EpiDISH" |
| DSA | Digital Sorting Analysis—a deconvolution algorithm that only needs the name of signature features for reference cell types rather than actual measurements themselves. This may be particularly platform-independent as a consequence | Zhong, Wan, Pang, Chow, and Liu (2013) Available on GitHub at the path "zhandong/DSA" |

## References

Amabile, A., Migliara, A., Capasso, P., Biffi, M., Cittaro, D., Naldini, L., et al. (2016). Inheritable silencing of endogenous genes by hit-and-run targeted epigenetic editing. *Cell*, *167*(1). https://doi.org/10.1016/j.cell.2016.09.006. 219-232.e214.

Andersson, R., Gebhard, C., Miguel-Escalada, I., Hoof, I., Bornholdt, J., Boyd, M., et al. (2014). An atlas of active enhancers across human cell types and tissues. *Nature*, *507*(7493), 455–461. https://doi.org/10.1038/nature12787.

Aryee, M. J., Jaffe, A. E., Corrada-Bravo, H., Ladd-Acosta, C., Feinberg, A. P., Hansen, K. D., et al. (2014). Minfi: A flexible and comprehensive bioconductor package for the analysis of Infinium DNA methylation microarrays. *Bioinformatics*, *30*(10), 1363–1369. https://doi.org/10.1093/bioinformatics/btu049.

Cancer Genome Atlas Research Network. (2017). Comprehensive and integrated genomic characterization of adult soft tissue sarcomas. *Cell*, *171*(4). https://doi.org/10.1016/j.cell.2017.10.014. 950-965.e928.

Carter, S. L., Cibulskis, K., Helman, E., McKenna, A., Shen, H., Zack, T., et al. (2012). Absolute quantification of somatic DNA alterations in human cancer. *Nature Biotechnology, 30*(5), 413–421. https://doi.org/10.1038/nbt.2203.

Chakravarthy, A., Furness, A., Joshi, K., Ghorani, E., Ford, K., Ward, M. J., et al. (2018). Pan-cancer deconvolution of tumour composition using DNA methylation. *Nature Communications, 9*(1), 3220. https://doi.org/10.1038/s41467-018-05570-1.

Du, P., Zhang, X., Huang, C. C., Jafari, N., Kibbe, W. A., Hou, L., et al. (2010). Comparison of beta-value and M-value methods for quantifying methylation levels by microarray analysis. *BMC Bioinformatics, 11*, 587. https://doi.org/10.1186/1471-2105-11-587.

Fortin, J. P., Labbe, A., Lemire, M., Zanke, B. W., Hudson, T. J., Fertig, E. J., et al. (2014). Functional normalization of 450k methylation array data improves replication in large cancer studies. *Genome Biology, 15*(12), 503. https://doi.org/10.1186/s13059-014-0503-2.

Fortin, J. P., Triche, T. J., Jr., & Hansen, K. D. (2017). Preprocessing, normalization and integration of the Illumina HumanMethylationEPIC array with minfi. *Bioinformatics, 33*(4), 558–560. https://doi.org/10.1093/bioinformatics/btw691.

Gujar, H., Weisenberger, D. J., & Liang, G. (2019). The roles of human DNA methyltransferases and their isoforms in shaping the epigenome. *Genes (Basel), 10*(2), 172. https://doi.org/10.3390/genes10020172.

Houseman, E. A., Accomando, W. P., Koestler, D. C., Christensen, B. C., Marsit, C. J., Nelson, H. H., et al. (2012). DNA methylation arrays as surrogate measures of cell mixture distribution. *BMC Bioinformatics, 13*, 86. https://doi.org/10.1186/1471-2105-13-86.

Houseman, E. A., Kile, M. L., Christiani, D. C., Ince, T. A., Kelsey, K. T., & Marsit, C. J. (2016). Reference-free deconvolution of DNA methylation data and mediation by cell composition effects. *BMC Bioinformatics, 17*, 259. https://doi.org/10.1186/s12859-016-1140-4.

Houseman, E. A., Molitor, J., & Marsit, C. J. (2014). Reference-free cell mixture adjustments in analysis of DNA methylation data. *Bioinformatics, 30*(10), 1431–1439. https://doi.org/10.1093/bioinformatics/btu029.

Jaffe, A. E., & Irizarry, R. A. (2014). Accounting for cellular heterogeneity is critical in epigenome-wide association studies. *Genome Biology, 15*(2), R31. https://doi.org/10.1186/gb-2014-15-2-r31.

Kundaje, A., Meuleman, W., Ernst, J., Bilenky, M., Yen, A., Heravi-Moussavi, A., et al. (2015). Integrative analysis of 111 reference human epigenomes. *Nature, 518*(7539), 317–330. https://doi.org/10.1038/nature14248.

Malta, T. M., Sokolov, A., Gentles, A. J., Burzykowski, T., Poisson, L., Weinstein, J. N., et al. (2018). Machine learning identifies stemness features associated with oncogenic dedifferentiation. *Cell, 173*(2). https://doi.org/10.1016/j.cell.2018.03.034. 338-354.e315.

Newman, A. M., Liu, C. L., Green, M. R., Gentles, A. J., Feng, W., Xu, Y., et al. (2015). Robust enumeration of cell subsets from tissue expression profiles. *Nature Methods, 12*(5), 453–457. https://doi.org/10.1038/nmeth.3337.

Newman, A. M., Steen, C. B., Liu, C. L., Gentles, A. J., Chaudhuri, A. A., Scherer, F., et al. (2019). Determining cell type abundance and expression from bulk tissues with digital cytometry. *Nature Biotechnology, 37*(7), 773–782. https://doi.org/10.1038/s41587-019-0114-2.

Onuchic, V., Hartmaier, R. J., Boone, D. N., Samuels, M. L., Patel, R. Y., White, W. M., et al. (2016). Epigenomic deconvolution of breast tumors reveals metabolic coupling between constituent cell types. *Cell Reports, 17*(8), 2075–2086. https://doi.org/10.1016/j.celrep.2016.10.057.

Shen-Orr, S. S., Tibshirani, R., Khatri, P., Bodian, D. L., Staedtler, F., Perry, N. M., et al. (2010). Cell type-specific gene expression differences in complex tissues. *Nature Methods*, 7(4), 287–289. https://doi.org/10.1038/nmeth.1439.

Smyth, G. K. (2004). Linear models and empirical Bayes methods for assessing differential expression in microarray experiments. *Statistical Applications in Genetics and Molecular Biology3.* , https://doi.org/10.2202/1544-6115.1027. Article 3.

Teschendorff, A. E., Breeze, C. E., Zheng, S. C., & Beck, S. (2017). A comparison of reference-based algorithms for correcting cell-type heterogeneity in epigenome-wide association studies. *BMC Bioinformatics*, *18*(1), 105. https://doi.org/10.1186/s12859-017-1511-5.

Yin, Y., Morgunova, E., Jolma, A., Kaasinen, E., Sahu, B., Khund-Sayeed, S., et al. (2017). Impact of cytosine methylation on DNA binding specificities of human transcription factors. *Science*, *356*(6337). https://doi.org/10.1126/science.aaj2239.

Yong, W. S., Hsu, F. M., & Chen, P. Y. (2016). Profiling genome-wide DNA methylation. *Epigenetics & Chromatin*, *9*, 26. https://doi.org/10.1186/s13072-016-0075-3.

Zheng, S. C., Breeze, C. E., Beck, S., & Teschendorff, A. E. (2018). Identification of differentially methylated cell types in epigenome-wide association studies. *Nature Methods*, *15*(12), 1059–1066. https://doi.org/10.1038/s41592-018-0213-x.

Zhong, Y., & Liu, Z. (2011). Gene expression deconvolution in linear space. *Nature Methods*, *9*(1), 8–9. author reply 9. https://doi.org/10.1038/nmeth.1830.

Zhong, Y., Wan, Y. W., Pang, K., Chow, L. M., & Liu, Z. (2013). Digital sorting of complex tissues for cell type-specific gene expression profiles. *BMC Bioinformatics*, *14*, 89. https://doi.org/10.1186/1471-2105-14-89.

CHAPTER FOUR

# Metabolomic approaches to study the tumor microenvironment

**Giuseppe Astarita[a,†], Suraj Dhungana[b,†], Bindesh Shrestha[b,†], Evagelia C. Laiakis[a,c,*,†]**

[a]Department of Biochemistry and Molecular & Cellular Biology, Georgetown University, Washington, DC, United States
[b]Waters Corporation, Milford, MA, United States
[c]Department of Oncology, Lombardi Comprehensive Cancer Center, Georgetown University, Washington, DC, United States
*Corresponding author: e-mail address: ecl28@georgetown.edu

## Contents



## Abstract

Tumors are characterized by metabolic dysregulation, reprogramming, and the presence of metabolites, which can act both as energy mediators and signaling messengers. Measuring the concentration and composition of metabolites in the tumor microenvironment can help to better understand the tumor pathology and might improve therapeutic treatments. Metabolomics can provide a description of the physiological and pathological status, as well as help to identify biomarkers of the disease. Additionally, mass spectrometry-based tissue imaging techniques can show the spatial distribution of metabolites. In this chapter we present protocols for the extraction and analysis of metabolites and lipids, with emphasis on liquid chromatography-mass spectrometry and mass spectrometry imaging.

† All authors contributed equally to the chapter.

*Methods in Enzymology*, Volume 636
ISSN 0076-6879
https://doi.org/10.1016/bs.mie.2019.07.037

## 1. Introduction

The tumor microenvironment (TME) is characterized by the dynamic interaction between malignant and non-transformed cells. Immune cells, vascular endothelial cells, cytokines acting in an autocrine or paracrine fashion, and alterations of cell elasticity create a highly active environment that can also be characterized by pro-inflammatory responses. In addition, small molecules <1 kDa (metabolites, lipids) and alterations in cellular metabolism have been closely linked to cancer progression and activation of signaling pathways, either in neighboring cells or intracellularly. Significant knowledge has been acquired on cancer cell metabolism and reprogramming over the years, with the classical Warburg effect being at the forefront of cancer metabolism. This theory defines cancer cells as relying on glycolysis for adenosine triphosphate (ATP) production, unlike normal cells that depend on the tricarboxylic acid (TCA) cycle (Kim, 2019). However, cancer energy metabolism may depend on cytosolic availability of molecules. Additionally, signaling based on small molecules through the TME may further dysregulate such processes, with heterocellular crosstalk between tumor and non-tumor cells being a hallmark of TME. Most notably, nutrient availability, oxidative stress, and pH can result in perturbations of biochemical processes followed by adaptive rewiring of microenvironment metabolomics (Hanahan & Weinberg, 2011; Makohon-Moore & Iacobuzio-Donahue, 2016). Furthermore, within the TME metabolites are being recognized to play a critical role in cellular signaling and serve as substrates to enzymes that regulate the epigenome (Lorendeau, Christen, Rinaldi, & Fendt, 2015; Pavlova & Thompson, 2016). Insights into the role metabolites play in the tumor development and progression are key to developing targeted cancer therapies and designing new drugs (Anastasiou, 2017; Murray, 2016).

In the recent years, metabolomics has been the major analytical tool to investigate metabolism in health and disease. There are two main metabolomics approaches: (1) the untargeted approach to screen in an unbiased fashion the metabolite profiles and (2) the targeted approach to monitor specific pathways or metabolites of choice. Such metabolomics approaches have allowed to describe both static and dynamic metabolic differences of the TME (Badur & Metallo, 2018; Zasada & Kempa, 2016).

Liquid chromatography (LC) coupled to mass spectrometry (MS) is the technique of choice for metabolomics. Although additional analytical approaches including gas chromatography (GC) and nuclear magnetic resonance (NMR) have been used for metabolomics, LC-MS has gained popularity given its sensitivity and selectivity, which allows to cover multiple metabolite classes with various degrees of concentration. Additionally, state-of-the-art MS-based techniques, such as MS imaging, allow to visualize spatial distribution of select metabolites in biological tissues, providing important information for better understanding TME. In this chapter we will discuss protocols regarding sample preparation for downstream metabolomic analysis using LC-MS or MS imaging approaches.

## 2. Untargeted metabolomics

Untargeted metabolomics aims to analyze a wide range of polar and nonpolar molecules providing a snapshot of the metabolic status of a tissue. Generally, tissues are rapidly collected and snap-frozen in liquid nitrogen. Metabolite composition is altered during tissue thawing at room temperature. Thus, to avoid cellular degeneration, tissue samples are cut and weighed while still frozen. The use of ultra high performance LC in combination with high resolution MS instruments can increase the resolving power allowing the separation of complex mixture of analytes. Here we report sample preparation procedures to extract polar metabolites and lipids from tissues, as metabolomic analysis has become an important tool in TME research (Kepp, Loos, Liu, & Kroemer, 2017; Park, Coffey, Limoges, & Le, 2018; Vitale, Manic, Coussens, Kroemer, & Galluzzi, 2019).

### 2.1 Tissue metabolomics

1. Weigh frozen tissues (~20 mg), record the weights and transfer them into low-bind Eppendorf tubes containing 300 μL of chilled methanol:water (v:v, 1:1) with internal standards, such as 30 μM 4-nitrobenzoic acid, 4 μM debrisoquine sulfate, and/or 5 μM chlorpropamide.
2. Homogenize with a bead-beater, such as MagNA Lyser (Roche). To each tube add 20 washed stainless-steel beads and use a 40 s pulse at 2000 rpm.
3. Remove 10 μL of the sample and measure protein levels, to be used for normalization purposes.
4. Centrifuge at 13,000 × *g* for 15 min at 4 °C.
5. Transfer the supernatant to a clean siliconated 1.5 mL tube.

6. Add an equal volume of chilled acetonitrile:water (v:v, 1:1) to the pellet from step 5 (300 μL), followed by rapid vortexing.
7. Incubate on ice for 15 min, and centrifuge at 13,000 × *g* for 15 min at 4 °C.
8. Combine the supernatant with the previously collected sample from step 5.
9. Vacuum dry the samples with no heat or evaporate under a gentle stream of $N_2$.
10. Resuspend the sample in methanol:water (v:v, 1:1) and vortex thoroughly. (*Note*: an extra step of filtering of the samples with a 0.2 μM size filter is sometimes warranted to remove floating particulates.)
11. Transfer the sample into an MS vial and store at −80 °C until processing on an LC-MS instrument.
12. Analysis can be conducted on a high resolution MS instrument such as time-of-flight (TOF) or Orbitrap (Astarita, Kendall, Dennis, & Nicolaou, 2015; Laiakis et al., 2017).

## 2.2 Tissue lipidomics

1. Weigh frozen tissues (~20 mg), record the weights and transfer them into low-bind tubes containing 1 mL of methanol with internal standards, such as SPLASH® LIPIDOMIX® Mass Spec Standard (Avanti Polar Lipids, Inc).
2. Homogenize with a bead-beater, such as MagNA Lyser (Roche). To each tube add 20 washed stainless-steel beads and use a 40 s pulse at 2000 rpm.
3. Add 2 mL of chloroform and vortex.
4. Add 0.75 mL of water and vortex.
5. Centrifuge at 1000 × *g* for 10 min at 4 °C to separate the mixture into two phases with a protein disk at their interface.
6. After centrifugation, collect the lower organic phase and transfer it into a new tube.
7. Evaporate the solvent to dryness using a gentle nitrogen stream.
8. Resuspend in 100 μL of methanol.
9. Transfer the sample into an MS vial and store at −80 °C until processing on an LC-MS instrument.
10. Analysis can be conducted on a high resolution MS instrument such as TOF or Orbitrap (Astarita, Stocchero, & Paglia, 2018; Paglia et al., 2016).

## 3. Targeted metabolomics

Targeted approaches are more sensitive and quantitative compared to an untargeted approach, primarily because of the instrument used for analysis (triple or tandem quadrupole MS instruments). Panels of metabolites of interest can be designed to encompass multiple targets from different metabolic pathways, or alternatively an assay can be designed to quantitate all the intermediates in a selected pathway. Storage and handling of tissue samples are critical steps to control for the precision of metabolite quantification. Other important considerations should include the use of appropriate internal standards, and the determination of limits of detection and quantitation, dynamic range and linearity of quantification. When in need to quantitate very low abundance metabolites, a cleanup method with solid phase extraction can be incorporated for enrichment of the metabolites of interest. Here we present selected protocols to prepare samples for the analysis of metabolites, which have been previously associated with TME and cancer phenotype (Anderson, Mucka, Kern, & Feng, 2018; Gomes, Felipe da Costa, & Colquhoun, 2018; Lyssiotis & Kimmelman, 2017; Norris & Serhan, 2018).

### 3.1 Eicosanoid analysis

Eicosanoids are bioactive metabolites of polyunsaturated fatty acids (PUFAs) generated through the action of lipoxygenases (LOXs), cyclooxygenases (COXs), and cytochrome P450 enzymes as well as non-enzymatic oxidative reactions. Eicosanoids are mediators of inflammation and angiogenesis. Here we describe a general extraction protocol for lipids, which can be applied to the analysis of the most abundant eicosanoid species.

1. Weigh frozen tissues (~20 mg), record the weight and transfer them into low-bind Eppendorf tubes containing 400 μL of methanol with internal standards [Deuterated Primary COX and LOX LC-MS Mixture (Cayman Chemicals, Ann Arbor, MI, USA)].
2. Homogenize the tissues with a bead-beater, such as MagNA Lyser (Roche). To each tube add 20 washed stainless-steel beads and use a 40 s pulse at 2000 rpm.
3. Centrifuge for 10 min at 21,000 × *g* at 4 °C.
4. Leave at −20 °C for 2 h to allow further precipitation of proteins.
5. Centrifuge for 20 min at 21,000 × *g* at 4 °C.
6. Transfer to an MS vial and store at −80 °C until analysis.

7. Targeted quantification can be conducted on a triple or tandem quadrupole MS system with select multiple reaction monitoring (MRM) transitions (Astarita et al., 2015; Laiakis et al., 2017).

## 3.2 Analysis of TCA cycle metabolites

Glucose and lactate metabolism have been shown to be significantly altered in different tumors types, as a result TCA cycle is a very well-studied pathway in the context of TME (Hensley et al., 2016; Kennedy et al., 2013; Sotgia et al., 2012). Sample preparation for the analysis of TCA cycle intermediates is outlined below.

1. Weigh the tumor tissue sample in a low-bind Eppendorf tube and record the weight.
2. To every 1 mg of tissue sample add 20 μL of acetonitrile:water (v:v, 1:1).
3. Homogenize the samples with a bead-beater such as MagNA Lyser (Roche). To each tube add 20 washed stainless-steel beads and use a 40 s pulse at 2000 rpm. Check to see if all the tissue is homogenized. If not, place the samples on the chilling rack for 1 min and repeat step.
4. Thoroughly mix on multiple tube vortexer for 2 min at 5000 rpm and transfer 200 μL of homogenate to a new pre-labeled low-bind Eppendorf tube.
5. Add 400 μL of ice cold acetonitrile:water (v:v, 1:1) with isotopically labeled internal standards (deuterium or $^{13}$C labeled versions of TCA cycle intermediates are commercially available. Isotopically labeled internal standards should be prepared in acetonitrile:water).
6. Thoroughly mix on multiple tube vortexer for 2 min at 5000 rpm.
7. Precipitate protein in a centrifuge operating at room temperature at 16,000 $\times g$ for 4 min.
8. Transfer 550 μL of the supernatant into a new pre-labeled 2.0 mL low-bind Eppendorf tube.
9. Dry the sample using a lyophilizer or a speed-vac to complete dryness. Lyophilization at −50 °C can be carried out overnight. Time needed to dry the sample on a Speed-vac at 40 °C can vary between 3 and 5 h depending on the strength of the vacuum.
10. If samples are not analyzed immediately, freeze the vials at −70 °C after they are dry.
11. For immediate analysis, add 50 μL of reconstitution buffer (water:methanol, v:v 95:5) to each sample.
12. Thoroughly mix on multiple tube vortexer for 10 min at 5000 rpm.

13. Centrifuge samples at room temperature and at 16,000 ×*g* for 4 min.
14. Transfer the supernatant to pre-labeled low-bind autosampler vials.
15. Inject 10 μL into LC-MS system for analysis (the injection volume may have to adjusted depending on the LC-MS system).
16. Targeted quantification can be conducted on a triple or tandem quadrupole MS system in multiple reaction monitoring (MRM) mode using the metabolite specific mass transitions (Barisón et al., 2017).

## 3.3 Analysis of amino acids

In certain cancers, such as pancreatic cancer, the amino acid alanine is excreted, which in turn outcompetes the glucose- and glutamine-derived carbon to fuel metabolism (Sousa et al., 2016). Cysteine is another amino acid that has an important role in maintaining the redox balance by maintaining glutathione homeostasis (Sullivan & Chandel, 2014). Similarly, other amino acids metabolism has also been shown to be significantly important in different cancer types (Anastasiou, 2017; Hanahan & Weinberg, 2011; Hensley et al., 2016; Kennedy et al., 2013; Lorendeau et al., 2015; Makohon-Moore & Iacobuzio-Donahue, 2016; Murray, 2016; Pavlova & Thompson, 2016). Sample preparation for the analysis of amino acids is outlined below.

1. Weigh the tumor tissue sample in a low-bind Eppendorf tube and record the weight.
2. To every 1 mg of tissue sample add 20 μL of acetonitrile:water (v:v, 1:1).
3. Homogenize the samples with a bead-beater such as MagNA Lyser (Roche). To each tube add 20 washed stainless-steel beads and use a 40 s pulse at 2000 rpm. Check to see if all the tissue is homogenized. If not place the samples on the chilling rack for 1 min and repeat step.
4. Thoroughly mix on multiple tube vortexer for 2 min at 5000 rpm and transfer 200 μL of homogenate to a new pre-labeled low-bind Eppendorf tube.
5. Add 400 μL of ice cold acetonitrile:water (v:v, 1:1) with isotopically labeled internals standards (deuterium or $^{13}C$ labeled versions of amino acids are commercially available. Isotopically labeled internals standards should be prepared in acetonitrile:water).
6. Thoroughly mix on multiple tube vortexer for 2 min at 5000 rpm.
7. Precipitate protein in a centrifuge operating at room temperature and at 16,000 × *g* for 4 min.
8. Transfer 550 μL of the supernatant into a new pre-labeled 2.0 mL low-bind Eppendorf tube.

9. Dry the sample using a lyophilizer or a speed-vac to complete dryness. Lyophilization at −50 °C can be carried out overnight. Time needed to dry the sample on a speed-vac at 40 °C can vary between 3 and 5 h depending on the strength of the vacuum.
10. If samples are not analyzed immediately, freeze the vials at −70 °C after they are dry.
11. For immediate analysis, the amino acids need to be derivatized. For derivatization, add 1 mL of acetonitrile to the AccQ·Tag Ultra reagent powder (Waters Corporation), vortex, and dissolve by heating at 55 °C (no longer than 15 min).
12. Add 70 μL of borate buffer (pH 8.6) to each sample. Thoroughly mix on multiple tube vortexer for 10 min at 5000 rpm.
13. Add 20 μL of the AccQ·Tag Ultra reagent solution prepared above (step 11). Vortex the samples at 55 °C for 10 min.
14. Dilute the sample 1:100 with Optima grade water for LC-MS analysis.
15. Transfer the sample to pre-labeled low-bind autosampler vials. Inject 10 μL into an LC-MS system for analysis (the injection volume may have to adjusted depending on the LC-MS system).
16. Targeted quantification can be conducted on a triple or tandem quadrupole MS system in multiple reaction monitoring (MRM) mode using the metabolite specific mass transitions (Gray et al., 2017).

## 4. MS imaging

MS imaging provides spatial distributions of specific lipid ions in the biological tissues (Amstalden van Hove, Smith, & Heeren, 2010; Gode & Volmer, 2013). Lipids can be ionized, detected, and mapped by several imaging ionization modalities, e.g., matrix-assisted laser desorption ionization (MALDI) (Berry et al., 2011; Jackson et al., 2007), laser ablation electrospray ionization (Shrestha et al., 2010), and desorption electrospray ionization (DESI) (Dill, Ifa, Manicke, Ouyang, & Cooks, 2009; Wiseman, Ifa, Song, & Cooks, 2006). During MS imaging, localized lipid content of tissue is ionized in a pixel-by-pixel manner. Each pixel along a two-dimensional grid will have MS scan with intensities of detected lipid species in that pixel or scan. A motorized translational stage is moved in two dimensions to obtain a unique mass spectrum at each pixel. Ion intensities in each pixel are rendered into false color heatmap images based on their intensities.

MS imaging requires a localized ionization source. Often ionization is performed by an analytical beam, such as focused laser irradiation in MALDI or electrospray jet in DESI, before detection of ions by a mass spectrometer (Škrášková et al., 2016). In some ionization source, often referred to as ambient ionization (Harris, Nyadong, & Fernandez, 2008; Huang, Yuan, Cheng, Cho, & Shiea, 2010), extraction is performed in concert with the ionization step removing often tedious sample preparation steps such as matrix deposition in MALDI. In ambient ionization source, ions are produced directly from the samples at atmospheric pressure or ambient conditions with as minimum sample preparation as technically possible. DESI is one of the first examples of ambient ionization source where extraction and ionization occur in tandem by pneumatically assisted electrospray jet. DESI has been employed to detect lipid species from microbial colonies (Zhang et al., 2011), cell cultures (Bodzon-Kulakowska, Drabik, Marszalek, Kotlinska, & Suder, 2014), mammalian tissue sections (Wiseman, Puolitaival, Takáts, Cooks, & Caprioli, 2005), etc.

Several types of MS detectors can be coupled with imaging ionization sources, each with their inherent advantages and disadvantages. For example, mass analyzers such as a quadrupole time-of-flight (Q-TOF) can provide a high mass accuracy for the tentative identification of molecules during discovery MS imaging, while tandem quadrupole are more suitable for targeted imaging of pre-defined analytes with higher sensitivity and quantitative profile. For lipid analysis, often these ion sources are coupled with quadrupole time-of-flight (Q-TOF) MS systems to take advantage of its speed and mass accuracy. Some Q-TOF systems are equipped with post-ionization ion mobility separation (IMS). For direct analysis such as imaging, IMS removes interfering background before MS detection, resulting in a better signal-to-noise ratio. Also, the drift time of IMS can be converted to collisional cross-section (CCS) or size information of lipid ions that can be searched against databases to support their identification.

In this protocol, strategies for lipid imaging using DESI will be demonstrated by showing the example of a few lipid species from rat brain section.

## 4.1 Tissue preparation for MS imaging

Sectioning, chemical processing, storing, or any tissue modification can affect the MS imaging results. Typically, MS imaging is performed on a fresh-frozen section of tissue and cannot use formalin-fixed paraffin-embedded (FFPE) tissue specimens. Avoid washing tissue section for lipid imaging because most of the washing protocols are designed to remove lipid from the tissue. The following tissue processing steps are recommended for obtaining optimal sections for DESI-MS imaging.

1. Use fresh or flash frozen tissue stored at −80 °C. Note: Fresh tissue needs to be frozen before cryo-sectioning.
2. Avoid the use of any medium to embed frozen tissue. Embedding medium, such as optimal cutting temperature (O.C.T.), contains polyethylene glycols, which interfere with the ion signal from lipids during direct analysis. Alternatively, use carboxymethylcellulose (CMC) or cellulose gum that does not ionize well. Prepare 2.5% CMC solution in deionized water.
3. To mount a tissue section, add a few drops of CMC at the bottom of the cryomold. If possible, use CMC as glue to hold a tissue on mounting chuck without completely covering the tissue with CMC. Note: Label the orientation (e.g., radial or transverse) of the mounted tissue specimen.
4. Snap freeze the mounted tissue specimen.
5. Remove the mounted tissue and attach to cryostat (e.g., M45010-HM525NX, Leica, Buffalo Grove, IL) for sectioning. Note: Check cryostat manufacturer manual for setting up and sectioning guideline for such tissue. Exact setting and temperature may depend on the type of tissue and model of the cryostat.

   Caution: Be cautious during sectioning as cryostat blades are very sharp.
6. The thickness of MS imaging tissue sections is similar to histological sectioning, between 5 and 30 μm. An exact sectioning thickness depends on the physical properties of tissue and the size of tissue. For example, a large tissue section such as whole-body sections needs to be thicker to avoid tearing. Typically, tissue sections for MS imaging are between 10 and 15 μm in thickness.
7. Sectioned tissues can be stored in an airtight slide holder inside a −80 °C freezer until further use.
8. Remove the frozen tissue sections and quickly transfer to a vacuum desiccator before analysis doing best to avoid any condensation. Note: Allow at least 15–30 min to defrost the frozen tissue section.

## 4.2 DESI imaging

A conceptual schematic of a DESI ion source coupled with Q-TOF MS with IMS is shown in Fig. 1. The figure depicts an electrospray emitter with

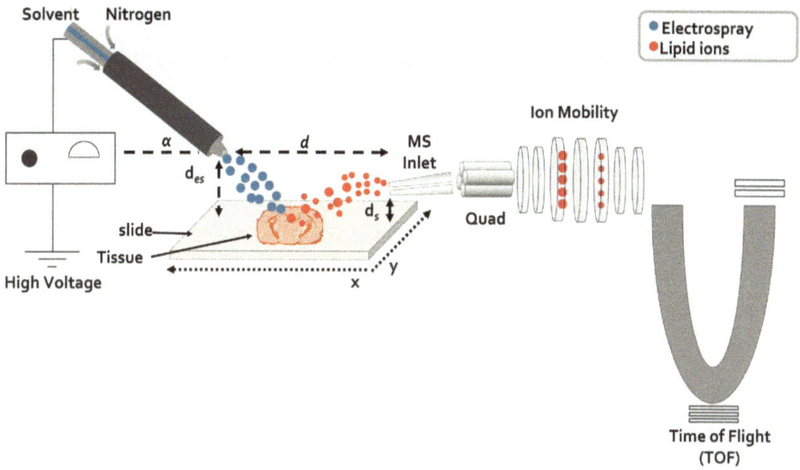


**Fig. 1** Schematic representation of DESI ion source coupled with Q-TOF mass spectrometer with ion mobility cell is shown. Briefly, pneumatically assisted electrospray solvent jet impinges on tissue surface desorbing and ionizing lipids. The ionized ions are sampled inside a mass spectrometer by an ion transfer capillary placed very close to the sample surface. MS imaging is done by moving the sample stage in a defined XY grid by a two-dimensional translational stage.

nitrogen gas flow assembly on a sample mounted on the two-dimensional translational stage. Ion transfer capillary of a mass spectrometer is placed very close to the sample surface. The spatial resolution of the DESI image depends on the size of the electrospray impact and the speed of the two-dimensional translational stage (step velocity). An optimization of DESI source, ideally performed with a similar sample such as adjacent serial section, is required for obtaining robust MS signal (Abbassi-Ghadi et al., 2015; Tillner et al., 2017). MS imaging software, e.g., High-definition imaging (HDI) software (Waters Corp, Milford, MA), interface with a mass spectrometer to acquire RAW data for MS imaging experiments. MS imaging software facilitates the acquisition of RAW mass spec data where each MS scan can be correlated to a single pixel. Here, we describe optimization strategies for commercial DESI source (Claude, Jones, & Pringle, 2017).

1. Tune the mass spectrometer (Xevo G2-XS or SYNAPT G2-S*i*, Waters Corporation, Milford, MA) according to the manufacturer's guidelines.
2. To optimize, continuously scan tissue and optimize DESI ion source parameters by monitoring ion intensities of the analyte of interest. Generally, DESI optimization means a geometric realignment of DESI electrospray assembly against sample surface and ion inlet. This includes electrospray incidence angle ($\alpha$), the angle of ion collecting inlet

capillary, the distance between emitter and sample surface ($d_{es}$), the distance between the electrospray emitter and MS inlet capillary (*d*), the distance between ion collection capillary and sample surface ($d_s$). In a new setup, adjustments may be needed to ensure correct microfluidic connections are made and the sprayer is assembled correctly. Start with the recommended values provided by the vendor and perform optimization according to their recommendation.

3. Electrospray composition for DESI imaging depends on the analyte of interest and the tissue. A typical composition of 95–98% methanol with 0.1% formic acid is recommended as a starting point for a positive ion mode DESI imaging.
4. Since the extraction of analyte depends on the solvent composition, method development is often done by changing the percent composition of aqueous or switching organic components, e.g., acetonitrile instead of methanol.
5. An internal standard (e.g., leucine enkephalin 250 pg/μL) added to DESI solvent is used as a lock mass to correct any mass drift during extended MS imaging runs. After applying lock mass, a low ppm mass accuracy can be obtained for higher confidence molecular annotation.
6. DESI-MS imaging integrated with ion mobility separation can provide collision cross section (CCS) of each ion. The CCS values can be used together with accurate mass to charge ratio ($m/z$) to increase the confidence during compound identification. DESI imaging is commercially coupled with traveling wave (T-Wave) ion mobility cell within a Q-TOF mass spectrometer (SYNAPT G2-Si, Waters Corp). Before obtaining CCS value, singly charged poly-DL-alanine oligomers are used as the calibrant. CCS calibration can be done using DESI or separately using electrospray ionization.
7. Define the mass spec acquisition mode (e.g., MS, MS/MS, ion mobility MS (HDMS), HDMS/MS, HDMSE), detector mode (sensitivity, resolution), and $m/z$ acquisition range. Refer to the manufacturer's instruction for additional parameters.
8. HDI software processes RAW mass spec files to an intensity table containing intensities for a defined number of ions, e.g., top 3000 $m/z$ peaks or a target $m/z$ list, for each pixel. Ions can also be defined based on $m/z$ ratio and drift times. Several image processing parameters such as $m/z$ window, lock-mass can be defined during the acquisition or later in post-acquisition.

**9.** HDI contains several MS imaging features such as multiple ion overlay, histological overlay, hierarchal clustering of ions based on spatial distribution, spatial correlation analysis on a single ion showing the ranking of all other ions based on spatial similarity, regions of interest (ROI's) analysis, normalization, etc. If additional data processing is required, the RAW data can be exported easily to the common imzML (Schramm et al., 2012) format to be used by other software such as Cardinal and SCiLS.

Successful DESI-MS imaging experiments using this protocol will result in simultaneous images of a large number of lipid ions in tissue in a single measurement. Fig. 2 shows an example of a sagittal rat brain section using DESI imaging on Q-TOF IMS mass spectrometer. Ion overlay of three ion species, e.g., *m/z* 369.3516 (red), 800.5512 (blue), 848.6522 (green), is shown top left, whereas other images in the panel of those ions separately. The image was generated after normalizing the ion intensities per peak against the doped internal standard, LeuEnk peak. Tentative molecular identification of lipid and accurate experimental *m/z* are water-loss protonated cholesterol, potassium adduct of PC (34:0), and water-loss PC (42:4).

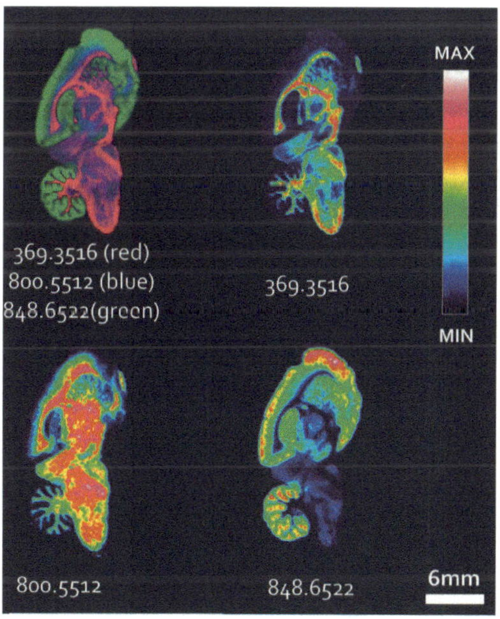


**Fig. 2** Multiple ion overlay of lipid species is presented on the top left, their individual ion intensity heatmaps are shown in the rest of the panel.

## 5. Summary

Investigators who wish to assess the metabolic profile of the TME will find themselves able to select from a myriad of options. LC-MS has proven to be a sensitive method for metabolomic profiling and for quantitative assessment of whole pathways or select biomarkers. MS imaging can highlight the spatial distribution of a metabolite of interest in biological tissues, allowing for multiple metabolite assessment and evaluation of co-localization, which can be used for assessment of metabolic alterations in TME.

## Acknowledgments

This work was supported by the National Institutes of Health National Institute of Allergy and Infectious Diseases (NIAID) grant U19 AI067773 (P.I. David J. Brenner, performed as part of Columbia University Center for Medical Countermeasures against Radiation) and pilot award AWD-7773347 (P.I. Evagelia C. Laiakis), NIAID R01AI101798 (P.I. Albert J. Fornace, Jr), and National Cancer Institute R01CA184168 (P.I. Albert J. Fornace, Jr, former P.I. Eliot M. Rosen).

## References

Abbassi-Ghadi, N., Jones, E. A., Veselkov, K. A., Huang, J., Kumar, S., Strittmatter, N., et al. (2015). Repeatability and reproducibility of desorption electrospray ionization-mass spectrometry (DESI-MS) for the imaging analysis of human cancer tissue: A gateway for clinical applications. *Analytical Methods*, 7, 71–80.

Amstalden van Hove, E. R., Smith, D. F., & Heeren, R. M. (2010). A concise review of mass spectrometry imaging. *Journal of Chromatography. A*, *1217*, 3946–3954.

Anastasiou, D. (2017). Tumour microenvironment factors shaping the cancer metabolism landscape. *British Journal of Cancer*, *116*, 277–286.

Anderson, N. M., Mucka, P., Kern, J. G., & Feng, H. (2018). The emerging role and targetability of the TCA cycle in cancer metabolism. *Protein & Cell*, *9*, 216–237.

Astarita, G., Kendall, A. C., Dennis, E. A., & Nicolaou, A. (2015). Targeted lipidomic strategies for oxygenated metabolites of polyunsaturated fatty acids. *Biochimica et Biophysica Acta*, *1851*, 456–468.

Astarita, G., Stocchero, M., & Paglia, G. (2018). Unbiased lipidomics and metabolomics of human brain samples. *Methods in Molecular Biology*, *1750*, 255–269.

Badur, M. G., & Metallo, C. M. (2018). Reverse engineering the cancer metabolic network using flux analysis to understand drivers of human disease. *Metabolic Engineering*, *45*, 95–108.

Barisón, M. J., Rapado, L. N., Merino, E. F., Furusho Pral, E. M., Mantilla, B. S., Marchese, L., et al. (2017). Metabolomic profiling reveals a finely tuned, starvation-induced metabolic switch in Trypanosoma cruzi epimastigotes. *The Journal of Biological Chemistry*, *292*, 8964–8977.

Berry, K. A., Hankin, J. A., Barkley, R. M., Spraggins, J. M., Caprioli, R. M., & Murphy, R. C. (2011). MALDI imaging of lipid biochemistry in tissues by mass spectrometry. *Chemical Reviews*, *111*, 6491–6512.

Bodzon-Kulakowska, A., Drabik, A., Marszalek, M., Kotlinska, J. H., & Suder, P. (2014). DESI analysis of mammalian cell cultures—Preparation and method optimisation. *Journal of Mass Spectrometry*, *49*, 613–621.

Claude, E., Jones, E. A., & Pringle, S. D. (2017). DESI mass spectrometry imaging (MSI). *Methods in Molecular Biology*, *1618*, 65–75.

Dill, A. L., Ifa, D. R., Manicke, N. E., Ouyang, Z., & Cooks, R. G. (2009). Mass spectrometric imaging of lipids using desorption electrospray ionization. *Journal of Chromatography. B, Analytical Technologies in the Biomedical and Life Sciences*, *877*, 2883–2889.

Gode, D., & Volmer, D. A. (2013). Lipid imaging by mass spectrometry—A review. *Analyst*, *138*, 1289–1315.

Gomes, R. N., Felipe da Costa, S., & Colquhoun, A. (2018). Eicosanoids and cancer. *Clinics (São Paulo, Brazil)*, *73*, e530s.

Gray, N., Zia, R., King, A., Patel, V. C., Wendon, J., McPhail, M. J., et al. (2017). High-speed quantitative UPLC-MS analysis of multiple amines in human plasma and serum via precolumn derivatization with 6-aminoquinolyl-N-hydroxysuccinimidyl carbamate: Application to acetaminophen-induced liver failure. *Analytical Chemistry*, *89*, 2478–2487.

Hanahan, D., & Weinberg, R. A. (2011). Hallmarks of cancer: The next generation. *Cell*, *144*, 646–674.

Harris, G. A., Nyadong, L., & Fernandez, F. M. (2008). Recent developments in ambient ionization techniques for analytical mass spectrometry. *Analyst*, *133*, 1297–1301.

Hensley, C. T., Faubert, B., Yuan, Q., Lev-Cohain, N., Jin, E., Kim, J., et al. (2016). Metabolic heterogeneity in human lung tumors. *Cell*, *164*, 681–694.

Huang, M. Z., Yuan, C. H., Cheng, S. C., Cho, Y. T., & Shiea, J. (2010). Ambient ionization mass spectrometry. *Annual Review of Analytical Chemistry (Palo Alto, California)*, *3*, 43–65.

Jackson, S. N., Ugarov, M., Egan, T., Post, J. D., Langlais, D., Albert Schultz, J., et al. (2007). MALDI-ion mobility-TOFMS imaging of lipids in rat brain tissue. *Journal of Mass Spectrometry*, *42*, 1093–1098.

Kennedy, K. M., Scarbrough, P. M., Ribeiro, A., Richardson, R., Yuan, H., Sonveaux, P., et al. (2013). Catabolism of exogenous lactate reveals it as a legitimate metabolic substrate in breast cancer. *PLoS One*, *8*, e75154.

Kepp, O., Loos, F., Liu, P., & Kroemer, G. (2017). Extracellular nucleosides and nucleotides as immunomodulators. *Immunological Reviews*, *280*, 83–92.

Kim, S. Y. (2019). Targeting cancer energy metabolism: A potential systemic cure for cancer. *Archives of Pharmacal Research*, *42*, 140–149.

Laiakis, E. C., Pannkuk, E. L., Chauthe, S. K., Wang, Y. W., Lian, M., Mak, T. D., et al. (2017). A serum small molecule biosignature of radiation exposure from total body irradiated patients. *Journal of Proteome Research*, *16*, 3805–3815.

Lorendeau, D., Christen, S., Rinaldi, G., & Fendt, S. M. (2015). Metabolic control of signalling pathways and metabolic auto-regulation. *Biology of the Cell*, *107*, 251–272.

Lyssiotis, C. A., & Kimmelman, A. C. (2017). Metabolic interactions in the tumor microenvironment. *Trends in Cell Biology*, *27*(11), 863–875.

Makohon-Moore, A., & Iacobuzio-Donahue, C. A. (2016). Pancreatic cancer biology and genetics from an evolutionary perspective. *Nature Reviews. Cancer*, *16*, 553–565.

Murray, P. J. (2016). Amino acid auxotrophy as a system of immunological control nodes. *Nature Immunology*, *17*, 132–139.

Norris, P. C., & Serhan, C. N. (2018). Metabololipidomic profiling of functional immunoresolvent clusters and eicosanoids in mammalian tissues. *Biochemical and Biophysical Research Communications*, *504*(3), 553–561.

Paglia, G., Stocchero, M., Cacciatore, S., Lai, S., Angel, P., Alam, M. T., et al. (2016). Unbiased Metabolomic investigation of Alzheimer's disease brain points to dysregulation of mitochondrial aspartate metabolism. *Journal of Proteome Research*, *15*, 608–618.

Park, J. K., Coffey, N. J., Limoges, A., & Le, A. (2018). The heterogeneity of lipid metabolism in Cancer. *Advances in Experimental Medicine and Biology, 1063*, 33–55.

Pavlova, N. N., & Thompson, C. B. (2016). The emerging hallmarks of cancer metabolism. *Cell Metabolism, 23*, 27–47.

Schramm, T., Hester, A., Klinkert, I., Both, J. P., Heeren, R. M., Brunelle, A., et al. (2012). imzML—A common data format for the flexible exchange and processing of mass spectrometry imaging data. *Journal of Proteomics, 75*, 5106–5110.

Shrestha, B., Nemes, P., Nazarian, J., Hathout, Y., Hoffman, E. P., & Vertes, A. (2010). Direct analysis of lipids and small metabolites in mouse brain tissue by AP IR-MALDI and reactive LAESI mass spectrometry. *Analyst, 135*, 751–758.

Škrášková, K., Claude, E., Jones, E. A., Towers, M., Ellis, S. R., & Heeren, R. M. (2016). Enhanced capabilities for imaging gangliosides in murine brain with matrix-assisted laser desorption/ionization and desorption electrospray ionization mass spectrometry coupled to ion mobility separation. *Methods, 104*, 69–78.

Sotgia, F., Martinez-Outschoorn, U. E., Howell, A., Pestell, R. G., Pavlides, S., & Lisanti, M. P. (2012). Caveolin-1 and cancer metabolism in the tumor microenvironment: Markers, models, and mechanisms. *Annual Review of Pathology, 7*, 423–467.

Sousa, C. M., Biancur, D. E., Wang, X., Halbrook, C. J., Sherman, M. H., Zhang, L., et al. (2016). Pancreatic stellate cells support tumour metabolism through autophagic alanine secretion. *Nature, 536*, 479–483.

Sullivan, L. B., & Chandel, N. S. (2014). Mitochondrial reactive oxygen species and cancer. *Cancer & metabolism, 2*, 17.

Tillner, J., Wu, V., Jones, E. A., Pringle, S. D., Karancsi, T., Dannhorn, A., et al. (2017). Faster, more reproducible DESI-MS for biological tissue imaging. *Journal of the American Society for Mass Spectrometry, 28*, 2090–2098.

Vitale, I., Manic, G., Coussens, L. M., Kroemer, G., & Galluzzi, L. (2019). Macrophages and metabolism in the tumor microenvironment. *Cell Metabolism, 30*, 36–50.

Wiseman, J. M., Ifa, D. R., Song, Q., & Cooks, R. G. (2006). Tissue imaging at atmospheric pressure using desorption electrospray ionization (DESI) mass spectrometry. *Angewandte Chemie (International Ed. in English), 45*, 7188–7192.

Wiseman, J. M., Puolitaival, S. M., Takáts, Z., Cooks, R. G., & Caprioli, R. M. (2005). Mass spectrometric profiling of intact biological tissue by using desorption electrospray ionization. *Angewandte Chemie (International Ed. in English), 44*, 7094–7097.

Zasada, C., & Kempa, S. (2016). Quantitative analysis of cancer metabolism: From pSIRM to MFA. *Recent Results in Cancer Research, 207*, 207–220.

Zhang, J. I., Talaty, N., Costa, A. B., Xia, Y., Tao, W. A., Bell, R., et al. (2011). Rapid direct lipid profiling of bacteria using desorption electrospray ionization mass spectrometry. *International Journal of Mass Spectrometry, 301*, 37–44.

CHAPTER FIVE

# Immunoscore assay for the immune classification of solid tumors: Technical aspects, improvements and clinical perspectives

Florence Marliot[a,b], Lucie Lafontaine[a], Jérôme Galon[a,*]
[a]INSERM, Laboratory of Integrative Cancer Immunology, Équipe Labellisée Ligue Contre le Cancer, Sorbonne Université, Sorbonne Paris Cité, Université Paris Descartes, Université Paris Diderot; Centre de Recherche des Cordeliers, Paris, France
[b]Laboratory of Immunology, AP-HP, European Hospital Georges Pompidou, Paris, France
*Corresponding author: e-mail address: jerome.galon@crc.jussieu.fr

## Contents



## Abstract

The past two decades witnessed the appreciation of the importance of specific tumor-infiltrating immune cells in influencing tumor evolution. The discovery that a favorable immune contexture is linked to a prolonged patients' survival, and more specifically that intratumoral cytotoxic T lymphocytes hold powerful prognostic value, provided the foundations for the development of the Immunoscore. Immunoscore is a digital

*Methods in Enzymology*, Volume 636
ISSN 0076-6879
https://doi.org/10.1016/bs.mie.2019.07.018

pathology, IHC-based immune assay measuring the densities of $CD3^+$ and $CD8^+$ T cells at different tumor locations, linking them with patients' clinical outcome. The present chapter provides a detailed overview of the assay development and procedure, from the bench to the data analysis, and of the internationally approved and validated guidelines to perform a *consensus* Immunoscore for colon cancer patients. Assay strengths and limitations are also discussed, as well as the clinical implications of its introduction in routine practice.

## 1. Introduction

The American Joint Committee on Cancer/Union for International Cancer Control (AJCC/UICC) TNM classification system based on tumor features is used for prognosis estimation and treatment recommendations in most cancers. However, the clinical outcome can vary significantly among patients within the same tumor stage, and TNM classification does not predict response to therapy. Much effort has been focused on the identification of new markers. Multiple tumor cell-based approaches have been proposed but very few have been translated into the clinic. It is now demonstrated with a considerable amount of data from animal models and human patients that the immune system can recognize and eliminate primary developing tumors (Dunn, Old, & Schreiber, 2004). For 30 years, several publications have evaluated with an increasing level of precision the quality, the density and the functional orientation of T cells infiltrating human tumors. There is now accumulated evidence showing a positive association between the densities of intratumoral lymphocyte infiltrates in solid tumors and increased patient survival (Fridman, Pages, Sautes-Fridman, & Galon, 2012). Human solid tumors are commonly infiltrated by cells from the innate (innate lymphoid cells, NK cells, NK-T cells, γδ T cells, macrophages, neutrophils, eosinophils and mast cells) and adaptive immunity (T and B lymphocytes, dendritic cells). Strikingly, this immune infiltrate influences the course of the disease despite the apparent insensitivity of the tumor cells at primary site to the immune attack. The recent demonstration of the essential role of the immune system in tumor progression has allowed great advance in the understanding of this complex disease and in the design of novel therapies (Chew, Toh, & Abastado, 2012; Emens, Butterfield, Hodi, Marincola, & Kaufman, 2016). In previous multiple studies, we characterized the subtypes of tumor-infiltrating immune cells (Bindea et al., 2013; Galon et al., 2006). This immune contexture (i.e., the type, functional orientation, density, and location of immune cells) of solid tumors could be determinant of clinical outcome (Galon et al., 2006; Galon, Fridman, & Pages, 2007). Thus, a biological test assessing the tumor immune infiltrate could become a central biomarker for prognostic and (immuno)therapy predictive purposes (Galon, Angell, Bedognetti, & Marincola, 2013; Galon & Bruni, 2019). These observations brought to suggest the development of an immune scoring system in order to facilitate the translation of this immune investigation into clinical practice. After the first Society for Immunotherapy of Cancer (SITC) meeting in 2012, a consensus Immunoscore has been defined and approved and this new test was developed (http://www.immunoscore.org). This test consists of assessing the density of $CD3^+$ and $CD8^+$ T cells effectors within the tumor and its invasive margin using specific standardized criteria. The prognostic value of Immunoscore has recently been validated with the support of the SITC by a major international multicenter retrospective study in stage I–III colon cancer (Pages et al., 2018). The Immunoscore could improve the patient's prognostic and the therapeutic management in clinical practice. Leveraging on this powerful concept, a new Immuno-Oncology diagnostic company, HalioDx, was founded and focused on the development of an Immunoscore Colon Kit® and its introduction in clinical practice. This test has been approved for clinical use in colorectal cancer (CE-IVD).

Here, we described the standardized procedures of the Immunoscore assay, the advantages, the inconveniences, and possible future applications.

## 2. Immunoscore® assay

The Immunoscore assay is the first standardized immune-based assay for the classification of cancer. The densities of $CD3^+$ T cells and $CD8^+$ cytotoxic T cells, within the CT and the IM were assessed by immunohistochemistry (IHC) on formalin-fixed, paraffin-embedded (FFPE) tissue sections. A dedicated software is used to quantify the effector T-cells in the CT and IM, and digitally assess the immune score. The hypothesis was that each tumor region provided complementary information that could affect the tumor progression and provide a prognostic indicator. In previous studies, a significant correlation was found for each marker of adaptive immunity studied between each tumor region (CT and IM) and the clinical outcome of the patient in disease-free survival (DFS) and overall survival (OS). The combination between immune markers, and these two regions (CT/IM) further enhanced prognostic precision and increased differences between patient groups (Galon et al., 2006).

Of note, previous studies relied on the use of the memory T cell marker CD45RO, instead of CD3, for Immunoscore evaluation (Galon et al., 2006; Mlecnik et al., 2011). However, the presence of staining background, as well as antigen loss under normal storage conditions, led to the decision to employ the T cell lineage marker CD3 as more technically exploitable, yet retaining equivalent prognostic power (Galon et al., 2014; Hermitte, 2016). Indeed, the prognostic value of these immune parameters was superior to all known biomarkers, including the TNM classification. Cox multivariate analysis showed that Immunoscore remained the most powerful and only significant parameter (Galon et al., 2006; Mlecnik et al., 2011). Similar results were found for early-stage colorectal cancer (Pages et al., 2009).

## 2.1 Sample requirements and preparation of tissue sections

The colon tissue piece is fixed in 10% formol saline for 24 h, as per routine clinical processing (Bancroft & Cook, 1985; Dey, 2018; Ramos-Vara, 2011). The tissue is embedded in paraffin using molds and an embedding station. The paraffin block is removed from the mold and trimmed on a rotary microtome. For each colon tumor, several FFPE tumoral blocks could be eligible for the Immunoscore assay. The pathologist selects the tumoral block from Hematoxylin Eosin Safran (HES) stained slides. The block must contain the tumor core, which should represent at least 5% of the overall tissue, and the surrounding healthy tissue, joined to the tumor at the IM. Initial studies, including the SITC-led international validation study, performed the Immunoscore® assay on the tissue block containing the highest immune infiltrate. More recent evidence demonstrated that any tumor block containing the IM could be used with similar results (manuscript in preparation). To minimize variation in tissue section thickness, the selected block is placed for 10 min at −20 °C on a cold plate (Myr EC350-2, Especialidades Medicas Myr, S.L., Spain), allowing the paraffin to cool and harden, and enabling an easier cut of a thin section. Serial sections are cut with a microtome (Microm HM340E, Thermo Fisher Scientific, USA), with the section thickness set to 4 μm. Each section is floated out on a non-albuminated distilled water drop put beforehand on a positively-charged glass slides (Super-Frost Plus, Menzel Glaser, Germany; DakoFlex, Agilent Technologies Inc., California,) placed on a hot-plate at 45 °C (Medax, Biosystems, Switzerland AG). After gently removing the excessed water localized between the glass slide and the tissue section, the latter adheres to the glass slide. The unstained sections are dried on a rack for 1 h at room temperature, and placed in a drying-oven for minimum 12h at 56 °C to ensure the section sticks to the slide. It is strongly recommended to use the tissue section in short delay (2 weeks) after cutting to avoid modifications in tissue section antigenicity. The delay between the cut and the immunohistochemistry should not exceed 4 months, with a storage at 4 °C immediately after cutting.

## 2.2 Optimization of CD3 and CD8 immunostaining for the Immunoscore assay in research procedure

To be applied in research setting, two adjacent tumor slides are stained with antibodies against CD3 and CD8, respectively, using the automated BenchMark XT immunostainer (Ventana Medical System, Arizona, USA), according to a standardized technique. The Benchmark XT enables a standard deparaffinization of the tissue sections, using EZ Prep buffer (Ventana, Arizona, USA) at 75 °C. Then, we defined optimal immunostaining protocols with the Benchmark XT immunostainer for CD3 and CD8 immunostaining: antigen retrieval is performed in Tris-EDTA base buffer, pH 8.5 (Cell Conditioning 1, Ventana, USA) for 60 min at 95 °C, and the endogenous peroxidase activity is quenched with 3% $H_2O_2$ for 10 min. Primary antibody against CD3 (2GV6, ready-to-use; Ventana, Arizona, USA) and primary antibody against CD8 (C8/144B, 3 μg/mL, Dako, Glostrup, Denmark) are incubated at 37 °C for 20 and 32 min, respectively. They are revealed in brown color with an UltraView Universal Diaminobenzidine (DAB) IHC Detection kit (Ventana, Arizona, USA). The tissue is counterstained with Mayer's Hematoxylin II and Bluing Reagent (Ventana, Arizona, USA) to visualize cell nuclei in blue. Slides are washed manually in four successive baths of distilled water with soap (Aniosafe, France) and then rinsed in four successive baths of distilled water. Then, the tissues are manually dehydrated in 70%, 90% and 100% successive alcohol baths for 4 min each, and slides are covered using an automatic coverslip-placing instrument (Microm CTM6, Thermo Scientific, USA) with Pertex glue (Histolab, Golthenburg, Sweden). Staining protocols were validated by the NordicQC organization (https://www.nordiqc.org/) using their external controls (Tonsil, Appendix, Tonsil, T-cell Lymphoma (CD8 negative), T-cell Lymphoma (CD8 positive)).

### 2.3 Immunoscore® assay for in vitro diagnostic test for clinical use

To be applied in clinical practice, the Immunoscore procedure and analysis was standardized and approved as an in vitro diagnostic test for clinical use (CE-IVD) (HalioDx, Marseille, France). An Immunoscore® colon kit (https://www.haliodx.com/) was proposed by HalioDX company with CD3 (HDX1, HalioDx, France) and CD8 (HDX2, HalioDx, France) pre-diluted monoclonal antibodies to standardized the immunostaining protocols. In first intention, these antibodies were validated with the BenchMark XT autostainer, and more recently with the BenchMark ULTRA system (both from Ventana, Arizona, USA). A staining protocol is recommended by HalioDx company following the autostainer used. As previously described, two adjacent tissue sections of 4 μM from FFPE tumoral block with the presence of tumor and invasive marge are recommended and immunostained using CD3 (HDX1, HalioDx, France) and CD8 (HDX2, HalioDx, France) monoclonal antibodies. Detection is performed with the ultraView Universal DAB Detection Kit (Ventana, Arizona, USA) followed by counterstaining with Mayer's Hematoxylin II and Bluing Reagent (Ventana, Arizona, USA). Immunostained slides are then washed, dehydrated and mounted on coverslips following the recommendation in previous chapter. Moreover, Immunoscore® colon kit proposes one separate control slide with three external controls: one negative tissue control (placenta) and two positive controls (one tissue: tonsil and a cell line pellet) are processed identically in each immunohistochemistry run, thereby enabling the monitoring of the quality of the staining.

## 3. Digitization of slides and analysis

In order to convert physical slides into digital images, slides are scanned using a Hamamatsu NanoZoomer (Hamamatsu Photonics, Japan). The instrument applies a 20 × scanning resolution mode (0.45 μm/pixel resolution) on a single focus plane, as automatically identified by the machine. Images are thereby created. After scanning, the operator validates for each slide the quality of the image by assessing the quality of the tissue section, the sharpness of the image, the presence of staining ($CD3^+$ and $CD8^+$ T-cells are colored in brown and the counterstain is blue, Fig. 1A), of background and of artifacts. This scanner is also the one approved for the CE-IVD Immunoscore® Assay in clinical practice.

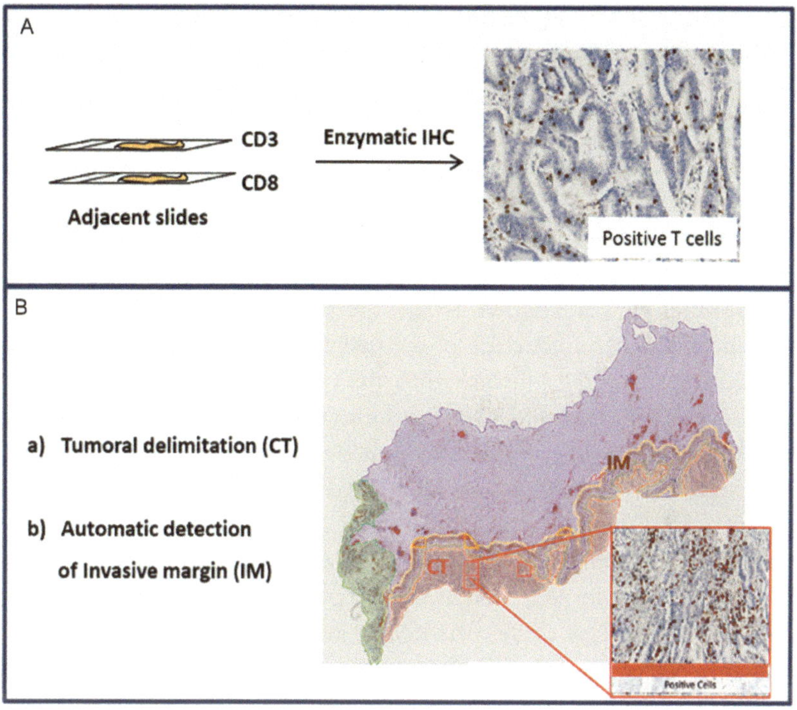


C **Immunoscore Classification**

| Percentiles | Immunoscore Classes | Patients IS groups |
|---|---|---|
| >95-100% | I4 | High |
| >70-95% | I3 | |
| >25-70% | I2 | Intermediate |
| >10-25% | I1 | Low |
| 0-10% | I0 | |

**Fig. 1** The Immunoscore assay. (A) Different steps of Immunoscore process with the standardized enzymatic immunohistochemistry in Diaminobenzdine (DAB) on two cutting adjacent colon cancer sections, as obtained with an automatic stainer. (B) Standardized tumoral classification using the Immunoscore Analyzer® software with automatic detection of tissue in blue, epithelium in red, tumor core (CT) in red and automatic detection of the tumor invasive margin (IM) in yellow; automatic quantification of brown stained cells, detected in red. (C) Immunoscore classifications with mean percentile categorization translated in Immunoscore classes and Immunoscore (IS) groups.

## 3.1 Immunoscore software and digital procedure for colon cancer

We first developed a dedicated Immunoscore module to quantify the immune cells in the colon tumor and its invasive margin (Immunoscore module, INSERM Paris France, under software Architect developed from software Developer XD, Definiens). This software was developed specifically for simple enzymatic immunohistochemistry revealed with Diaminobenzidine (DAB) in brown color and for the assessment of staining cell density in the CT and IM. Images are uploaded into the software; the tissue and histological structure of the tissue are automatically detected by the software. Using the software tools, the user defines a tumoral classification: the tumor (adenocarcinoma), healthy tissue (sub-mucosa, muscularis propria, serosa) and the epithelium (mucosa) are selected (Fig. 1B); areas of necrosis, areas of abscess and artifacts (bubbles folds, torn areas, background) are excluded. The tumoral emboli and the metastasis lymph nodes are also excluded of the classification. The IM with predetermined and fixed width is automatically proposed by the software at the junction between the tumor and the healthy tissue, oriented inside of the colon wall tissue. The annotation can be corrected by the user at any time before starting the analysis. To detect and quantify correctly immunostained cells, a threshold is determined and fixed. A quality assessment of the intensity of the staining is performed by the software to ensure proper automatic counting of the positive immune cells. Thus, the intensity of all detected cells is monitored to provide a distribution of the intensity of the stained cells. The mean and median intensity for each analyzed slide is calculated. A histogram with the distribution of the positive T-cells staining intensity is shown in order to validate the quality of the staining for each slide. During the analysis, the software divides the predefined tumor regions in fields (tiles). The number of positive cells is calculated, the tissue area and the density of immunostained cells are assessed in each field. At the end of the analysis with the software, a validation step by a pathologist ensures the correct classification of the tumor, the quality of the staining intensity and that a minimum of 70–80% of positive T-cells in the regions of interest (CT and IM) are detected by the software.

## 3.2 The Immunoscore® analyzer software

To facilitate the process of the assay and its introduction in clinical practice, an analysis software dedicated to the routine Immunoscore assessed in colon tumor (Immunoscore® Analyzer, HalioDx, France) has also been developed, improving the previous version of the Immunoscore software. Sets of two digital images, showing, respectively, the CD3 and CD8 marker in immunostained colon tissue, are both imported into the Immunoscore® Analyzer software and automatically processed for the detection of the whole tissue, the tumor area, the healthy tissue, the epithelium, and for the detection of $CD3^+$ and $CD8^+$ T-cells in the whole tissue. The detection of the stained cells implements the same algorithm as the previous version of the Immunoscore software. The stained cells are detected and quantified applying a fixed threshold. If required, the regions can be manually modified and abscess area, necrosis and artifacts can be deleted by the user. Then, the IM is automatically displayed, having a width spanning 360 μm into the healthy tissue and 360 μm into the tumor, calculated from the border between these two tissue types. As previously described, a pathologist validates the correct tumoral classification, the quality of the staining intensity based on the mean intensity of the histogram and the correct detection of the signal (70–80% minimum of T-cells detected in CT and IM regions) (Fig. 1B). The Immunoscore® Analyzer generates a printable report which contains a capture of all quality control results as well as the analysis results. The Immunoscore® Analyzer is the software approved for the CE-IVD Immunoscore® Assay in clinical practice.

## 3.3 Score calculation

The Immunoscore allows to obtain an immune score from the density of the $CD3^+$ and $CD8^+$ effector T-cell within the tumor and its invasive margin. After image processing validation, densities of $CD3^+$ and $CD8^+$ positive lymphocytes in the two regions of interest (ROIs), CT or IM, are reported. The mean density of each marker is determined in each region by the image analysis program. According to an automated calculation, the density level of each marker in each region is translated to density percentile defined in a large international retrospective validation study (>3500 Stage I–III colon cancer patients; see following chapter) (Pages et al., 2018). The mean density percentile is then calculated. Based on the average density percentile classes, the Immunoscore is categorized into five classes, from Immunoscore 0 (IS-0) to Immunoscore 4 (IS-4). Immunoscore classified from 0 to 1 corresponds to low $CD3^+$ and $CD8^+$ lymphocyte infiltrate (IS Low), IS-2 to a moderate $CD3^+$ and $CD8^+$ lymphocyte infiltrate (IS Int.), and IS-3-4 to a high $CD3^+$ and $CD8^+$ lymphocyte infiltrate (IS High) (Fig. 1C).

To be applied in clinical setting, Immunoscore can be performed in Clinical Laboratory Improvement Amendments (CLIA)-certified laboratories in Europe (HalioDx, Marseille, France) and in the United States (HalioDx Inc., Richmond, VA).

## 4. Advantages and limitations of Immunoscore assay

Immunoscore is a novel method for quantification of immunohistochemical staining which combines whole slide imaging with image analysis. Several steps are needed for the Immunoscore assay: immunohistochemistry of tissue, scanning of slides to obtain digital images and quantification of positive markers ($CD3^+$ and $CD8^+$) in regions of interest (CT and IM). Traditionally, IHC staining of tissue has always been a fundamental technique in tissue diagnosis and research. Clinical pathology laboratories typically perform IHC on FFPE tissue sections by standard methods, making potentially available hundreds of millions of specimens for subsequent study. Therefore, validated IHC assays with standardized procedures may be readily implemented into clinical practice. Actually, for the analysis of the immunostaining on the tissue in clinical practice, pathologists still rely on visual assessments. Pathologists' visual scoring without quantification of positive markers with analysis image software is a qualitative method fraught with problems due to subjectivity in interpretation. Automated IHC measurements promise to overcome these limitations. Whole-slide imaging systems are widely available to convert immunohistochemistry of FFPE tissues mounted on glass slides into diagnostic quality digital images. Automated IHC measurements are precise in ranges of staining that are too weak for ocular assessment (Rimm, 2006) and produce continuous data (Rimm et al., 2007). Moreover, when automated IHC measurements are provided to a pathologist during visual scoring, computer aided IHC analysis substantially improves both intra- and interobserver agreement (Gavrielides, Gallas, Lenz, Badano, & Hewitt, 2011). Automated analysis is objective and temporally linear, regardless of the number of the analyzed cases (Ong et al., 2010). Nevertheless, this technique suffers from significant variability depending on the intensity of the staining, and strict standardized procedures and methods of quality control are required in final data analysis to exclude tissue areas with artifacts such as tissue folds, bubbles, or high background. The quality of stain is known to be affected by section thickness and the storage in long-term of the tissue sections (Jacobs, Prioleau, Stillman, & Schnitt, 1996). The variability in immunohistochemical staining within and between laboratories has implications for diagnostic accuracy and quality control in current laboratory practice, but also for the repeatability and reproducibility of image analysis algorithms for diagnostic use in future. This is why a strict standardized operating procedure (SOP), as described in the previous chapter, has been developed for the consensus Immunoscore Assay. It is demonstrated that the Immunoscore assay is simple, robust and repeatable but its assessment requires certain constraints (Pages et al., 2018). To achieve reproducible Immunoscore measurements, the cut of the tissue and the immunohistochemistry have to be performed according to a stringently standardized process. Our experience showed us that it is very important to avoid folded, torn tissues, and background on the tissue to obtain a proper Immunoscore. Long-term storage of unstained slides is not recommended, due to decreased staining intensity. The development of the Immunoscore® colon kit (CE-IVD, HalioDx, France) and the additional availability of the test in CLIA certified environment guarantee its performance. The image acquisition performed with different scanners (reproducibility measure) shows significant variations due to different resolutions in the virtual slide image, either due to illumination or some difference in performance of the image capture system. The choice of the resolution and the absence of scanner calibration could bring different signal intensities of RGB colors. This has implications for the use of image analysis in digital pathology—care must be taken to ensure that intensity measurements are repeatable over time for a same device and between devices. So, today, the Immunoscore assay is recommended with a fixed resolution (0.45 μm/pixel), and a regular calibration of the instrument is strongly recommended. Immunoscore® Colon CE-IVD workflow is also compatible with the Philips IntelliSite Pathology Solution. A future novel version of the software is being currently optimized to yield high performance on most high-resolution scanners.

Our current software allows the automated assessment of the staining intensity level to control the quality of the IHC. The analysis of the immune quantification using the image analysis software relies on a semi-automatic method where the operator contributes to the classification and validation steps. Moreover, the operator must have an expertise in digital pathology and the anatomical pathology of the type of tumor treated. Guidelines to direct different classifications of the tumor with the image analysis software will be have to be produced to obtain the best standardization and comparability of the Immunoscore validation in the future. Finally, if the Immunoscore assay is processed in clinical practice, the analysis of

Immunoscore will have to be validated by pathologists or authorized person, like any other analysis in clinical practice. The digital information extracted from the Immunoscore analysis reports the internal control of the stain intensity in the colon tissue, the different densities of each region (CT and IM) in each marker ($CD3^+$ and $CD8^+$), the delimitation of the tumor and the invasive margin and the Immunoscore classification of the patient. Further information can be achieved as a spatial and numerical analysis which may be combined in visualizations, such as heat maps, to indicate areas of higher or lower immune T-cells quantification. Routine pathology diagnosis with a microscope is rapid, simple and cost-effective, but is also a semi-quantitative evaluation, highly subjective and much less reproducible than digital quantification (Pages et al., 2018). Implementing a fully digital pathology-based workflow would require significant expenditure on high-resolution scanners, computer servers and workstations for pathologists and medical displays. The cost could be justified in term of its benefits on quality, reproducibility or income (Henricks, 2010). Alternative options, with the use of remotely secured computing computers and Immunoscore software, would not require a sophisticated IT infrastructure, and would be more rapid, user-friendly and cost-effective.

## 5. Analytical performance and prognostic validation of the Immunoscore

To introduce the Immunoscore in clinical practice and validate its prognostic value, the test needed to be robust, reproducible, quantitative, standardized, pathology-based and powerful. Indeed, the Immunoscore fulfilled all these key aspects (Pages et al., 2018). Nonetheless, the complexity of immunohistochemistry, in conjunction with protocol variability in each laboratory, might contribute to the variability of the results obtained, thus requiring a standardized consensus method. Large-scale assay harmonization efforts have already been witnessed, and conducted for commonly-used immunological assays of peripheral blood immune cell populations (van der Burg et al., 2011). It was therefore essential to pursue assay uniformity to reduce these limitations. A clinical validation of the Immunoscore with standardized procedures was necessary to reach clinical applicability for individual patients (Pages et al., 2018). We investigated multiple Immunoscore quality controls to test the accuracy and repeatability of the method. Software was used to determine the mean staining intensities of each slide, which allowed the reliability of each staining to be assessed, and avoided underestimation of the total cell count. We first observed that the automated cell-counting method achieved a very good level of correlation with the optical counting for CD3 and CD8 immunostainings and an excellent reproducibility of the count of stained cells. Moreover, a strong interobserver reproducibility was found for the determination of mean cell densities in each tumor region ($r=0.97$ for tumor; $r=0.97$ for invasive margin; $P<0.0001$) (Pages et al., 2018). The reproducibility of the results and the prognostic performance of the Immunoscore were compared with that of a visual assessment of the density of tumor-infiltrating T cells in tumor tissue stained with hematoxylin and eosin (H&E). For TIL evaluation on H&E slides, only 4% of cases were concordant between all observers (Pages et al., 2018). Moreover, an evaluation of a specific subset of immune cells is generally not possible with H&E-stained sections. Lymphocytes with opposite functions such as $CD4^+$ T cells with $T_H1$ or $T_H2$ orientation, or immune cells with regulatory functions ($T_{REG}$ cells), NK cells, NKT cells, B cells or cytotoxic $CD8^+$ T cells are indistinguishable without proper marker evaluation. A deep assessment of the analytical performance of Immunoscore has been carried out (manuscript in preparation). All the quality control tests demonstrated the reliability of the Immunoscore.

Numerous data collected on large cohorts of colorectal cancer demonstrated that high $CD3^+$ and $CD8^+$ T lymphocyte densities observed in the tumor and the invasive margin are favorable predictors for time to recurrence and overall survival, conferring the Immunoscore a prognostic value independent to that provided by the AJCC-UICC TNM classification system (Galon et al., 2006; Mlecnik et al., 2011). The TNM classification was never surpassed in multivariate analysis by alternative methods such as immunohistochemistry for tumor biomarkers, flow-cytometry for DNA content, molecular signatures or genetic features. However, it was shown that the analysis of the specific type of intratumoral immune response, as that provided by the Immunoscore, was indeed surpassing the TNM classification in multivariate analysis (Galon et al., 2006; Mlecnik et al., 2011). Thus, tumor progression has now to be considered as the result of a balance between an invasive tumor process and a defense system, whose major component is constituted by the host immune response. The Immunoscore may add to the significance of the current AJCC/UICC TNM classification. An international consortium of 14 centers in 13 countries, led by the SITC, has been established to validate and promote the Immunoscore in routine clinical settings (Galon et al., 2014). The Immunoscore classification, demonstrating the prevalence of immune infiltrates, was confirmed to have a prognostic significance superior to that of the AJCC/UICC TNM classification system (Pages et al., 2018). Multivariate Cox analysis on stage I/II/III CRC patients revealed that only the immune criteria remained highly significantly associated with prognosis (time-to-recurrence, TRR; DFS and OS); in contrast, the histopathological staging system was no longer significant (Mlecnik et al., 2016). Of note, such demonstration that the Immunoscore remained the only significant criterion over the classical AJCC/UICC TNM classification for DFS and OS led to an editorial entitled "TNM staging in colorectal cancer: T is for T cell and M is for memory" accompanying the publication by Mlecnik et al. (Broussard & Disis, 2011; Mlecnik et al., 2011). Given the power of the pre-existing immunity (Immunoscore) to stratify patients, the results of the international Consensus Immunoscore consortium should result in the implementation of the Immunoscore as a new component for the classification of cancer, designated TNM-I (TNM-Immune) (Pages et al., 2018). We also investigated the Immunoscore combining with other biomarkers among stage III colon cancer patients treated with adjuvant FOLFOX (Sinicrope et al., 2018). Tumors with a high versus low Immunoscore were significantly associated with lower T stage, right-sided location, and Mismatch Repair Deficient (dMMR) profile. An increasing Immunoscore was associated with a progressive improvement in 3-year DFS rates. Moreover, a high versus low Immunoscore was associated with significantly better DFS independent of other prognostic factors, with a similar hazard ratio when excluding dMMR tumors. Among predefined low (T1-3N1) and high risk (T4 or N2) patient subsets, the Immunoscore was significantly prognostic within each risk stratum (Sinicrope et al., 2018). In numerous studies, the Immunoscore was shown to be very powerful, for instance, in CRC patients with clinically localized colorectal cancer, with no detectable tumor spreading to lymph nodes or distant organs (Galon et al., 2019; Pages et al., 2009).

## 6. Clinical applications of the Immunoscore

The first clinical application of the Immunoscore could be performed in colorectal cancer due to the initial collection of compiled data on large cohorts of this cancer over several years. The discovery and ultimate validation of the prognostic power of Immunoscore is accompanied by several clinically relevant implications. First, the proposed TNM-I joint classification, incorporating the immune (I) component to the classical AJCC-UICC-provided TNM staging (Galon et al., 2014), could be instrumental in directing the therapeutic choices. Of note, no clear consensus has been reached yet over the decision to administer (neo)adjuvant therapies, i.e., before or after surgical removal of the primary tumor, which is the most adopted therapeutic intervention in non-metastatic colorectal cancer. An Immunoscore-based stratification of patients enables to distinguish subgroups of patients with very high risk of recurrence (low Immunoscore) and one characterized by very low risk of recurrence (high Immunoscore) (Galon et al., 2006; Mlecnik et al., 2011; Pages et al., 2018). Thus, the adoption of such fine-tuned stratification within the same TNM-stage could at once: direct the reinforcement of a stricter patients' surveillance or the administration of (neo)adjuvant treatments for the former, low Immunoscore group; the avoidance of these therapeutic strategies, and inherent unnecessary side effects (in case of envisaged (neo) adjuvant therapies) for the latter, high Immunoscore group (Galon & Bruni, 2019; Pages et al., 2018). Indeed, such "Watch-and-Wait" strategy approach has been suggested as a better option for rectal cancer patients who had a complete response to pre-operative chemoradiation therapy (Bernier, Balyasnikova, Tait, & Brown, 2018; Smith et al., 2019).

In clinical routine, most often the decision to administer neoadjuvant treatment in conjunction or not with a primary tumor or metastatic resection is based on various parameters, including the evaluation of an individual biopsy. In order to meet such clinical needs, a modified version of the Immunoscore has been specifically designed for biopsies, thus permitting to evaluate the Immunoscore status of patients easily on these specimens and just as accurately, without the need for more invasive surgery. Hence, both the prognostic aspect (risk of recurrence) and the therapeutic management of patients can be assessed on a single biopsy. This conclusion was supported by a study demonstrating that the predictive accuracy of the Immunoscore from a single biopsy was not only high, but also superior to that of PD-L1 (Van den Eynde et al., 2018).

The same study showed that the Immunoscore measure on the least infiltrated metastasis is the one with the strongest association with patient outcome then that on other metastases (Van den Eynde et al., 2018). It was then demonstrated that even when performed on a random biopsy (which is the case in clinical setting), the Immunoscore retained high specificity, with lowly infiltrated patients correctly identified in over 90% of the cases. Nonetheless, the accurate identification of highly infiltrated patients required the assessment of multiple biopsies and/or metastases, due to the high inter-metastatic heterogeneity (Van den Eynde et al., 2018).

The prognostic performance and the feasibility of the Immunoscore assay is being currently assessed in several ongoing multicentric prospective studies (Immucol_1 and Immucol_2, national PHRC and translational INCA, and clinical trials NCT01688232 and NCT02274753).

Immune analysis of the primary tumor can help predict the occurrence and development of both early and distant metastases (Mlecnik et al., 2016; Pages et al., 2005). A study analyzing the main possible tumor or immune determinants associated with metastatic propagation revealed that immune-, rather than tumor-related parameters are the ones associated with dissemination to distant metastasis. Indeed, parameters such as chromosomal instability and genomic mutations had a minimal impact on the presence of metastases, while high Immunoscore was significantly associated with the absence of metastases. Furthermore, when combined to lymphatic vessel density and the expression specific immune contexture parameters, the Immunoscore accurately predicted the presence of synchronous metastases (Mlecnik et al., 2016). In a metastatic setting, high immune infiltrates, Immunoscore and T and B cells score were significantly associated with prolonged patient survival and decrease relapse incidence. The immune parameters remained the only statically significant prognostic factor associated with DFS and OS. A more recent study tracing the spatiotemporal evolution of metastases found the Immunoscore to be one of the four parameters associated with metastatic dissemination, together with the immunoediting, the distance between $CD3^+$ T-cells and proliferating tumor cells, and the size of the parent metastasis (Angelova et al., 2018). Overall, it became clear that the assessment of immune infiltrates, whether in primary tumors or metastatic settings, and whether on fully resected tumor specimens or tumor biopsies, provides a great deal of prognostic information, which can be leveraged by the means of Immunoscore to predict patient survival and clinical managements of colorectal cancer patients. In view of extending the use of the Immunoscore to other solid tumors, a Pan Cancer study evaluating the prognostic value of Immunoscore in multiple solid malignancies is currently ongoing. Moreover, an adaptation of Immunoscore to rectal biopsies (ISb) is also currently investigated in order to predict the response to neoadjuvant treatment and delineate clinical responders that would benefit from a "Watch-and-Wait" strategy with acceptable outcomes (Anitei et al., 2014).

Apart from being a prognostic tool, the Immunoscore could also serve as a biomarker to predict the response to most recent anti-tumor immunotherapies, such as the anti-PD-1, PD-L1 and CTLA-4 (Gnjatic et al., 2017; Tumeh et al., 2014). The evaluation of the immune contexture and Immunoscore could also be used to select patients eligible for specific therapies characterized by a limited or variable percentage of responders. A four-group, Immunoscore-based stratification of patients has been proposed (Camus et al., 2009; Galon & Bruni, 2019), and includes: inflamed, immune-hot tumors (high Immunoscore); non-inflamed, immune-cold tumors (low Immunoscore), with absent $CD3^+$ and $CD8^+$ T cells infiltration; and two immune-altered (intermediate Immunoscore) categories, with either excluded ($CD3^+$ and $CD8^+$ T cells at the tumor edges, but not at the tumor core) and immunosuppressed (low infiltrating $CD3^+$ and $CD8^+$ T cells) phenotypes (Galon & Bruni, 2019). Each category has been associated with possible therapeutic strategies, or associations thereof, most likely eliciting clinical benefit. Hence, such Immunoscore-based group provides a rationale on which to most accurately select patients' clinical management (Galon & Bruni, 2019).

## 7. Concluding remarks

The consensus Immunoscore has proved to be a valuable, unprecedented immune-based tool providing prognostic information beyond the classic TNM-based staging system in colorectal cancer. In this chapter, we presented the Immunoscore, its improvement and discussed in details the assay procedure and the subsequent data analysis, at the same time providing strong recommendations to perform the analysis properly and obtain a correct classification of Immunoscore. The strengths and current limitations of the assay, as well as its clinical implications and future developments are also discussed. We believe the implementation of the Immunoscore in clinical routine could relieve the need for an improved patients' stratification, with evident benefit on the consequent therapeutic management.

## Declaration of interests

J.G. has patents associated with immune prognostic biomarkers and is co-founder of HalioDx biotech company. Immunoscore® a registered trademark owned by the National Institute of Health and Medical Research (INSERM).

## References

Angelova, M., Mlecnik, B., Vasaturo, A., Bindea, G., Fredriksen, T., Lafontaine, L., et al. (2018). Evolution of metastases in space and time under immune selection. *Cell, 175*(3). https://doi.org/10.1016/j.cell.2018.09.018. 751-765.e716.

Anitei, M. G., Zeitoun, G., Mlecnik, B., Marliot, F., Haicheur, N., Todosi, A. M., et al. (2014). Prognostic and predictive values of the immunoscore in patients with rectal cancer. *Clinical Cancer Research, 20*(7), 1891–1899. https://doi.org/10.1158/1078-0432.CCR-13-2830.

Bancroft, J. D., & Cook, H. C. (1985). *Manual of histological techniques*. Edinburgh, London, Melbourne, New York: Churchill Livingstone.

Bernier, L., Balyasnikova, S., Tait, D., & Brown, G. (2018). Watch-and-Wait as a therapeutic strategy in rectal cancer. *Current Colorectal Cancer Reports, 14*(2), 37–55. https://doi.org/10.1007/s11888-018-0398-5.

Bindea, G., Mlecnik, B., Tosolini, M., Kirilovsky, A., Waldner, M., Obenauf, A. C., et al. (2013). Spatiotemporal dynamics of intratumoral immune cells reveal the immune landscape in human cancer. *Immunity, 39*(4), 782–795. https://doi.org/10.1016/j.immuni.2013.10.003.

Broussard, E. K., & Disis, M. L. (2011). TNM staging in colorectal cancer: T is for T cell and M is for memory. *Journal of Clinical Oncology, 29*(6), 601–603. https://doi.org/10.1200/Jco.2010.32.9078.

Camus, M., Tosolini, M., Mlecnik, B., Pages, F., Kirilovsky, A., Berger, A., et al. (2009). Coordination of intratumoral immune reaction and human colorectal cancer recurrence. *Cancer Research, 69*(6), 2685–2693. https://doi.org/10.1158/0008-5472.CAN-08-2654.

Chew, V., Toh, H. C., & Abastado, J. P. (2012). Immune microenvironment in tumor progression: Characteristics and challenges for therapy. *Journal of Oncology, 2012*, 12. https://doi.org/10.1155/2012/608406.

Dey, P. (2018). *Basic and advanced laboratory techniques in histopathology and cytology*. Springer.

Dunn, G. P., Old, L. J., & Schreiber, R. D. (2004). The three Es of cancer immunoediting. *Annual Review of Immunology, 22*, 329–360. https://doi.org/10.1146/annurev.immunol.22.012703.104803.

Emens, L. A., Butterfield, L. H., Hodi, F. S., Jr., Marincola, F. M., & Kaufman, H. L. (2016). Cancer immunotherapy trials: Leading a paradigm shift in drug development. *Journal for Immunotherapy of Cancer, 4*, 42. https://doi.org/10.1186/s40425-016-0146-9.

Fridman, W. H., Pages, F., Sautes-Fridman, C., & Galon, J. (2012). The immune contexture in human tumours: Impact on clinical outcome. *Nature Reviews Cancer, 12*(4), 298–306. https://doi.org/10.1038/nrc3245.

Galon, J., Angell, H. K., Bedognetti, D., & Marincola, F. M. (2013). The continuum of cancer immunosurveillance: Prognostic, predictive, and mechanistic signatures. *Immunity, 39*(1), 11–26. https://doi.org/10.1016/j.immuni.2013.07.008.

Galon, J., & Bruni, D. (2019). Approaches to treat immune hot, altered and cold tumours with combination immunotherapies. *Nature Reviews Drug Discovery, 18*, 197–218. https://doi.org/10.1038/s41573-018-0007-y.

Galon, J., Costes, A., Sanchez-Cabo, F., Kirilovsky, A., Mlecnik, B., Lagorce-Pages, C., et al. (2006). Type, density, and location of immune cells within human colorectal tumors predict clinical outcome. *Science, 313*(5795), 1960–1964. https://doi.org/10.1126/science.1129139.

Galon, J., Fridman, W. H., & Pages, F. (2007). The adaptive immunologic microenvironment in colorectal cancer: A novel perspective. *Cancer Research, 67*(5), 1883–1886. https://doi.org/10.1158/0008-5472.CAN-06-4806.

Galon, J., Hermitte, F., Mlecnik, B., Marliot, F., Bifulco, C., Lugli, A., et al. (2019). Immunoscore clinical utility to identify good prognostic colon cancer stage II patients with high-risk clinico-pathological features for whom adjuvant treatment may be avoided. *Journal of Clinical Oncology, 37*(4_suppl), 487. https://doi.org/10.1200/JCO.2019.37.4_suppl.487.

Galon, J., Mlecnik, B., Bindea, G., Angell, H. K., Berger, A., Lagorce, C., et al. (2014). Towards the introduction of the 'Immunoscore' in the classification of malignant tumors. *The Journal of Pathology, 232*(2), 199–209. https://doi.org/10.1002/path.4287.

Gavrielides, M. A., Gallas, B. D., Lenz, P., Badano, A., & Hewitt, S. M. (2011). Observer variability in the interpretation of HER2/neu immunohistochemical expression with unaided and computer-aided digital microscopy. *Archives of Pathology & Laboratory Medicine, 135*(2), 233–242.

Gnjatic, S., Bronte, V., Brunet, L. R., Butler, M. O., Disis, M. L., Galon, J., et al. (2017). Identifying baseline immune-related biomarkers to predict clinical outcome of immunotherapy. *Journal for Immunotherapy of Cancer, 5*, 44. https://doi.org/10.1186/s40425-017-0243-4.

Henricks, W. H. (2010). *Not so fast? Concrete considerations for digital move. In Advancing practice, instruction, and innovation through informatics*. CAP TODAY Available at: http://www.captodayonline.com/Archives/0110/0110ha_not_so_fast.html.

Hermitte, F. (2016). Biomarkers immune monitoring technology primer: Immunoscore® Colon. *Journal for ImmunoTherapy of Cancer, 4*(1), 57. https://doi.org/10.1186/s40425-016-0161-x.

Jacobs, T. W., Prioleau, J. E., Stillman, I. E., & Schnitt, S. J. (1996). Loss of tumor marker-immunostaining intensity on stored paraffin slides of breast cancer. *Journal of the National Cancer Institute, 88*(15), 1054–1059.

Mlecnik, B., Bindea, G., Kirilovsky, A., Angell, H. K., Obenauf, A. C., Tosolini, M., et al. (2016). The tumor microenvironment and Immunoscore are critical determinants of dissemination to distant metastasis. *Science Translational Medicine, 8*(327), 327ra326. https://doi.org/10.1126/scitranslmed.aad6352.

Mlecnik, B., Tosolini, M., Kirilovsky, A., Berger, A., Bindea, G., Meatchi, T., et al. (2011). Histopathologic-based prognostic factors of colorectal cancers are associated with the state of the local immune reaction. *Journal of Clinical Oncology, 29*(6), 610–618. https://doi.org/10.1200/JCO.2010.30.5425.

Ong, C. W., Kim, L. G., Kong, H. H., Low, L. Y., Wang, T. T., Supriya, S., et al. (2010). Computer-assisted pathological immunohistochemistry scoring is more time-effective than conventional scoring, but provides no analytical advantage. *Histopathology, 56*(4), 523–529. https://doi.org/10.1111/j.1365-2559.2010.03496.x.

Pages, F., Berger, A., Camus, M., Sanchez-Cabo, F., Costes, A., Molidor, R., et al. (2005). Effector memory T cells, early metastasis, and survival in colorectal cancer. *The New England Journal of Medicine, 353*(25), 2654–2666. https://doi.org/10.1056/NEJMoa051424.

Pages, F., Kirilovsky, A., Mlecnik, B., Asslaber, M., Tosolini, M., Bindea, G., et al. (2009). In situ cytotoxic and memory T cells predict outcome in patients with early-stage colorectal cancer. *Journal of Clinical Oncology, 27*(35), 5944–5951. https://doi.org/10.1200/JCO.2008.19.6147.

Pages, F., Mlecnik, B., Marliot, F., Bindea, G., Ou, F. S., Bifulco, C., et al. (2018). International validation of the consensus Immunoscore for the classification of colon cancer: A prognostic and accuracy study. *Lancet, 391*(10135), 2128–2139. https://doi.org/10.1016/S0140-6736(18)30789-X.

Ramos-Vara, J. A. (2011). Principles and methods of immunohistochemistry. *Methods in Molecular Biology, 691*, 83–96. https://doi.org/10.1007/978-1-60,761-849-2_5.

Rimm, D. L. (2006). What brown cannot do for you. *Nature Biotechnology, 24*(8), 914–916. https://doi.org/10.1038/nbt0806-914.

Rimm, D. L., Giltnane, J. M., Moeder, C., Harigopal, M., Chung, G. G., Camp, R. L., et al. (2007). Bimodal population or pathologist artifact? *Journal of Clinical Oncology*, *25*(17), 2487–2488. https://doi.org/10.1200/Jco.2006.07.7537.

Sinicrope, F. A., Shi, Q., Hermitte, F., Heying, E. N., Benson, A., Gill, S., et al. (2018). Immunoscore to provide prognostic information in low-(T1-3N1) and high-risk (T4 or N2) subsets of stage III colon carcinoma patients treated with adjuvant FOLFOX in a phase III trial (NCCTG N0147; Alliance). *Journal of Clinical Oncology*, *36*(4), 614. https://doi.org/10.1200/JCO.2018.36.4_suppl.614.

Smith, J. J., Strombom, P., Chow, O. S., Roxburgh, C. S., Lynn, P., Eaton, A., et al. (2019). Assessment of a watch-and-wait strategy for rectal cancer in patients with a complete response after Neoadjuvant therapy. *JAMA Oncology*, e185896. https://doi.org/10.1001/jamaoncol.2018.5896.

Tumeh, P. C., Harview, C. L., Yearley, J. H., Shintaku, I. P., Taylor, E. J., Robert, L., et al. (2014). PD-1 blockade induces responses by inhibiting adaptive immune resistance. *Nature*, *515*(7528), 568–571. https://doi.org/10.1038/nature13954.

Van den Eynde, M., Mlecnik, B., Bindea, G., Fredriksen, T., Church, S. E., Lafontaine, L., et al. (2018). The link between the multiverse of immune microenvironments in metastases and the survival of colorectal cancer patients. *Cancer Cell*, *34*(6). https://doi.org/10.1016/j.ccell.2018.11.003. 1012-1026.e103.

van der Burg, S. H., Kalos, M., Gouttefangeas, C., Janetzki, S., Ottensmeier, C., Welters, M. J., et al. (2011). Harmonization of immune biomarker assays for clinical studies. *Science Translational Medicine*, *3*(108), 108ps144. https://doi.org/10.1126/scitranslmed.3002785.

CHAPTER SIX

# Transforming growth factor beta (TGF-β) activity in immuno-oncology studies

**Gabriel Gallo-Oller[a], Marianna Di Scala[b], Fernando Aranda[c], Javier Dotor[d,*]**

[a]Group of Translational Research in Child and Adolescent Cancer, Vall d'Hebron Research Institute, Barcelona, Spain
[b]VIVEbiotech, San Sebastián, Spain
[c]Immunoreceptors of the Innate and Adaptative System, Institut d'Investigacions Biomèdiques August Pi i Sunyer, Barcelona, Spain
[d]Facultad de Ciencias Biosanitarias, Universidad Francisco de Vitoria, Madrid, Spain
*Corresponding author: e-mail address: javier.dotor@ufv.es

## Contents



## Abstract

Transforming growth factor beta (TGF-β) is a potent pleiotropic polypeptide cytokine, with a complex and context dependent control of its activation, signaling and effector functions. This cytokine is pivotal in the regulation of immunological responses, tumor initiation and development, stromal homeostasis and all their intricate related interactions. Last decade advances in cancer immunotherapy have reactivated the clinical interest on potential drug with TGF-β inhibition effect, combined with immunomodulating enhancer drugs. The correct quantification of the *in vitro* and *in vivo* biological activity of this cytokine is essential to understand the intrinsic underlying biological mechanisms and TGF-β role in the immune system, tumor and stromal

*Methods in Enzymology*, Volume 636
ISSN 0076-6879
https://doi.org/10.1016/bs.mie.2019.06.008

codevelopment, modulation and interactions. There is a wide variety of available procedures to quantify TGF-β activity, which includes different methodological approximations like ELISA, Bioassays including reporter gene assays, Flow cytometry (FC), Western blotting (WB), immunochemical/fluorescence microscopy, among others. Here, we detail available methods for TGF-β biological activity analysis, together with their applicability and suitability for each experimental setting, in order to get a complete analytical perspective and more comprehensive information along the development and design of combined antitumor immunotherapies, which include the inhibition of TGF-β biological activity.

## 1. Introduction

Transforming growth factor β (TGF-β) is a pleiotropic dimeric cytokine (2 × 112 aa, 25 kDa) belonging to a wide cytokine superfamily that includes over 30 members comprising regulatory molecules such as bone morphogenetic proteins, activins, nodals, inhibins, growth and differentiation factors among others (Massagué, 2012; Weiss & Attisano, 2013). TGF-β exerts its activity through intracrine, autocrine, paracrine and endocrine action modes and presents three different isoforms expressed in mammalian tissues: TGF-β1, 2 and 3, encoded by genes located in different chromosomes but containing highly conserved regions. These isoforms present high sequence homology (70–80%) but functional differences, isoform 1 and 2 are closer in terms of biological effect, whereas isoform three exerts some opposite activities. Despite the presence of three isoforms, TGF-β1 is the most abundant and ubiquitously expressed isoform in the context of biological regulation, and strongly induced in tumor and inflammation context. TGF-β2 was first described in human glioblastoma and initially defined as "glioblastoma-derived T-cell suppressor factor" and is also expressed in many cells types and tissues, in particular neurons and astroglial cells, lung, smooth muscle and sexual gland/tissues. It is also important in tumor growth enhancing cell proliferation in an autocrine manner and/or reducing immune-surveillance of tumor development. TGF-β3 was isolated from a rhabdomyosarcoma cell line (Kubiczkova, Sedlarikova, Hajek, & Sevcikova, 2012). In addition, each isoform presents a high interspecies homology along mammals, greater than 97%, reaching even more than 99% among humans and experimental rodent species (Massagué, 1992).

The expression of TGF-β and its receptors is considered ubiquitous. In most cells, three types of cell surface proteins mediate TGF-β signaling:

TGF-β receptor I (TβRI), II (TβRII), and III (TβRIII: betaglycan or endoglin). As other members of this superfamily, TGF-β ligands are secreted as dimers that bind to heterodimeric receptor complexes consisting of TβRII and TβRI with serine/threonine kinase domains (Derynck & Zhang, 1996). Both receptors are subject to reversible posttranslational modifications that regulate their stability and availability, but phosphorylation is the key activator element of TGF-β signaling pathway. First, the ligand could be presented by TβRIII to TβRII or binds directly to TβRII, followed by subsequent phosphorylation of the regulatory region within TβRI. This leads to incorporation of TβRI and formation of a large ligand-receptor complex: dimeric TGF-β ligand and two pairs of TβRI and TβRII (Kubiczkova et al., 2012; Massagué, 2012).

The TGF-β pathway directly transduces extracellular stimuli from the cell-surface transmembrane receptors to the nucleus through Smad (suppressor of mothers against decapentaplegic) intracellular mediators proteins that includes the receptor-regulated Smad (R-Smad: Smad2 and 3); the common Smad (Co-Smad: Smad4); and the inhibitory Smad (I-Smad: Smad7) (Kubiczkova et al., 2012; Massagué, 2012; Weiss & Attisano, 2013). After TGF-β binding to receptors complex, TβRI recognizes and phosphorylates R-Smad, a pivotal event in the initiation of TGF-β signal. After ligand binding to the receptors complex, activation is facilitated by Smad anchor for receptor activation (SARA), initiating a canonical intracellular signal that involves recruitment and phosphorylation of two C-terminal serine residues of the R-Smads, in particular Smad2 or 3 that dimerized with Smad4 (Heifetz et al., 1990; Miyazono, Katsuno, Koinuma, Ehata, & Morikawa, 2018). Smad2 phosphorylation decreases its affinity to SARA and induces further dissociation. Afterward, phosphorylated complex of Smad2/3 forms a higher-order complex with Smad4 and moves to the nucleus. The Smad2/3-4 complex is retained in the nucleus by interactions with additional and diverse protein binding partners and DNA sequences. Each Smad-partner combination targets a particular subset of genes and recruits either transcriptional coactivators or corepressors, depending on the cell type and in a context-dependent manner (Bayin, Modrek, & Placantonakis, 2014; Kubiczkova et al., 2012; Weiss & Attisano, 2013; Ying et al., 2014). In addition, TGF-β activates also noncanonical pathways, including extracellular signal-regulated kinase (Erk) 1 and 2, c-Jun N-terminal kinase (JNK), p38 mitogen-activated protein (MAP) kinase pathways, phosphoinositide 3-kinase (PI3K)-Akt pathway, Src tyrosine kinase pathway and Rho GTPase pathway (Pannu et al., 2007; Zhang, 2017). Both, canonical and noncanonical activated pathways interacts mediating a specific and context dependent transcriptional activity regulation in the nucleus (Fig. 1).

The three isoforms of TGF-β are synthetized as preproteins with a large amino-terminal prodomain called the latency associated peptide (LAP), which is required for proper folding and dimerization of the mature isoform. This complex is called small latent complex (SLC) and biologically inactive. After folding and dimerization, TGF-β dimer is cleaved from its propeptides in trans-Golgi but it remains strongly associated with LAP through noncovalent interactions, this SLC likewise binds latency TGF-β binding proteins (LTBPs), a 120–140 kDa glycoproteins generating the large latent complex (LLC). Most cultured cell types release latent TGF-β into extracellular matrix (ECM) as LLC which includes LTBPs. The LTBP participates in the regulation of latent TGF-β bioavailability by addressing its final latent storage through transglutaminases mediated crosslinking reactions with ECM proteins, like fibronectin and fibrillin (Fig. 1). Nonactive TGF-β stays in ECM and the further activation is a critical step in the regulation of its activity (Kubiczkova et al., 2012; Stover, Bierie, & Moses, 2007).

TGF-β activation is a complex and critical process that release active TGF-β from the different latent ECM storage forms (Koli, Saharinen, Hyytiäinen, Penttinen, & Keski-Oja, 2001). The latent TGF-β physiological activation is a multifactorial process robustly controlled, that could be Alpha(V) integrin dependent through some matrix metalloproteases (e.g., MMP14) mediated proteolysis, or through mechanical conformation changes implying actin cytoskeletal attachment (Fig. 1). Moreover, there are also integrins-independent activation mechanisms based on hydroxyl radicals from reactive oxygen species (ROS), MMP2/9, thrombospondin and plasminogen/plasmin involving mannose-6-phosphate/type II insulin-like growth factor receptor (M6P/IGFII-R) with the urokinase plasminogen activator (uPA) and its receptor (uPA-R) (Fortunel, Hatzfeld, & Hatzfeld, 2000). *In vitro* TGF-β activation could be also achieved by extreme pH (1.5 or 12), freeze thaw cycles, heat (5 min at 80°C) and chaotropic agents such as SDS and urea (Areström, Bartek, Theresa, & Niklas, 2012). Once released, extracellular active TGF-β could be inhibited by the activity of different proteins: noggin, chordin, follistatin and CAN family among others. TGF-β type III membrane receptors (betaglycan and endoglin) enhance the TGF-β and type II receptor interaction, but block the same interaction when released as soluble forms able to attach with high affinity free TGF-β (Gatza, Oh, & Blobe, 2010) (Fig. 1). All these molecular processes generate

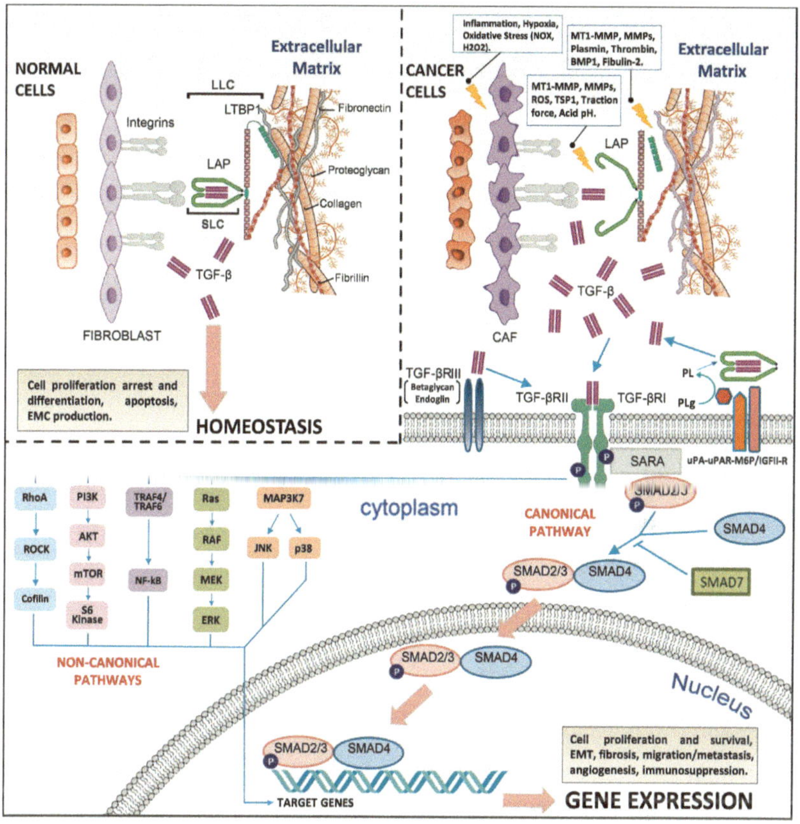


**Fig. 1** Extracellular TGF-β activation and subsequent signaling in the context of homeostasis and tumor progression. In normal tissues, basal levels of TGF-β activity maintains cell proliferation and differentiation under control, and extracellular matrix (EMC) synthesis and degradation balanced, *LAP*, latency associated peptide; *SLC*, small latent complex; *LLC*, large latent complex; *LTBP*, latent TGF-β binding protein. In tumor microenvironment tumor cells induce fibroblast activation (CAF: cancer associated fibroblast) inducing TGF-β secretion and activation through inflammatory process (oxidative stress) and hypoxia. Latent TGF-β is activated through several pathways including integrin dependent (*MT1-MMP*, membrane type 1 matrix metalloproteinase; *MMP*, matrix metalloproteinase; physical traction, acid pH, etc.) and independent mechanisms (*ROS*, reactive oxygen species; *TSP-1*, thrombospondin 1; *BMP1*, bone morphogenetic protein 1; *Plg/PL*, plasminogen/plasmin; *uPA-uPAR-M6P/IGFII-R*, urokinase plasminogen activator and receptor-mannose-6-phosphate/insulin like growth factor 2-receptor membrane complex). TGF-β exerts its biological activity through Canonical (Smad) and noncanonical TGF-β signaling pathways (*MAPKs*, mitogen-activated protein kinases; *PI3K*, phosphatidylinositide 3-kinase; *TRAF4/6*, TNF receptor-associated factor 4/6, *NF-κB*, nuclear factor-κB; *mTOR*, mammalian target of rapamycin, Rho and Ras families of small GTPases and other transcription factors) inducing the expression of protumoral genes involve in cell survival and proliferation, epithelial to mesenchymal transition (EMT), angiogenesis, fibrosis, immunosuppression, cell migration and metastasis.

a TGF-β free active form with a short half-life of about 2 min, while latent forms could reach a half-life ≥90 min. Even after the initiation of intracellular signaling of TGF-β, several intracellular factors like Smad6, Smad7, Smurf, STRAP, decorin and nephrocan, could inhibit the activated cascade pathway (Pang, Ryan, Baxevanis, & Martindale, 2011). The pleiotropic biological activity of TGF-β provides animal cells with tightly controlled developmental programs and cell behavior, exerting a wide variety of cellular processes including cell proliferation control, differentiation, apoptosis, morphogenesis, motility and invasion, extracellular matrix production, and immune cells homeostasis among others. Accordingly, TGF-β generates and modulates several tissular and systemic processes like extracellular matrix remodeling, *in vivo* angiogenesis, immunosuppression, chronic inflammation, fibrosis, tumor progression, metastasis and many others (Fig. 1). Consistent with its ubiquitous activity, aberrant TGF-β signaling is associated with a wide range of human pathologies including immunological, cardiovascular, and fibrotic diseases, as well as many aspects of cancer development (Kubiczkova et al., 2012; Massagué, 2012; Weiss & Attisano, 2013).

Regarding TGF-β role in tumorigenesis, tumors frequently present a TGF-β signaling dysregulation with pivotal implications in tumor initiation, development and metastasis. The role of the TGF-β pathway as a tumor-promoter or suppressor is still a matter of debate, due to its differential cell type dependent effects at the early and late stages of carcinogenesis and dependency on the previous presence or absence of chronic inflammation. At a microenvironment level, the TGF-β pathway enhances advanced tumor progression and metastasis throughout all the steps of carcinogenesis including the induction of angiogenesis, immune evasion, and stromal remodeling. Cancer cells often evade growth inhibition effects of TGF-β, while leaving intact TGF-β-mediated cellular responses that promote tumor progression (Kubiczkova et al., 2012; Massagué, 2012). TGF-β-mediated suppressive effects become more dramatic in premalignant state related with oncogenic mutations. In some tumor types, this pressure selects mutations over TGF-β signaling that implies partial or complete abrogation, and/or aberrant pathway activation (Massagué, 2012). This paradox is reflected in clinic, where in early stage cancers high levels of TGF-β are positively associated with a favorable prognosis. But in advanced malignancies, high microenvironment TGF-β levels are positively associated with tumor size and invasiveness, making TGF-β a useful prognostic biomarker and predictor of recurrence after initial or failed therapy (Principe et al., 2014). There is a growing list of different identified tumoral markers directly related with

TGF-β signaling. Among them, Ski and SnoN are two examples of negative regulators of TGF-β signaling described in several tumors, acting as transcriptional corepressors that block TGF-β-mediated cell cycle arrest when overexpressed (Seoane, 2006; Stolfi, Marafini, De Simone, Pallone, & Monteleone, 2013). Ski and SnoN can interact with Smad2, 3, and 4 blocking the interaction with Smad-binding element of the TGF-β-responsive promoters (Luo et al., 1999; Nomura et al., 1989) through the recruitment of corepressor complexes. Many human tumors express high levels of Ski and/or SnoN escaping from TGF-β tumor-suppressive function (Seoane, 2006). Other relevant markers, like SOX4 (Ikushima et al., 2009) or LIF (Hau, Jachimczak, Schlaier, & Bogdahn, 2011; Penuelas et al., 2009), directly induced by TGF-β, are implicated with stem-like phenotype maintenance. Thus, several studies indicate a pivotal role of TGF-β family signaling in maintenance and differentiation of somatic and cancer stem cells (CSC) in several tumor types (Sakaki-Yumoto, Katsuno, & Derynck, 2013). Other specific TGF-β targets are involved in tumoral invasion and metastasis of tumors, like HDM2, a downstream target of TGF-β with a pivotal function in cancer progression and evaluated as therapeutic target in late-stage cancer (Drabsch & ten Dijke, 2011). Metastatic genes directly induced by TGF-β also include CTGF (Kang et al., 2005), MMP-1 (Dunn et al., 2009) and MMP-9 (Jenkins, 2008; Wipff & Hinz, 2008).

In the context of fibrosis, TGF-β is the master factor in regulation and induction of fibrotic processes in multiple organs (Biernacka et al., 2011; Leask & Abraham, 2004; Meng, Nikolic-Paterson, & Lan, 2016). TGF-β plays a well stablished and described key role in fibrosis (Meng et al., 2016; Pohlers et al., 2009; Verrecchia & Mauviel, 2007), that is defined as the abnormal connective tissue deposition involving the dysregulation of the extracellular matrix balance. TGF-β induces collagen and fibronectin expression (Ignotz & Massagué, 1986; Ohji, SundarRaj, & Thoft, 1993) and many other fibrosis and ECM related markers (Table 3), such as CTGF, induced by TGF-β via a functional Smad3 binding site in the CTGF promoter, which in turn stimulates myofibroblast differentiation, collagen synthesis and ECM mechanical retraction (Duncan et al., 1999). CTGF acts as mediator of the profibrotic functions of TGF-β (Ihn, 2002) together with α-SMA (Desmoulière, Geinoz, Gabbiani, & Gabbiani, 1993).

TGF-β also regulates immune response in several key regulation points (Travis & Sheppard, 2014). For instance, G protein-coupled receptor GPR15 is induced by TGF-β in T regulatory cells (Tregs) promoting its homing to the large intestine (Kim et al., 2013). TGF-β can also downregulate the differentiation of Th2 cells by modulation of GATA3, a key transcription factor for the Th2 cell induction (Gorelik, Fields, & Flavell, 2000). In addition, the suppression of T cell responses by Forkhead box P3 (FoxP3) positive Tregs is regulated by TGF-β through Smad2/3 (Chen et al., 2003; Zheng, Wang, Wang, Gray, & Horwitz, 2007). The regulation of CD8+ T cells by two transcription factors (ThPOK and Runx3) in a TGF-β-specific manner has been stablished (Mucida et al., 2013; Reis, Rogoz, Costa-Pinto, Taniuchi, & Mucida, 2013). Additionally, in the last decade numerous studies have proposed an important role for Treg in the maintenance of peripheral tolerance. Tregs (FC markers: CD4+CD25+-FoxP3+ mice cells, CD4+CD25+CD127− human cells) modulate CD8+ T cell activity by suppressing the production of IFN-γ, increasing the secretion of IL-4, IL-10 and TGF-β, and also by proliferation arrest. Tregs, physiologically suppress aberrant immune response against self-antigens, but also suppress antitumor immune response. Infiltration of a large number of Treg cells into tumor tissues is often associated with poor prognosis. There is accumulating evidence that the removal of Treg cells is able to evoke and enhance antitumor immune response (Tanaka & Sakaguchi, 2017).

To further complicate this matter, TGF-β in the presence of IL-6 promote the differentiation of Th17 CD4+ T cells lineage, while its survival and expansion are dependent on IL-21 and IL-23. Th17 is a subset of potent proinflammatory T helper cells mediated by RORγT and Tbet transcription factors, defined by induction of IL-17 secretion and with mutual antagonism with Tregs (Zhang, 2018). Th17 cells play an important role in mucosal microbial clearance, allergy, autoimmunity and inflammatory disorders. Its role in cancer is controversial and could display beneficial versus detrimental effects in tumor growth and immunity upon context (Chen & Feng-Hou, 2019).

Immunosuppression is thus a pivotal effect of TGF-β in the tumor progression context, primarily by suppressing the T-cell immune surveillance capacity based on the well stablished inhibitory effect of TGF-β on T cell proliferation, activation, and effector functions. Moreover, TGF-β subverts T cell immunity by induction of Treg differentiation reinforcing immunosuppression within tumor microenvironments (Drabsch & ten Dijke, 2012). The profile of its biological activities defined the TGF-β signaling inhibition as an emerging strategy for cancer therapy, and promoted the development of strategies to block TGF-β activity, as monotherapy and in combination with other adjuvant therapy to restore antitumor immune response. Indeed, TGF-β inhibition strategy, as demonstrated in different cancer clinical trials, has failed when used as a single and isolated target. Interestingly, in the last decade, the development and application of cancer immunotherapies have reinforced the clinical interest on TGF-β inhibition, based on the potential additive or synergistic effects of different antitumor therapies combination. These therapies include antigen based vaccination, cell therapy, immuno-checkpoint blockers such as cytotoxic T-lymphocyte antigen 4 (CTLA-4), programmed cell death 1 (PD-1), and PD ligand 1 (PD-L1); among other immunotherapy strategies under preclinical or clinical study (Emens et al., 2017).

There is a growing development of *in vitro, in vivo* and analytical procedures, technics and models related to immunological, carcinogenesis and immunotherapy studies at basic, applied and clinical research level. In this entire knowledge environment TGF-β biological activity evaluation is relevant, but this activity measurement implies challenges related with several factors: TGF-β is generally produced in moderate quantities in the site of action; activity could implies nonsoluble forms (membrane to membrane TGF-β presentation); soluble active state presents a very short half-life with high binding affinity to different extracellular proteins, lab material and tubes surfaces; not to forget its clear pleiotropic profile and complexity through activation and signaling regulation. The precise and contextual quantification of TGF-β is crucial to obtain the optimal and relevant information for efficient clinical TGF-β related therapeutics development. In the present chapter, we intend to guide the selection of adequate methodology for the TGF-β biological activity quantification, including a description of these procedures and key methodological aspects to avoid missing information, irrelevant or nonaccurate biological data for further clinical development of tumor immunotherapy.

## 2. Detection of TGF-β

This chapter will address the many options for quantification of TGF-β biological activity, considering for this purpose the following categories of molecular targets:

**(1)** Direct quantification of free TGF-β that could be executed quantifying unaltered samples that contains active TGF-β, or samples after TGF-β release through activation of latent TGF-β complexes (bound to LAP-LTBP-ECM). Total TGF-β represents active plus latent states of the cytokine. Latent TGF-β value is the result of the difference between total and free TGF-β. This value enables the estimation of the free/

latent ratio that reflects the activation degree of TGF-β signaling and could be correlated with tumor development stage and inflammation status. Under physiological conditions, in the absence of inflammation, latent TGF-β in blood represents more than 98% of the total TGF-β (Walther, Jallow, Jeffries, & Walther, 2009) plus any free bioactive TGF-β potentially present in the samples prior to the dissociation. Problems connected with the samples handling are of pivotal relevance in the context of TGF-β quantification. It is advisable to divide the sample into different aliquots in order to run the quantification assay in duplicate, one a subset without activation, that allow the quantification of the free TGF-β and the other subset after activation, that allow the quantification of total TGF-β. The difference between these two values will determine the amount of latent TGF-β in the sample, and both values generate a ratio of activation. This ratio range can varies significantly depending on the type of tissue, organ or in pathological situations like inflammation or cancer where the activation rate of this cytokine can be considerably increased (Khan, Joyce, & Tsuda, 2012). This option is generally addressed by ELISA and cell bioassays.

**(2)** Intracellular TGF-β canonical signaling pathway, focused on the quantification of phosphorylated Smad2 and Smad3 proteins, mainly addressed by WB.

**(3)** Surrogate factors (EM proteins and regulators, inflammatory and tumoral transcription factors, epithelial-mesenchymal markers, cell cycle regulators, etc.), which expression is directly induced or repressed after the TGF-β effector activity. This strategy is the best choice in cases where no clear TGF-β activity is detected, due to challenges above mentioned that implies the accelerated molecular kinetics of TGF-β. This option could be faced with almost all experimental approximations, but WB and FC are the most commonly used.

TGF-β mechanism of action behaves with an extremely controlled but rapid ligand presentation. This kinetic significantly complicates the direct quantification of active TGF-β dimer or mRNA levels, due to its low half-life that provides underestimation of real TGF-β biological activity, while quantification of downstream effector signaling factors (phosphorylated Smad2/3) and induced/repressed proteins expression allow the perception of a comprehensive view of TGF-β activity. Accordingly, in many experimental conditions it is highly recommended to evaluate and quantify the effector molecules under the induction impact of TGF-β, identified in each specific context. This approach gives an indirect and cumulative quantification of active TGF-β presence, but more informative and realistic than direct measurement of TGF-β protein or mRNA. Previous publications afford data that suggest low reliable subrogation between TGF-β mRNA transcript and protein levels, and in some conditions active protein levels do not correlates proportionally with real signaling activity (Flanders et al., 2016), especially in cancer initial and intermediate progression stages.

## 2.1 Biological samples processing

Since measurement of plasmatic TGF-β1 can be confounded by *in vitro* release of TGF-β1 from platelets, as many works have shown in mouse models, developed and validated procedures for human or mice blood drawing and plasma preparation, are required in order to prevent unintended release or degradation of TGF-β during the preanalytical phase. In addition, the same processing particular approximations are applied for other biological samples (culture supernatants, tissue homogenates, etc.); in order to avoid the unintended activation, as well as to induce intentional activation to release all TGF-β dimers from its latent form (Brown, Wakefield, Levinson, & Sporn, 1990; Mancini et al., 2018).

### *2.1.1 Blood processing protocol for TGF-β quantification*

1. Collect blood samples from healthy volunteers/patients into syringes containing sodium citrate (0.38% final concentration) or commercial vacuum tubes (Vaccutainer; BD BioSciences, Franklin Lakes, NJ) containing 1 of 4 different anticoagulants, sodium citrate, ethylenediaminetetraacetic acid, heparin, or a combination of citrate, theophylline, adenosine, and dipyridamole, but avoid EDTA tubes since this compound does not prevent TGF-β activation and further degradation (Walther et al., 2009).
2. Centrifuge blood at either 3000*g* for 15 min at RT or at 12,000*g* for 5 min at 4°C, either immediately or 2h after blood collection. In some cases, the platelet release reaction inhibitor $PGE_1$ (1 μM) must be added to the blood collection tube to prevent the release of platelet granules content.
3. For total TGF-β assessment, *in vitro* release of latent TGF-β from serum, plasma, cell culture supernatants, tumor/tissues lysates, etc; could be assess through different optional procedures:
   a. *Acidic activation*: (i) Slow: Add 30 μL of serum or protein lysate (100–300 μg protein) in 200 μL of MEM/bovine serum albumin (BSA) and add 10 μL of 4N HCl. Keep in ice for 1 h at 4°C. Acidic

activation is neutralized with addition of 10 μL of 4N NaOH. (ii) Rapid: Add 20 μL of 1N HCl to 100 μL (cell culture supernate) or 40 μL (serum/plasma), mix properly and incubate 10 min at RT. Neutralize the acidified sample by adding 20 μL of 1.2 NaOH/0–5 M HEPES, mix properly. Final pH should range 7.2–7.6. Activated serum and plasma samples may be stored for up to 24 h at 2–8°C before use. Activated cell culture supernatants/urine samples must be assayed immediately after activation. Final concentration values must be multiplied by the dilution factor.

**b.** *Freeze-thaw cycles*: Subject a sample aliquot to 3–5 × rapid freeze thaw cycles in dry ice or liquid nitrogen/37°C water bath. Number of cycles for optimal activation depends on sample nature and total protein composition/concentration. For this reason, previous protocol optimization could be required.

**c.** *Heat activation*: Subject a sample aliquot to heat (5 min at 80°C) and avoid temperatures above 90°C that could induce TGF-β1 thermal denaturation.

**d.** *Chaotropic activation*: Subject an aliquot to final concentration of 0.02% sodium dodecyl sulfate (SDS) or urea 8 M. This option requires O/N standard dialysis to avoid protein denaturizing properties of chaotropic agents.

**e.** *Mechanical friction*: Subject an aliquot of whole blood collected in 0.38% sodium citrate from patients to a shear rate of $1800\,s^{-1}$ for 10 min in a cone and plate device.

**4.** Centrifuge samples at 12,000 *g* for 5 min at 4°C, and separate plasma and subsequently subject to a second step of activation, in the case of mechanical friction, with a shear rate of $1800\,s^{-1}$ for another 2 h.

**5.** Total and free TGF-β1 could be then measured by ELISA, immunochemistry, immunoblotting, luminex or using high-sensitivity quantification by Singulex Erenna SMC (single-molecule counting) assay technology, among others.

Acid-ethanol extraction, originally used for TGF-β isolation and purification from tissues is an optimized method that minimizes proteolytic activity and extracts high yield of acid-stable proteins (Frolik, Dart, Meyers, Smith, & Sporn, 1983; Roberts et al., 1980). Acid-ethanol extraction is a labor-intensive multistep protocol, but is the method of election since generates the better yield in TGF-β extraction. One handicap of this method is related to the induction of protein precipitation during homogenization of frozen tissues.

As different tissues must be assayed at very different dilutions to fall within range of the standard curve, acid-ethanol extraction from RIPA homogenates, followed by dialysis for solvent exchange, is the recommended procedure since stabilizes the TGF-β value, making results independent of sample dilution. Since TGF-β levels differ widely between tissues, the acid-ethanol step is necessary for accurate TGF-β quantitation, avoiding matrix effects (low samples dilutions) that could have a disproportional impact on samples with low TGF-β expression.

Cause platelets are the *in vivo* most concentrated source of TGF-β1 (approximately 40–100 times more TGF-β1 content than most cells) (Assoian, Komoriya, Meyers, Miller, & Sporn, 1983), residual blood within some highly vascularized tissue samples, could contribute significantly to TGF-β levels, (liver and some type of tumors). In the case of animal models, a previous general or organ/tissue perfusion is recommended to avoid this contribution.

### 2.1.2 Tissue samples (including tumors) processing protocol for TGF-β quantification

1. Acid-ethanol solution composition: 60 mL 95% ethanol, 30 mL $H_2O$, 1.2 mL conc. HCl, 6 mg phenylmethylsulfonyl fluoride (Sigma-Aldrich #P7627) and 0.3 mg pepstatin A (Sigma-Aldrich #P5318).
2. Frozen tissues (at least 20 mg, optimally ~100 mg) must be weighed, placed into Precellys® reinforced tubes (2 mL) containing 2.8 mm ceramic beads (CK28-R) (Bertin Technologies #KT-03961-1-002.2) and put on dry ice until use. For fresh tissue or RIPA extracts samples put on ice and transferred to 15 mL tubes containing 6 mL acid-ethanol/mL RIPA extract or 6 mL acid-ethanol/gram of fresh tissue. Samples must be rocked overnight at 4°C.
3. Add 6 mL/g tissue of RIPA lysis buffer (Millipore #20-188) containing HALT protease inhibitor cocktail (ThermoScientific #78440) and homogenize samples in a Precellys® 24 homogenizing system with Cryolys cooling at 4°C at 5500 rpm (2 × 20 s). The Precellys® system allows reproducibly high efficiency and rapid simultaneous homogenization of multiple samples under controlled temperature conditions.
4. Centrifuge extracts at 3500 *g* for 10 min and transfer the supernatants to pre-rinsed MWCO 3500 Spectra/Por6 dialysis tubing (Spectrum Labs #132590). Dialyzed samples against 4 mM HCl (3 × 100 vol) at 4°C for 72 h and then transfer to tubes and centrifuge at 3500 *g* for 10 min.

5. Transfer supernatants to polypropylene 15 mL conical tubes pretreated with the siliconizing agent Sigmacote (Sigma Aldrich #SL2-100) to minimize losses of TGF-β to the tube walls, and freeze.
6. Lyophilize frozen samples to dryness, dissolve in 400 μL assay buffer (4 mM HCl/150 mM NaCl/0.1% crystalline BSA (Sigma-Aldrich cat # 05470)), transfer to a 1.5 mL microfuge tube pretreated with Sigmacote and store at −80°C.

## 2.2 ELISA for TGF-β measurement

TGF-β levels can be measured using ELISA assays that are sensitive and reproducible. This method can detect low levels of TGF-β in cell culture conditioned medium, serum, plasma, urine, protein extraction samples from tissues or cells lysates, etc. The assay employs a quantitative sandwich enzyme immunoassay technique. The wells of a microplate are coated with either anti-TGF-β antibody or TGF-β soluble receptor type II protein. This represents the immobilized component of the ELISA assay that captures TGF-β from the samples analyzed. After applying the secondary antibody conjugated with reporter enzyme, the substrate is added, color is developed, and the intensity of the color is measured and quantified. Because protocols from various companies vary slightly, the protocol described here is based on R&D System's Quantikine® TGF-β1 immunoassay:

1. Prepare all reagents, samples, and standards; activate TGF-β in the samples by acid or heat treatment (serum present in conditioned media contains high levels of latent TGF-β that can result in readings above those of the standards. Dilutions of the samples or use of serum free media can circumvent this problem).
2. Add 200 μL of sample or standard in appropriate diluent per well. Cover with adhesive strip and incubate 3 h at RT (incubation time differs depending on the system used).
3. Aspirate each well and wash with 400 μL wash buffer three times. Removal of the wash buffer after each wash improves the performance of the procedure. After the last wash, aspirate wash buffer completely or blot the plate against paper towel.
4. Add 200 μL of TGF-β antibody conjugate in appropriate diluent to each well. Cover with a new adhesive strip and incubate 1.5 h at RT.
5. Repeat the aspiration/wash procedure as in step 3.
6. Add 200 μL of substrate solution to each well and incubate 20 min at RT away from light.

7. Add 50 μL of stop solution to each well and tap the plate to ensure proper mixing.
8. Determine the intensity of color in a plate reader at 450 nm. Correction readings should be taken at 540 and 570 nm to compensate for plate interferences. These readings should be subtracted from the 450 nm reading. In the absence of this correction, the readings at 450 nm are higher and less accurate.

There are also available ELISA kits for LAP and LTBP (latent TGF-β states elements) quantification.

## 2.3 Bioassays for TGF-β quantification

There are several available bioassays that allow the quantitative determination of TGF-β biological activity or potency in samples, by testing the molecular/phenotypic response of a cell culture to this cytokine. Consequently, these assays allow not only the determination of TGF-β content in any biological sample but also the titration of any factor with inhibitory, enhancer, antagonist or agonist effect over the TGF-β biological activity.

Bioassays implementation requires a standard response curve to known TGF-β concentrations and specific inhibitors of TGF-β (e.g., monoclonal antibodies, inhibitor peptides, specific kinase inhibitors, etc.), to properly validate the bioassay mechanism. These bioassays are able to quantify free or total (postactivation) TGF-β present in tumor cell supernatants, serum, plasma, tissular protein extracts or even direct secretion or cell surface presentation in coculture assays. Based on the wide range of biological activities exerted by TGF-β, many experimental designs are available and could be compatible with specific *ad hoc* bioassays that implies cells of interest (tumoral, mesenchymal or immune cell lines or primary cultures), with special applicability in the case of wound healing and transwell migration assay. Table 1 summarizes standardized available bioassay to address TGF-β quantification as target itself or as methods for screening of TGF-β blocker drugs with potential application in cancer immunotherapy. Besides Table 1 content, there are other available specific but complex bioassays that measure TGF-β biological activity, such as determination of specific TGF-β ability to dose-dependent inhibition of IL-4 induced cell proliferation of mouse HT-2 cells (BALB/c spleen activated by sheep erythrocytes in the presence of IL-2) with a sensitivity from 10 to 500 pg/mL for human TGF-β1 (Ruegemer et al., 1990). Other optional sensitive bioassay is based on TGF-β1/2 ability to inhibit the interleukin-5 induced cell proliferation of

**Table 1** Main available bioassays for direct or indirect measurement of active TGF-β presence in biological samples and related biological activity quantification, including assay based in primary cultures, standard and gene reporter transfected cell lines.

| Assay | | Cell type | TGF-β effect (response parameter) | Quantification | Specificity | Sensitivity (linear response) | References |
|---|---|---|---|---|---|---|---|
| Mv 1 Lu | | Mink lung epithelial cells (NBL7, ATCC: CCL-64™) | Inhibition of cell proliferation | T3H, MTT, XTT, WST-1, Resazurine, cell count | TGF-β1, 2, 3 | 5–300 pg/mL | Garrigue-Antar, Barbieux, Lieubeau, Boisteau, and Grégoire (1995) |
| Wound Healing | | Tumor and mesenchymal primary cultures and cell lines | Induction of 2D cell migration | Microscopy images, ImageJ | Low, many growth factors including TGF-β1, 2, 3 | Variable, depending on cell type and cell culture conditions | Lampugnani (1999) |
| Transwell migration assay | | Tumor primary cultures and cell lines | Induction of 3D cell migration | Colorimetry (lysate OD or ImageJ) | Low, many factors including TGF-β1, 2, 3 | Variable, depending on cell type and cell culture conditions | Marshall (2011) |
| SACF assay/Spheroid assay | | Tumor primary cultures and cell lines | Anchorage-independent cell growth | Number of colonies/ invasiveness image analysis | Low, many growth factors including TGF-β1, 2, 3 | Variable, depending on cell type and cell culture conditions | Naber, Wiercinska, Ten Dijke, and van Laar (2011), Borowicz et al. (2014), and Berens, Holy, Riegel, and Wellstein (2015) |
| Reporter gene assays | Mv 1 Lu Luciferase | PAI-1 Reporter (Luc)—Mv1 Lu | Luciferase expression | Luminescence | TGF-β1, 2, 3 | 5 to >750 pg/mL | Abe et al. (1994) |
| | 293T-CD4 T cells | 293T cells (caga)12 SBE luciferase CD32 (Fc receptor) and CD86 | Luciferase expression | Luminescence | TGF-β1, 2, 3 | 5–160 pg/mL | Oida and Weiner (2011) |
| | MFB-F11 | TGF-$\beta 1^{-/-}$ embrionic mice fibroblast (SBE-SEAP) | Alkaline phosphatase expression | Colorimetry | TGF-β1, 2, 3 | TGF-β1 from 1 pg/mL to 10 ng/mL | Tesseur, Zou, Berber, Zhang, and Wyss-Coray (2006) |

*Mv 1 Lu*, mink vison lung epithelial cell; *OD*, optical density; *SACF*, soft agar colony formation; *PAI*, plasminogen activator inhibitor; *SBE*, Smad binding element; *SBE-SEAP*, Smad-binding elements coupled to

the erythroleukemia cell line TF-1, with response range below 500 fg/mL of TGF-β1, and 5–10 pg/mL TGF-β2 (Randall, Wadhwa, Thorpe, & Mire-Sluis, 1993). As last example, another screening assay measures TGF-β mediated secretion of IL-11 by A549 lung carcinoma cells, which is part of a profibrotic response in lung fibroblasts or epithelial cells, and also implicated in metastatic processes (Rapoza, Fu, & Sendak, 2006). This bioassay present specificity to TGF-β1, 2 and 3 isoforms, and display a linear sensitivity from 10 pg/mL to10 ng/mL.

## 2.4 Immunohistochemistry and immunofluorescence TGF-β quantification

TGF-β can be traced in cell culture or tissue sections, including tumors biopsies, using different methods that cover immunofluorescence or immunohistochemistry (ICH) on paraffin-embedded tissues. Many of the reagents required are also available in commercial kit form (Vectastain ABC, Vector Labs, Burlingame, CA). The following method has been successfully utilized to localize abnormal TGF-β1 protein deposition in human glioblastoma tissue samples (Frei et al., 2015). For alternative indirect immunofluorescence staining, secondary antibody must be replace by fluorochromes labeled antibodies and detected with fluorescence microscopy.

**Protocol for TGF-β ICH of paraffin-embedded tissue sections**

1. Heat paraffin sections (4 μm thick) to 60°C and deparaffinize by soaking in three exchanges of xylene (30 min each).
2. Subsequently rehydrate sections through a graded series of alcohol washes (100% ethanol, 95%, 70%, 50%, and 30% ethanol in deionized water, and deionized water; 20 min for each exchange).
3. Boil in 10 mmol/L citrate buffer, pH 6.0 for 15 min in a steamer for heat-induced proteins antigenic epitopes retrieval.
4. After washing the sections three times for 10 min each in phosphate-buffered saline (PBS), the sections are incubated for 15 min in 1% hydrogen peroxide to quench endogenous peroxidase activity.
5. Wash the sections three times for 10 min each in PBS, next incubate in tris-buffered saline (TBS) containing 10% swine serum, 0.2% Triton and 2% bovine serum albumin (BSA) for 30 min at RT in a humid chamber to reduce nonspecific binding of antibodies.
6. After washing again, incubate sections with primary rabbit polyclonal anti human TGF-β1 (G1221, Promega, WI, USA) 1/50 for 4 h at RT.
7. Rinse sections, as described earlier, and subsequently incubate with secondary anti-rabbit antibody 1/200 (414141F Nichirei Biosciences,

Japan) from Histofine Simple stain Max PO® Universal Immuno-Peroxidase Polymer, for 30 min at RT.

8. Next, wash sections three times and treat with 3,3-diaminobenzidine tetrahydrochloride (Dako, Glostrup, Denmark) during 30 min at RT for visualization.
9. Following three 10 min washes in PBS, demonstrate bounded antibodies using 3-amino-9-ethylcarbazole substrate (AEC substrate, Dako) in the presence of 3% hydrogen peroxide for 15 min.
10. Following substrate development, rinse sections and counterstain by soaking for 1 min in Smith's hematoxylin stain (Sigma).
11. After mounting, samples may be randomized, coded, examined by a blinded observer, and scored with a subjective semiquantitatively manner or with quantitative image analysis software.

## 2.5 TGF-β signaling pathway analysis by Western blot

Western blot is the first choice technique for the analysis of cellular signaling pathways through specific protein detection. Based in mono/polyclonal antibodies technology, this technic allows the identification of gene expression changes due to the regulation of translation including posttranslational modifications, through a multistep method that detects and/or quantify a protein of interest from different samples (Taylor, Berkelman, Yadav, & Hammond, 2013). The possibility to detect protein modifications, allows a deep analysis of cellular processes and status, such as inhibition, activation, or the detection of regulation mechanisms. To obtain clear and reproducible results with this technique, a rigorous performing of different and interdependent technical steps is required, from cells or tissue samples processing to final chemiluminescence quantification (Bass et al., 2016; Taylor et al., 2013).

TGF-β signaling elements, for WB quantification, could be classified in ligands and previous latent forms, receptors, cytoplasmic signaling factors (Table 2) and specific target genes (Table 3).

For ligand detection, some particularities must be taken into account including temporal ligands secretion, tissue specificity, intracellular or extracellular detection, latent precursors and, despite its high homology among species, the detection of specific species isoforms must be considered (Kubiczkova et al., 2012; Weiss & Attisano, 2013).

Several technical aspects and optimization/calibration steps must be considered to obtain reliable results due to the multistep nature of Western blot

| TGF-β signaling level | Target protein | Observations | References |
|---|---|---|---|
| Ligands | TGF-β | Expression | D'Arpino, Fuchs, Sánchez, and Honoré (2018), Yuan and Lin (2018), Wu et al. (2018), Wang et al. (2016), Tian et al. (2015), Fang et al. (2016), Li, Wang, Li, et al. (2016), Li, Wang, Liu, et al. (2016), Du et al. (2012), Maurya et al. (2013), and Coste et al. (1998) |
| | LAP | Expression | Li, Wang, Li, et al. (2016), Li, Wang, Liu, et al. (2016), Maurya et al. (2013), Kast et al. (2014), and Mazzieri et al. (2005) |
| | SLC | Levels | Maurya et al. (2013) |
| | LLC | Levels | Maurya et al. (2013) and Chen et al. (2007) |
| | LTBP | Expression | Flaumenhaft et al. (1993), Maurya et al. (2013), Zilberberg et al. (2012), Kast et al. (2014), Chen et al. (2007), and Mazzieri et al. (2005) |
| Receptors | TβRII | Expression | Nyati et al. (2011), Baugé, Cauvard, Leclercq, Galéra, and Boumédiene (2011), Semlali et al. (2008), Hou, Zhang, Wang, Xiong, and Mi (2017), Chen, Ruan, et al. (2014), Chen, Wang, et al. (2014), D'Arpino et al. (2018), Wu et al. (2018), Wang et al. (2016), Wrighton, Lin, and Feng (2008), Bandyopadhyay et al. (2011), and Du et al. (2012) |
| | TβRI | Expression | Nyati et al. (2011), Baugé et al. (2011), Chandra et al. (2012), Semlali et al. (2008), Giroux, Delisle, O'Brien, Hébert, and Perreault (2010), Hou et al. (2017), Chen, Ruan, et al. (2014), Chen, Wang, et al. (2014), D'Arpino et al. (2018), Yuan and Lin (2018), Wu et al. (2018), Wang et al. (2016), Wrighton et al. (2008), Bandyopadhyay et al. (2011), and Du et al. (2012) |
| Cytoplasmic factors | SARA | Expression | Runyan, Hayashida, Hubchak, Curley, and Schnaper (2009), Bakkebø et al. (2012), Semlali et al. (2008), and Kong et al. (2016) |
| | Smad2 | Expression | Gallo-Oller et al. (2016), Nyati et al. (2011), Baugé et al. (2011), Runyan et al. (2009), Bakkebø et al. (2012), Pierreux, Nicolás, and Hill (2000), Giroux et al. (2010), Hou et al. (2017), Chen, Ruan, et al. (2014), Chen, Wang, et al. (2014), Wrighton et al. (2008), and Ai et al. (2015) |

*Continued*

**Table 2** Protein elements list of TGF-β signaling pathway, detectable by Western blot.—cont'd

| TGF-β signaling level | Target protein | Observations | References |
|---|---|---|---|
| | | Phosphorylation | de Galarreta et al. (2016), Nyati et al. (2011), Baugé et al. (2011), Runyan et al. (2009), Bakkebø et al. (2012), Pierreux et al. (2000), Giroux et al. (2010), Hou et al. (2017), Chen, Ruan, et al. (2014), Chen, Wang, et al. (2014), Wrighton et al. (2008), and Ai et al. (2015) |
| | | Translocation | de Galarreta et al. (2016), Chapnick and Liu (2010), Runyan et al. (2009), Pierreux et al. (2000), and Hou et al. (2017) |
| | Smad3 | Expression | Nyati et al. (2011), Baugé et al. (2011), Runyan et al. (2009), Bakkebø et al. (2012), Huang et al. (2011), Semlali et al. (2008), Pierreux et al. (2000), Giroux et al. (2010), Hou et al. (2017), Chen, Ruan, et al. (2014), Chen, Wang, et al. (2014), Wrighton et al. (2008), and Ai et al. (2015) |
| | | Phosphorylation | de Galarreta et al. (2016), Nyati et al. (2011), Baugé et al. (2011), Runyan et al. (2009), Bakkebø et al. (2012), Huang et al. (2011), Semlali et al. (2008), Giroux et al. (2010), Hou et al. (2017), Chen, Ruan, et al. (2014), Chen, Wang, et al. (2014), Wrighton et al. (2008), and Ai et al. (2015) |
| | | Translocation | de Galarreta et al. (2016), Chapnick and Liu (2010), Runyan et al. (2009), Pierreux et al. (2000), and Hou et al. (2017) |
| | Smad2/Smad3 complex | Complex detection (total and phosphorylated) | de Galarreta et al. (2016), Chandra et al. (2012), D'Arpino et al. (2018), Tian et al. (2015), Bandyopadhyay et al. (2011), and Cho et al. (2015) |
| | | Translocation | de Galarreta et al. (2016) and Ahn et al. (2016) |
| | Smad4 | Expression | Pierreux et al. (2000), Ai et al. (2015), and Ahn et al. (2016) |
| | | Translocation | Chapnick and Liu (2010), Pierreux et al. (2000), and Ahn et al. (2016) |
| | Smad7 | Expression | Gallo-Oller et al. (2016), Nyati et al. (2011), Semlali et al. (2008), Giroux et al. (2010), Ai et al. (2015), Li, Wang, Li, et al. (2016), Li, Wang, Liu, et al. (2016), and Du et al. (2012) |
| | | Translocation | Miyake, Alli, and McDermott (2010) |

*LAP*, latency associated peptide; *LLC*, large latent complex; *LTBP*, latent TGF-β binding protein; *SARA*, Smad anchor for receptor activation; *SLC*, small latent complex; *Smad*, suppressor of mothers against decapentaplegic; *TβR*, TGF-beta receptor; *TGF-β*, Transforming Growth Factor-beta.

**Table 3** Western blot affordable proteins with altered expression/stability pattern, as a result of TGF-β biological activity, in the context of tumoral development, stromal activation and effector activity of immune system.

| Context | Protein | Observations | References |
|---|---|---|---|
| Tumor cells | Ski | Expression | Gallo-Oller et al. (2016), Javelaud et al. (2011), Le Scolan et al. (2008), Li et al. (2010), Jacob, Grabner, Atanasoski, and Suter (2008), Yang, Zhang, Li, and Liu (2003), and Macias-Silva, Li, Leu, Crissey, and Taub (2002) |
| | SnoN | Expression | Javelaud et al. (2011), Li et al. (2010), Yang et al. (2003), Macias-Silva et al. (2002), and Zhu et al. (2016) |
| | SOX4 | Expression | Jang et al. (2013), Beekman et al. (2012), Chen, Ruan, et al. (2014), Chen, Wang, et al. (2014), Cheng et al. (2017), and Dai et al. (2017) |
| | LIF | Expression | De Breuck, Baeyens, and Bouwens (2006), Ruan et al. (2006), Hu, Huang, Wang, and Wu (2015), and Bauer et al. (2003) |
| | HDM2 | Expression | Yang et al. (2009), Ling et al. (2014), Araki et al. (2010), Chen, Lin, et al. (2015), and Chen, Zhu, et al. (2015) |
| | PDGF-B | Expression | Liu et al. (2017) and Jiang et al. (2017) |
| | MMP-1 | Expression | Milde-Langosch et al. (2004), Liu et al. (2013), Liu et al. (2016), and Cho, Il, and Lee (2013) |
| | MMP-9 | Expression | Chen, Ruan, et al. (2014), Chen, Wang, et al. (2014), Cheng et al. (2017), and Milde-Langosch et al. (2004) |
| Fibrosis | Collagen | Expression | Zhou et al. (2015), Cho et al. (2013), Schnoor et al. (2008), Lee, Chung, Kim, and Kim (2013), Xie et al. (2017), Li et al. (2016), Li et al. (2016), Chen, Lin, et al. (2015), Chen, Zhu, et al. (2015), Fu et al. (2006), Manojlovic, Blackmon, and Stefanovic (2013), and Riquelme-Guzmán, Contreras, and Brandan (2018) |

*Continued*

**Table 3** Western blot affordable proteins with altered expression/stability pattern, as a result of TGF-β biological activity, in the context of tumoral development, stromal activation and effector activity of immune system.—cont'd

| Context | Protein | Observations | References |
|---|---|---|---|
| | | Secretion | Manojlovic et al. (2013) and Kang et al. (2011) |
| | Fibronectin/EDA | Expression | Yang et al. (2003), Kim et al. (2012), Nalvarte, Damdimopoulos, Rüegg, and Spyrou (2015), Tiwari, Kumar, Ram, Sharma, and Luthra-Guptasarma (2016), and Riquelme-Guzmán et al. (2018) |
| | CTGF | Expression | Kang et al. (2011), Liu et al. (2011), Wickert, Chatain, Kruschinsky, and Gressner (2007), and Riquelme-Guzmán et al. (2018) |
| | α-SMA | Expression | Yang et al. (2003), Fu et al. (2006), Tiwari et al. (2016), and Jeon et al. (2006) |
| Immune system | GPR15 | Expression | Pan et al. (2017) and Pan et al. (2017) |
| | GATA3 | Expression | Gibson et al. (2013), Lin et al. (2017), Maneechotesuwan et al. (2007), Chen, Osada, Santamaria-Babi, and Kannagi (2006), and Izzo et al. (2014) |
| | FoxP3 | Expression | Gibson et al. (2013), Ramos et al. (2012), Lee, Gao, and Fang (2008), and Hinz et al. (2007) |
| | ThPOK | Expression | Gera et al. (2017) and Li et al. (2013) |
| | RUNX3 | Expression | Gera et al. (2017), Lee, Jung, Choi, and Lee (2015), Zhu et al. (2017), and Gan, Hao, Idell, and Tang (2015) |

*CTGF*, connective tissue growth factor; *EDA*, extra domain A (fibronectin alternative splicing); *FoxP3*, Forkhead box P3; *GATA*, specific DNA binding sequence; *GPR*, G protein regulator; *HDM*, human double minute; *LIF*, leukemia inhibitory factor; *MMP*, metalloproteinase; *PGF-B*, placenta growth factor B; *RUNX3*, runt related transcription factor 3; *SKI*, Sloan-Kettering Institute; *SMA*, smooth muscle actin; *SnoN*, Ski-related novel protein N; *SOX4*, *ThPOK*, T-helper-inducing POZ/Krüppel-like factor.

for each specific protein. Specific protocol details that correspond with WB figures included in this chapter (P-Smad2/Smad2) are described in Gallo-Oller et al. (2016).

### 2.5.1 General aspects of WB procedure

**(1)** *Sample preparation*: Lysis buffer selection and lysis conditions depend on sample characteristics (type of tissues, tumors, cell lines, primary cultures, etc.) and cellular localization of target protein. RIPA buffer is the most commonly used but other predefined buffers could be used according to specifics sample requirements. Presence of proteases and phosphatases inhibitors in the lysis buffer is mandatory. The target protein will drive the inhibitors selection that encompasses several options: $Na_4O_7P_2$ (sodium pyrophosphate), $NaVO_3$ (sodium orthovanadate), glycerolphosphate, NaF (sodium fluoride), DTT (dithiothreitol), PMSF (phenylmethane sulfonyl fluoride), and EDTA (ethylenediaminetetraacetic acid). Commercial inhibitors cocktails are also available with different composition and applications. During preparation/obtaining, samples must be maintained at 4°C and avoid the use of trypsin in cell cultures processing to prevent protein degradation, both facts can affect the final protein concentration and stability. Short sonication steps and mechanical homogenization are recommended to optimize the lysis process.

**(2)** *Protein quantification*: Comparable results acquisition requires the same amount of total protein in the loading samples. To assure this, different commercial kits are available for protein content quantification, generally based on colorimetric biochemical reactions between a reactive molecule and specific amino acids contained in the proteins lysates. Some compounds that could affect these chemical reactions must be avoided or controlled.

**(3)** *Electrophoresis*: Specific protein detection requires proteins sample spatial segregation. For this, samples' running though acrylamide gels under a uniform electric field is the method of choice. There are several options in the gel composition, acrylamide percentage, precast gels, buffer composition and separation methods that significantly decrease the time and improves the separation quality of the sample. As in DNA electrophoresis, the use of commercial molecular weight markers is mandatory in this step, and its selection should fit the weight range of the targeted proteins. Once selected, a test of weight marker accuracy is recommended. Additionally, loading buffers required some considerations. These buffers consist in a colorant that indicates the running front, a thickening agent (usually glycerol), a buffer solution to maintain the sample integrity and a reducing agent (e.g., 2-mercaptoethanol). Loading buffer does not admit significant variations, but could require some optimization.

**(4)** *Transference to membrane*: After electrophoretic separation, protein bands must be transferred to a membrane for further detection. The choice of wet, semidry or even capillary transfer system will depend on needs and available sources. Membrane material selection is also relevant, nitrocellulose and PVDF (polyvinylidene difluoride) membranes are available. PVDF presents higher protein affinity that improves transference but could increase the background signal. For both membrane materials, 0.2 μm membrane pore size is recommended to avoid small proteins lost during the transference. Transferring system, type of gel and membrane will define the final buffer composition to be used in the protocol.

**(5)** *Blotting*: Once proteins are separated and transferred to a membrane, specific detection is mediated through antibodies recognition. Previously, a membrane blocking step must be carried out to avoid unspecific antibody binding to the membrane. Usual blocking compounds are nonfat milk powder and BSA, employed at different concentrations, or even combined. Blotting step requires special optimization needs including: primary and secondary antibody selection and dilution, Tween or other detergent percentage included in the TBS buffer, blocking agent concentration and type, blocking incubation time and temperature.

**(6)** *Bands detection and quantification*: Protein bands detection is based on signal generated by different reagents and detected through films or CCD cameras. The reagents that trigger the signal usually include a substrate (of an enzyme attached to the secondary antibody) and a buffer that potentiates the signal intensity. There are several commercial options and the choice will depends basically on the signal quantification equipment and previous experience of each lab. For an affordable, accurate and reproducible approximation of signal quantification, background subtraction and Western blot densitometry band quantification through image analysis software are pivotal (Gallo-Oller, Ordoñez, & Dotor, 2018).

#### 2.5.1.1 Western blot of Smad2/P-Smad2 WB (cell culture)

1. Extract total proteins using RIPA buffer (50 mM TRIS-HCl pH 8.0, 150 mM NaCl, 0.5% Triton X-100 and 0.5% sodium deoxycholate) supplemented with PMSF 1 mM, DTT 10 mM and Halt™ Protease and Phosphatase Inhibitor Cocktail (Thermo Scientific, Rockford, IL, USA).
2. Quantify protein content with BCA Protein Assay Kit (Thermo Scientific, Rockford, IL, USA) following manufacturer's instructions.
3. Resolve 20 μg of total protein with reducing SDS-PAGE using Tris-Glycine gels and transfer onto a nitrocellulose membrane (GE Healthcare, Buckinghamshire, UK) at constant amperage (300 mA for 1 h).
4. After transfer, block membranes with 5% nonfat milk in TBS-Tween 0.1% buffer.
5. Incubate with primary antibody anti-Smad2 (86F7) dilution 1:4000 anti-Phospho Smad2 (Ser465/467) (138D4) dilution 1:2000, both rabbit mAb from Cell Signaling (Danvers, MA, USA) cat. Nos. #3122 and #3108, respectively, at 4°C during an O/N incubation. After primary antibody incubation, and washes with TBS-Tween (three washes for 10 min), incubate with secondary antibody anti-Rabbit IgG-H&L (Abcam, Cambridge, UK Cat No ab6721) at 1/10,000 dilution for 1 h at RT.
6. Perfom chemiluminescence detection with Amersham™ ECL™ Prime Western Blotting Detection Reagent (GE Healthcare, Buckinghamshire, UK) and Amersham Hyperfilm™ ECL™ (GE Healthcare, Buckinghamshire, UK).
7. Quantify bands intensity with ImageJ (Java-based image-processing and analysis software), normalizing data with GAPDH bands value. For this purpose, scan developed films with a standard office scanner at least with 600 pixels per inches (ppi). Convert images to 8-bit and apply "Subtract Background" function, stablishing the rolling ball radius value to visualize the background exclusively (with the option "create background—don't subtract" activated). The "background" image must be saved and then subtracted to the original 8-bit image with the "Image Calculator" function. Finally, the obtained image (negative) must be inverted in order to generate an image with detectable bands and a uniformly subtracted background (Gallo-Oller et al., 2018). Fig. 2 shows an example of Smad2/P-Smad2 WB analysis, in two different human

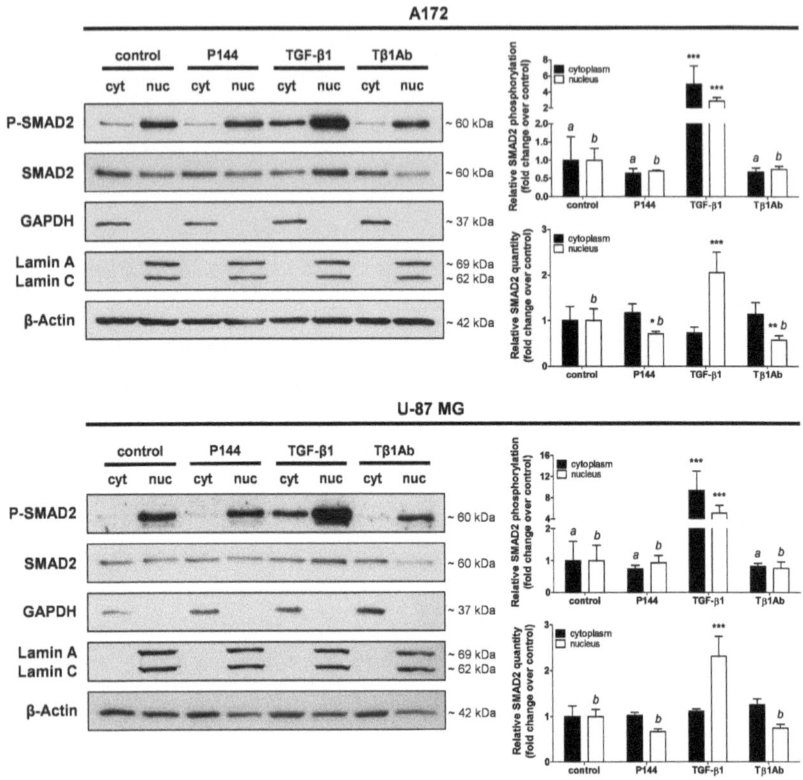


**Fig. 2** Western blot analysis of cytoplasmic (cyt) Smad2 phosphorylation and its nuclear translocation (nuc), in two human glioblastoma cell lines (A172 and U-87 MG), in the presence or absence of TGF-β1 and two different TGF-β1 inhibitors (P144, inhibitor peptide; Tb1Ab, anti TGF-β1 monoclonal antibody). Left panel shows representative Western blot images from three independent experiments. Right panel shows relative quantification of total Smad2 and P-Smad2. Statistical test: Kruskal-Wallis and Tukey's multiple comparisons. Statistical significance: $^{*}P < 0.05$; $^{**}P < 0.01$; and $^{***}P < 0.001$. (a) $P < 0.05$ in TGF-β1 group vs the other treatments in cytoplasmic and (b) nuclear fraction. Cytoplasmic marker: GAPDH; nuclear marker: Lamin A/C.

glioblastoma cell lines, in the presence or absence of recombinant human TGF-β1 and two different TGF-β1 inhibitors: P144: (Ezquerro et al., 2003) and anti-TGF-β1 antibody (Cat No. AB-101-NA, R&D Systems, Abingdon, UK).

### 2.5.1.2 TGF-β target genes Western blotting

Besides the determination of the signaling pathway status (receptors, Smad proteins and canonical regulators), it is important to consider the determination of TGF-β signaling pathway target genes according to research context, and briefly described in Section 1. Table 3 includes a brief review of most studied markers under TGF-β control in the context of tumor development, fibrosis induction and immune response regulation. This objective is conveniently addressed by WB due to its versatility.

## 2.6 Cytometry/luminex

Other methods, based on fluorescent labeled antibodies, can also quantitatively elucidate the presence of TGF-β in cell surface or intracellularly (FC) or in biological samples (Luminex). Analysis of total TGF-β requires *in vitro* release from the latent TGF-β, as previously stated, however, in FC analysis context this approach can indirectly induce cells to produce TGF-β and it determines fictitious altered results.

ELISA or bioassays that evaluate the antiproliferative effects of TGF-β on various cell lines are the most characterized standard methods to measure this cytokine, however, these methods do not allow the identification of the concrete cells that are effectively producing TGF-β, for this objective it would be better to perform a FC analysis.

In the case of TGF-β quantification through FC is important to consider that both free and latent states of this cytokine could be found as membrane-bound (Ahn, Lee, Sung, & Heo, 2009) or intracellularly.

TGF-β determination using FC provides the opportunity to detect TGF-β or LAP with standard surface membrane staining. In this case is desirable to increase the staining time at 4°C for 2h (Andersson et al., 2008).

In the case of intracellular TGF-β staining, after incubation with antibodies against surface markers, cells must be fixed with intracellular commercial fixation and permeabilization buffer (FacsFix, Cytofix/Cytoperm, etc.) or with 2–4% paraformaldehyde, followed by a Fc receptors required blocking and finally an incubation with antibodies anti-TGF-β or anti-LAP (Garba & Frelinger, 2001). However, the majority of TGF-β detection by FC targets LAP due to its stability while the active form of TGF-β has a very short half-life that complicates its detection (Joly et al., 2014; Nakamura, Kitani, & Strober, 2001). It is important to consider that the intracellular staining process can also induce an unspecific activation of TGF-β. For this reasons, the analysis of TGF-β activation is preferably quantified through intracellular staining of TGF-β downstream effector molecules.

### *2.6.1 Flow cytometry TGF-β measurement*

#### 2.6.1.1 Splenocyte or tumor infiltrating lymphocytes isolation

1. Homogenize spleens or tumors in 30–40mL of DPBS (Gibco/Life Technologies Corp, Grand Island, NY) by mechanical disruption and filter with a 70μm cell strainer (BD biosciences, San Jose, CA).

2. After 5 min of centrifugation at 300 *g* and 4°C, discard the supernatant and lysate the pellet for 3 min with 1–3 mL of ACK lysing buffer depending on the pellet size. Then, complete with DPBS to a volume of 40 mL and centrifuge again 5 min at 300 *g* and 4°C.
3. Discard supernatant and resuspend the pellet in 30 mL of DPBS and filtered again with the cell strainer.
4. Repeat point 2 centrifugation step, and resuspend the pellet in 1–5 mL of PBS, depending on the pellet size, and count cells at the Cell Coulter (Beckman coulter Inc, Fullerton, CA).

#### 2.6.1.2 LAP (TGF-β) staining

1. Harvest Tissue, Cells or Blood sample. Obtain desired sample to study TGF-β (e.g., tumor, lymph node, bone marrow, blood) and prepare a single cell suspension in standard cell staining buffer.
2. Add cell staining buffer up to ~15 mL and centrifuge at 350 × *g* for 5 min, discard supernatant.
3. Lyse red cells. If necessary (e.g., spleen, tumor), dilute 10 × red blood cell (RBC) lysis buffer and resuspend pellet in 3 mL 1 × RBC Lysis Buffer. Incubate on ice for 5 min.
4. Stop cell lysis by adding 10 mL cell staining buffer to the tube. Centrifuge for 5 min at 350 × *g* and discard supernatant.
5. Repeat wash as in step 2.
6. Count viable cells and resuspend in cell staining buffer at 5–10 × $10^6$ cells/mL and distribute 100 μL/tube of cell suspension (5–10 × $10^5$ cells/tube) into 12 × 75 mm plastic tubes.
7. Block Fc-Receptors incubation. Reagents that block Fc receptors may be useful for reducing nonspecific immunofluorescent staining. Human samples, antihuman CD16 (clone 3G8), CD32 (clone FUN-2), and CD64 (clone 10.1) antibodies. Block Fc receptors by preincubating cells with 0.25 μg antihuman CD16, CD32 and CD64 antibodies per $10^6$ cells in a 100 μL volume for 5–10 min on ice.
8. Cell-Surface Staining with Human LAP Antibody (R&D System, Clone # 27232). Resuspend pellet in residual buffer and add anti-LAP antibody 10 μL per $10^6$ cells in cell staining buffer, and incubate on ice in the dark for 15–20 min.
9. Repeat wash as in step 2.
10. Resuspend pellet with 200 μL per $10^6$ cells with cell staining buffer, and analyze samples using a flow cytometer.

#### 2.6.1.3 Whole blood flow cytometry (mouse/human)

1. After anesthesia, obtain mice blood samples from retro-orbital plexus in a heparinized 150 mm glass pasteur pipette and extract single-cell suspensions with FcR-Block (anti-CD16/32, BD Bioscience-Pharmigen). For human blood samples use vacutainer heparin tubes through venipuncture.
2. Stain cells surface markers for the identification of the specific population through addition of FC antibodies, after a vortex, and incubate the mixture for 10 min at 4°C.
3. Add 1 mL of FACS Lysing Solution (BD Biosciences, San Jose, CA) in each tube and incubate for 5 min in darkness and at RT.
4. Centrifuge mixtures for 4 min at 400 *g* and 4°C.
5. Remove supernatants and add 1 mL of FACS buffer per tube for washing.
6. Repeat centrifuge as in step 4.
7. Remove supernatants and resuspend pellet in 200 μL of FACS buffer for subsequent acquisition in a flow cytometer.

#### 2.6.1.4 FC regulatory T cells detection (Fig. 3)

Regulatory T cells Staining could be asses with "Human or mice Regulatory T cell Staining Kit" from eBioscience (San Diego, CA) following the manufacturer's instructions. Briefly, this consists on an intracellular staining performed as described above, where CD4 and CD25 are the surface antigens and FoxP3 is the intracellular antigen for mice Tregs detection, and CD127 surface antigen for human Tregs since CD127 inversely correlates with FoxP3 expression and has been identified as a more precise negative marker for human Tregs (Yu et al., 2012).

#### 2.6.1.5 FC Th17 cells detection (Fig. 3)

Functional analysis research of Th17 subpopulation requires a complex *in vitro* differentiation and amplification procedure, starting with a larger CD4+ T cell population in order to provide increased numbers and facilitate the studies (CellXVivo Human Th17 Cell Differentiation Kit, R&D Systems, Minneapolis, MN).

**For direct or poststimulation step quantification of Treg/Th17**

1. Plate $1 \times 10^6$ T cells per well in a round bottom 96-well plate.
2. Wash cells twice with 200 μL ice cold FACS buffer and centrifuge plate for 5 min at 300 × *g* and 4°C. Discard supernatants.

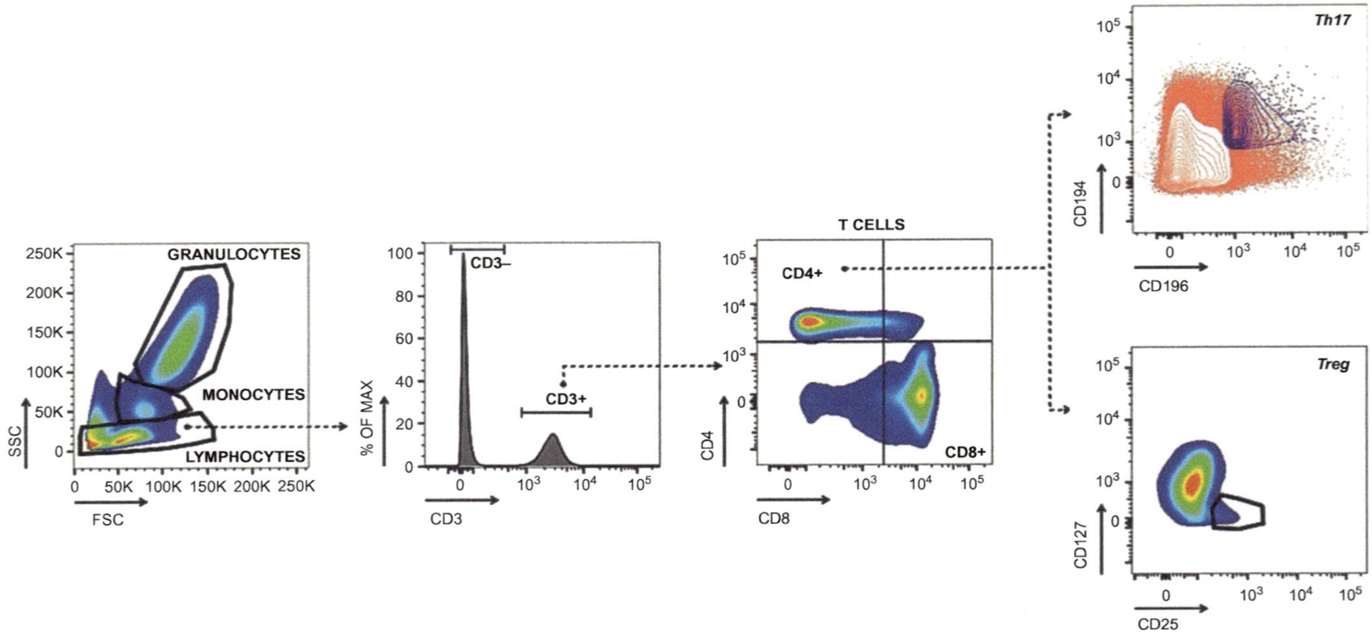


**Fig. 3** Diagram of flow cytometric (pseudo-color/smooth) plots showing a gating strategy example, used to define lymphocytes subpopulations Th17 (CD3+ CD4+ CD8 − CD194+ CD196+) and Tregs (CD3+ CD4+ CD8 − CD25+high, CD127 low), in tumor-derived infiltrating and peripheral blood lymphocytes. *FSC*, forward scatter; *SSC*, side scatter.

3. Resuspend cells in 100 μL ice cold FACS buffer containing the proper fluorochrome labeled antibodies (Treg: anti-CD3, CD4, CD35, and CD127; Th17: anti-CD3, CD4, CD194, and CD196, for human surface cell staining) at the appropriate dilution in FACS buffer at a final volume of 50 μL per well. Incubate plates for 10 min at 4°C in darkness.
4. Finally, wash plates twice in FACS buffer and resuspended in 200 μL of this buffer for acquisition at the flow cytometer.

**Flow cytometry analysis of intracellular antigens (FoxP3)**

1. Activate culture of splenocytes or tumor infiltrating lymphocytes with the appropriate stimulus (example: PMA/IONOMICIN).
2. After 1 h, add Brefeldin A to block the exocytosis of the intracellular cytokines (follow instructions of Golgi Stop/Golgi Plug BD).
3. Subsequently, perform the staining with surface antibodies as previously described. Wash cells, after this first staining, adding 100 μL of the BD Cytofix/Cytoperm buffer (BD Biosciences, San José, CA) and incubate the plate for 15 min at 4°C in darkness.
4. After centrifugation and removal of the fixation/permeabilization buffer, wash cells in permeabilization buffer.
5. Then, add the mouse Fc-Block antibody (BD Biosciences) in a final volume of 50 μL to each well (in permeabilization buffer) and incubate for 10 min at 4°C in darkness.
6. Once the incubation is finished, directly add the mixture of intracellular antibodies diluted in permeabilization buffer again in a final volume of 50 μL. Incubate the plate for 15 min at 4°C in darkness.
7. Washed cells twice in FACS buffer and then resuspend in 200 μL of FACS buffer for acquisition in a flow cytometer.

### *2.6.2 Luminex assay*

Luminex assay is a recent technological approach that allows the simultaneous quantification of many different analytes, including TGF-β three different isoforms (Flanders et al., 2016; Montoya et al., 2017). Luminex is based on both the principles of FC and sandwich-ELISA. Briefly, the sample is added to a mixture of color-coded beads, precoated with analyte-specific capture antibodies. The antibodies bind to the analytes of interest. Biotinylated detection antibodies specific to the analytes of interest are added and form a microspheres antibody-antigen sandwich. Phycoerythrin (PE) conjugated streptavidin is added and binds the biotinylated detection antibodies. Polystyrene beads are read on a dual-laser flow-based detection instrument (Luminex® 100™/200™ or Bio-Rad® Bio-Plex® analyzer).

One laser classifies the bead and determines the analyte that is being detected. The second laser determines the magnitude of the PE-derived emission signal, which is in direct proportion to the amount of analyte bound. Luminex technology could analyze samples as serums, supernatants and also tissues lysates (Flanders et al., 2016). These samples could analyze unaltered (TGF-β Free State) or previously acid-activated (Total TGF-β), and then incubated with TGF-β capture antibodies-coupled fluorescent microsphere for the three different TGF-β isoforms simultaneously. Many other TGF-β surrogate factors can be quantified with this technological platform.

## 3. Concluding remarks: Studying TGF-β as "Philosopher's Stone," the *status quo* in cancer diseases

Targeting TGF-β in cancer immunotherapies is critical to improve the antitumor immune responses in cancer patients. TGF-β has been shown to support the evasion of cancer cells from immunosurveillance and to contribute to the subversion of the immune cells. In this regard, TGF-β in the tumor microenvironment acts as a potent immunosuppressor (especially in effector immune cells such as CD8 and NK) being a promoter of malignant growth and metastasis. On the other hand, TGF-β is related to the proliferation of immunosuppressive immune cells such as Tregs, MDSC, and Macrophages M2. The development and progression of cancer cells themselves are also markedly impacted by the effects of TGF-β. Considering all these facts, TGF-β is classified as one of the most relevant molecular therapeutic targets in patients to determine the malignancy grade or select the best immunotherapy in cancer diseases, including combinations with TGF-β activity blockers. Mariathasan et al. (2018) have found, using a mouse model, that therapeutic administration of a TGF-β blocking antibody together with anti-PD-L1 reduced TGF-β signaling in stromal cells, facilitated T cell penetration into the tumor, and provoked vigorous antitumor immunity and tumor regression. Integration of these three independent biological features provides the best basis for understanding outcome in this setting and suggests that TGF-β shapes the tumor microenvironment to restrain antitumor immunity by restricting T cell infiltration. Furthermore, reduced circulating TGF-β1 seems to be a determinant of poor disease prognosis in patients–bearing breast cancer. This finding was studied when stratified by TGF-β1 levels, patients exhibiting low circulating levels of this cytokine (down 20 pg/mL) tended to a reduced overall survival when compared to those that presented high TGF-β1 levels (up 40 pg/mL) (Panis, Herrera, Victorino, Aranome, & Cecchini, 2013).

In similar works with pancreatic cancer, hepatocellular carcinoma and lung cancer, patients with high TGF-β1 plasma levels had worse prognosis. In these tumor types, treatment with Galunisertib, a TGFβRI inhibitor, resulted in a reduction of plasma TGF-β1 and a better outcome for patients. Galunisertib is currently under clinical development in combination with other checkpoint inhibitors.

Importantly, the identification of reliable and feasible predictive biomarkers to identify patients who are most likely to benefit from immunotherapy treatments is a foreseeable and critical issue. Expression levels of biomarkers related to the TGF-β pathway can be a valuable tool for this purpose, as may be the evaluation of the antitumor immune response. Through this, different approaches to precisely determine the quantity and activity of TGF-β are essential to assess the following steps in cancer treatment or predict cancer diseases by the Oncologist´s community.

## 4. Notes

- High homology among species of each mammalian isoform must be taking into account, because cross inter-species reactivity is a fact. In case of fetal bovine serum usage in cell cultures, adequate controls are required in order to subtract the background induction/signal. Absence or quantitative presence of FBS will modify sensitivity rate of TGF-β bioassays.
- The reduced half-life of TGF-β may generate data that underestimate the real presence and biological activity of this cytokine. In many experimental situations could be necessary the verification of P-Smad2/3 levels or the variation of induced surrogate proteins.
- It is recommended resuspension of recombinant TGF-β in solution with BSA, acidic pH and the usage of polypropylene test tubes, or other low binding material, to avoid free TGF-β binding to internal vials surface.
- To avoid unintended activation of TGF-β latent states in biological samples, slow freeze steps for storage and slow thawing prior analysis are recommended (RT → 4°C → −20°C → −80°C → −20°C → 4°C → RT). When compatible with assays logistics, it is recommended the use of fresh samples stored at 4°C within 24 h of collection to avoid partial activation of latent TGF-β.

- The establishment of several *in vitro* assays based in different cellular phenotypes/responses is crucial for a deeper understanding of TGF-β molecular mechanism. With this priority, and besides the available bioassays, we encourage the research groups to develop new and creative bioassays with intrinsic elements of each laboratory research: cell lines, cocultures, supernatants, etc. TGF-β generates powerful cellular responses that are relatively easy to quantify: proliferation, migration, apoptosis, secretion of GF, cytokines and ECM proteins, induction/inhibition of subrogated genes, etc. Analysis of presence or activity of TGF-β or potential inhibitor drugs can be established through bioassays combined with tumor extracts, cocultures, supernatants, etc.

## Acknowledgment

Fernando Aranda is supported by Sara Borrell's Program (CD15/00016) from Instituto de Salud Carlos III.

## References

Abe, M., Harpel, J. G., Metz, C. N., Nunes, I., Loskutoff, D. J., & Rifkin, D. B. (1994). An assay for transforming growth factor-beta using cells transfected with a plasminogen activator inhibitor-1 promoter-luciferase construct. *Analytical Biochemistry*, *216*(2), 276–284.

Ahn, Y.-O., Lee, J.-C., Sung, M.-W., & Heo, D. S. (2009). Presence of membrane-bound Tgf-beta1 and its regulation by IL-2-activated immune cell-derived IFN-gamma in head and neck squamous cell carcinoma cell lines. *Journal of Immunology (Baltimore, Md.: 1950)*, *182*(10), 6114–6120.

Ahn, J., Son, M. K., Jung, K. H., Kim, K., Kim, G. J., Lee, S. H., et al. (2016). Aminoacyl-tRNA synthetase interacting multi-functional protein 1 attenuates liver fibrosis by inhibiting TGFβ signaling. *International Journal of Oncology*, *48*(2), 747–755.

Ai, J., Nie, J., He, J., Guo, Q., Li, M., Lei, Y., et al. (2015). GQ5 hinders renal fibrosis in obstructive nephropathy by selectively inhibiting TGF-β-induced Smad3 phosphorylation. *Journal of the American Society of Nephrology*, *26*(8), 1827–1838.

Andersson, J., Tran, D. Q., Pesu, M., Davidson, T. S., Ramsey, H., O'Shea, J. J., et al. (2008). CD4+ FoxP3+ regulatory T cells confer infectious tolerance in a TGF-beta-dependent manner. *The Journal of Experimental Medicine*, *205*(9), 1975–1981.

Araki, S., Eitel, J. A., Batuello, C. N., Bijangi-Vishehsaraei, K., Xie, X. J., Danielpour, D., et al. (2010). TGF-beta1-induced expression of human Mdm2 correlates with late-stage metastatic breast cancer. *The Journal of Clinical Investigation*, *120*(1), 290–302.

Areström, I., Bartek, Z., Theresa, B., & Niklas, A. (2012). Measurement of human latent transforming growth factor-B1 using a latency associated protein-reactive ELISA. *Journal of Immunological Methods*, *379*(1–2), 23–29.

Assoian, R. K., Komoriya, A., Meyers, C. A., Miller, D. M., & Sporn, M. B. (1983). Transforming growth factor-beta in human platelets. Identification of a major storage site, purification, and characterization. *The Journal of Biological Chemistry*, *258*, 7155–7160.

Bakkebø, M., Huse, K., Hilden, V. I., Forfang, L., Myklebust, J. H., Smeland, E. B., et al. (2012). SARA is dispensable for functional TGF-β signaling. *FEBS Letters*, *586*(19), 3367–3372.

Bandyopadhyay, B., Han, A., Dai, J., Fan, J., Li, Y., Chen, M., et al. (2011). TbetaRI/Alk5-independent TbetaRII signaling to ERK1/2 in human skin cells according to distinct levels of TbetaRII expression. *Journal of Cell Science, 124*(Pt. 1), 19–24.

Bass, J. J., Wilkinson, D. J., Rankin, D., Phillips, B. E., Szewczyk, N. J., Smith, K., et al. (2016). An overview of technical considerations for Western blotting applications to physiological research. *Scandinavian Journal of Medicine & Science in Sports, 27*, 4–25.

Bauer, S., Rasika, S., Han, J., Mauduit, C., Raccurt, M., Morel, G., et al. (2003). Leukemia inhibitory factor is a key signal for injury-induced neurogenesis in the adult mouse olfactory epithelium. *The Journal of Neuroscience, 23*(5), 1792–1803.

Baugé, C., Cauvard, O., Leclercq, S., Galéra, P., & Bouméddiene, K. (2011). Modulation of transforming growth factor beta signaling pathway genes by transforming growth factor beta in human osteoarthritic chondrocytes: Involvement of Sp1 in both early and late response cells to transforming growth factor beta. *Arthritis Research & Therapy, 13*(1), R23.

Bayin, N. S., Modrek, A. S., & Placantonakis, D. G. (2014). Glioblastoma stem cells: Molecular characteristics and therapeutic implications. *World Journal of Stem Cells, 6*(2), 230–238.

Beekman, J. M., Vervoort, S. J., Dekkers, F., van Vessem, M. E., Vendelbosch, S., Brugulat-Panès, A., et al. (2012). Syntenin-mediated regulation of Sox4 proteasomal degradation modulates transcriptional output. *Oncogene, 31*(21), 2668–2679.

Berens, E. B., Holy, J. M., Riegel, A. T., & Wellstein, A. (2015). A cancer cell spheroid assay to assess invasion in a 3D setting. *Journal of Visualized Experiments*(105), e53409.

Biernacka, A., Dobaczewski, M., & Frangogiannis, N. G. (2011). TGF-β signaling in fibrosis. *Growth Factors (Chur, Switzerland), 29*(5), 196–202.

Borowicz, S., Van Scoyk, M., Avasarala, S., Karuppusamy Rathinam, M. K., Tauler, J., Bikkavilli, R. K., et al. (2014). The soft agar colony formation assay. *Journal of Visualized Experiments, 92*, e51998.

Brown, P. D., Wakefield, L. M., Levinson, A. D., & Sporn, M. B. (1990). Physicochemical activation of recombinant latent transforming growth factor-beta's 1, 2, and 3. *Growth Factors, 3*, 35–43.

Chandra, M., Zang, S., Li, H., Zimmerman, L. J., Champer, J., Tsuyada, A., et al. (2012). Nuclear translocation of type I transforming growth factor β receptor confers a novel function in RNA processing. *Molecular and Cellular Biology, 32*(12), 2183–2195.

Chapnick, D. A., & Liu, X. (2010). Analysis of ligand dependent nuclear accumulation of Smads in TGF-beta signaling. *Methods in Molecular Biology, 647*, 95–111.

Chen, C., & Feng-Hou, G. (2019). Th17 cells paradoxical roles in melanoma and potential application in immunotherapy. *Frontiers in Immunology, 10*, 187.

Chen, W. J., Jin, W. W., Hardegen, N., Lei, K. J., Li, L., et al. (2003). Conversion of peripheral CD4+CD25− naive T cells to CD4+CD25+ regulatory T cells by TGF-β induction of transcription factor Foxp3. *The Journal of Experimental Medicine, 198*, 1875–1886.

Chen, S. Y., Lin, J. S., Lin, H. C., Shan, Y. S., Cheng, Y. J., & Yang, B. C. (2015). Dependence of fibroblast infiltration in tumor stroma on type IV collagen-initiated integrin signal through induction of platelet-derived growth factor. *Biochimica et Biophysica Acta, 1853*(5), 929–939.

Chen, G. Y., Osada, H., Santamaria-Babi, L. F., & Kannagi, R. (2006). Interaction of GATA-3/T-bet transcription factors regulates expression of sialyl Lewis X homing receptors on Th1/Th2 lymphocytes. *Proceedings of the National Academy of Sciences of the United States of America, 103*(45), 16894–16899.

Chen, X., Ruan, A., Wang, X., Han, W., Wang, R., Lou, N., et al. (2014). miR-129-3p, as a diagnostic and prognostic biomarker for renal cell carcinoma, attenuates cell migration and invasion via downregulating multiple metastasis-related genes. *Journal of Cancer Research and Clinical Oncology, 140*(8), 1295–1304.

Chen, Q., Sivakumar, P., Barley, C., Peters, D. M., Gomes, R. R., Farach-Carson, M. C., et al. (2007). Potential role for heparan sulfate proteoglycans in regulation of transforming growth factor-beta (TGF-beta) by modulating assembly of latent TGF-beta-binding protein-1. *The Journal of Biological Chemistry, 282*(36), 26418–26430.

Chen, X., Wang, H., Liao, H. J., Hu, W., Gewin, L., Mernaugh, G., et al. (2014). Integrin-mediated type II TGF-β receptor tyrosine dephosphorylation controls SMAD-dependent profibrotic signaling. *The Journal of Clinical Investigation, 124*(8), 3295–3310.

Chen, J., Zhu, H., Zhang, Y., Cui, M. H., Han, L. Y., Jia, Z. H., et al. (2015). Low expression of phosphatase and tensin homolog in clear-cell renal cell carcinoma contributes to chemoresistance through activating the Akt/HDM2 signaling pathway. *Molecular Medicine Reports, 12*(2), 2622–2628.

Cheng, Q., Wu, J., Zhang, Y., Liu, X., Xu, N., Zuo, F., et al. (2017). SOX4 promotes melanoma cell migration and invasion though the activation of the NF-κB signaling pathway. *International Journal of Molecular Medicine, 40*(2), 447–453.

Cho, J. W., Il, K. J., & Lee, K. S. (2013). Downregulation of type I collagen expression in silibinin-treated human skin fibroblasts by blocking the activation of Smad2/3-dependent signaling pathways: Potential therapeutic use in the chemoprevention of keloids. *International Journal of Molecular Medicine, 31*(5), 1148–1152.

Cho, W., Kim, Y., Kim, J., Park, S., Park, D., Kim, B. C., et al. (2015). Suppressor of cytokine signaling 1 is a positive regulator of TGF-β-induced prostaglandin production in human follicular dendritic cell-like cells. *Journal of Immunology, 194*(9), 4287–4297.

Coste, A., Lefaucheur, J. P., Wang, Q. P., Lesprit, E., Poron, F., Peynegre, R., et al. (1998). Expression of the transforming growth factor beta isoforms in inflammatory cells of nasal polyps. *Archives of Otolaryngology—Head & Neck Surgery, 124*(12), 1361–1366.

Dai, W., Xu, X., Li, S., Ma, J., Shi, Q., Guo, S., et al. (2017). SOX4 promotes proliferative signals by regulating glycolysis through AKT activation in melanoma cells. *The Journal of Investigative Dermatology, 137*(11), 2407–2416.

D'Arpino, M. C., Fuchs, A. G., Sánchez, S. S., & Honoré, S. M. (2018). Extracellular matrix remodeling and TGF-β1/Smad signaling in diabetic colon mucosa. *Cell Biology International, 42*(4), 443–456.

De Breuck, S., Baeyens, L., & Bouwens, L. (2006). Expression and function of leukaemia inhibitory factor and its receptor in normal and regenerating rat pancreas. *Diabetologia, 49*(1), 108–116.

de Galarreta, M. R., Navarro, A., Ansorena, E., Garzón, A. G., Mòdol, T., López-Zabalza, M. J., et al. (2016). Unfolded protein response induced by Brefeldin A increases collagen type I levels in hepatic stellate cells through an IRE1α, p38 MAPK and Smad-dependent pathway. *Biochimica et Biophysica Acta, 1863*(8), 2115–2123.

Derynck, R., & Zhang, Y. (1996). Intracellular signaling: The mad way to do it. *Current Biology, 6*(10), 1226–1229.

Desmoulière, A., Geinoz, A., Gabbiani, F., & Gabbiani, G. (1993). Transforming growth factor-beta 1 induces alpha-smooth muscle actin expression in granulation tissue myofibroblasts and in quiescent and growing cultured fibroblasts. *The Journal of Cell Biology, 122*(1), 103–111.

Drabsch, Y., & ten Dijke, P. (2011). TGF-β signaling in breast cancer cell invasion and bone metastasis. *Journal of Mammary Gland Biology and Neoplasia, 16*(2), 97–108.

Drabsch, Y., & ten Dijke, P. (2012). TGF-beta signaling and its role in cancer progression and metastasis. *Cancer Metastasis Reviews, 31*(3–4), 553–568.

Du, J. X., Sun, M. Y., Du, G. L., Li, F. H., Liu, C., Mu, Y. P., et al. (2012). Ingredients of Huangqi decoction slow biliary fibrosis progression by inhibiting the activation of the transforming growth factor-beta signaling pathway. *BMC Complementary and Alternative Medicine, 12*, 33.

Duncan, M. R., Frazier, K. S., Abramson, S., Williams, S., Klapper, H., Huang, X., et al. (1999). Connective tissue growth factor mediates transforming growth factor beta-induced collagen synthesis: Down-regulation by CAMP. *FASEB Journal: Official Publication of the Federation of American Societies for Experimental Biology*, *13*(13), 1774–1786.

Dunn, L. K., Mohammad, K. S., Fournier, P. G., McKenna, C. R., Davis, H. W., Niewolna, M., et al. (2009). Hypoxia and TGF-beta drive breast cancer bone metastases through parallel signaling pathways in tumor cells and the bone microenvironment. *PLoS One*, *4*(9), e6896.

Emens, L. A., Ascierto, P. A., Darcy, P. K., Demaria, S., Eggermont, A. M. M., Redmond, W. L., et al. (2017). Cancer immunotherapy: Opportunities and challenges in the rapidly evolving clinical landscape. *European Journal of Cancer*, *81*, 116–129.

Ezquerro, I. J., Lasarte, J. J., Dotor, J., Castilla-Cortázar, I., Bustos, M., Peñuelas, I., et al. (2003). A synthetic peptide from transforming growth factor beta type III receptor inhibits liver fibrogenesis in rats with carbon tetrachloride liver injury. *Cytokine*, *22*(1–2), 12–20.

Fang, L., Zhao, J., Wang, D., Zhu, L., Wang, J., & Jiang, K. (2016). Jumonji AT-rich interactive domain 1B overexpression is associated with the development and progression of glioma. *International Journal of Molecular Medicine*, *38*(1), 172–182.

Flanders, K. C., Yang, Y.-A., Herrmann, M., Chen, J. Q., Mendoza, N., Mirza, A. M., et al. (2016). Quantitation of TGF-β proteins in mouse tissues shows reciprocal changes in TGF-B1 and TGF-B3 in normal vs neoplastic mammary epithelium. *Oncotarget*, 7(25), 38164–38179.

Flaumenhaft, R., Abe, M., Sato, Y., Miyazono, K., Harpel, J., Heldin, C. H., et al. (1993). Role of the latent TGF-beta binding protein in the activation of latent TGF-beta by co-cultures of endothelial and smooth muscle cells. *The Journal of Cell Biology*, *120*(4), 995–1002.

Fortunel, N. O., Hatzfeld, A., & Hatzfeld, J. A. (2000). Transforming growth factor-beta: Pleiotropic role in the regulation of hematopoiesis. *Blood*, *96*(6), 2022–2036.

Frei, K., Gramatzki, D., Tritschler, I., Schroeder, J. J., Espinoza, L., Rushing, E. J., et al. (2015). Transforming growth factor-β pathway activity in glioblastoma. *Oncotarget*, *6*(8), 5963–5977.

Frolik, C. A., Dart, L. L., Meyers, C. A., Smith, D. M., & Sporn, M. B. (1983). Purification and initial characterization of a type beta transforming growth factor from human placenta. *Proceedings of the National Academy of Sciences of the United States of America*, *80*, 3676–3680.

Fu, P., Liu, F., Su, S., Wang, W., Huang, X. R., Entman, M. L., et al. (2006). Signaling mechanism of renal fibrosis in unilateral ureteral obstructive kidney disease in ROCK1 knockout mice. *Journal of the American Society of Nephrology*, *17*(11), 3105–3114.

Gallo-Oller, G., Ordoñez, R., & Dotor, J. (2018). A new background subtraction method for Western blot densitometry band quantification through image analysis software. *Journal of Immunological Methods*, *457*, 1–5.

Gallo-Oller, G., Vollmann-Zwerenz, A., Melendez, B., Rey, J. A., Hau, P., Dotor, J., et al. (2016). P144, a transforming growth factor beta inhibitor peptide, generates antitumoral effects and modifies SMAD7 and SKI levels in human glioblastoma cell lines. *Cancer Letters*, *381*, 67–75.

Gan, H., Hao, Q., Idell, S., & Tang, H. (2015). Transcription factor Runx3 is induced by influenza A virus and double-strand RNA and mediates airway epithelial cell apoptosis. *Scientific Reports*, *5*, 17916.

Garba, M. L., & Frelinger, J. A. (2001). Intracellular cytokine staining for TGF-Beta. *Journal of Immunological Methods*, *258*(1–2), 193–198.

Garrigue-Antar, L., Barbieux, I., Lieubeau, B., Boisteau, O., & Grégoire, M. (1995). Optimisation of CCL64-based bioassay for TGF-beta. *Journal of Immunological Methods*, *186*(2), 267–274.

Gatza, C. E., Oh, S. Y., & Blobe, G. C. (2010). Roles for the type III TGF-beta receptor in human cancer. *Cellular Signalling*, *22*(8), 1163–1174.

Gera, R., Singh, V., Mitra, S., Sharma, A. K., Singh, A., Dasgupta, A., et al. (2017). Arsenic exposure impels CD4 commitment in thymus and suppress T cell cytokine secretion by increasing regulatory T cells. *Scientific Reports*, 7(1), 7140.

Gibson, H. M., Mishra, A., Chan, D. V., Hake, T. S., Porcu, P., & Wong, H. K. (2013). Impaired proteasome function activates GATA3 in T cells and upregulates CTLA-4: Relevance for Sézary syndrome. *The Journal of Investigative Dermatology*, *133*(1), 249–257.

Giroux, M., Delisle, J. S., O'Brien, A., Hébert, M. J., & Perreault, C. (2010). T cell activation leads to protein kinase C theta-dependent inhibition of TGF-beta signaling. *Journal of Immunology*, *185*(3), 1568–1576.

Gorelik, L., Fields, P. E., & Flavell, R. A. (2000). Cutting edge: TGF-β inhibits Th type 2 development through inhibition of GATA-3 expression. *Journal of Immunology*, *165*, 4773–4777.

Hau, P., Jachimczak, P., Schlaier, J., & Bogdahn, U. (2011). TGF-beta2 signaling in high-grade gliomas. *Current Pharmaceutical Biotechnology*, *12*(12), 2150–2157.

Heifetz, S., Hernandez, H., Laiho, M., ten Dijke, P., Iwata, K. K., & Massague, J. (1990). Distinct transforming growth factor-β (TGF-β) receptor subsets as determinants of cellular responsiveness to three TGF-bisoforms. *The Journal of Biological Chemistry*, *265*, 20533–20538.

Hinz, S., Pagerols-Raluy, L., Oberg, H. H., Ammerpohl, O., Grüssel, S., Sipos, B., et al. (2007). Foxp3 expression in pancreatic carcinoma cells as a novel mechanism of immune evasion in cancer. *Cancer Research*, *67*(17), 8344–8350.

Hou, X., Zhang, J., Wang, Y., Xiong, W., & Mi, J. (2017). TGFBR-IDH1-Cav1 axis promotes TGF-β signaling in cancer-associated fibroblast. *Oncotarget*, *8*(48), 83962–83974.

Hu, Q., Huang, C., Wang, Y., & Wu, R. (2015). Expression of leukemia inhibitory factor in the rat retina following acute ocular hypertension. *Molecular Medicine Reports*, *12*(5), 6577–6583.

Huang, C., Du, R., Zhang, P., Meng, H., Jia, H., Song, Y., et al. (2011). Expression, purification, and functional characterization of recombinant PTD-SARA. *Acta Biochimica et Biophysica Sinica Shanghai*, *43*(2), 110–117.

Ignotz, R. A., & Massagué, J. (1986). Transforming growth factor-beta stimulates the expression of fibronectin and collagen and their incorporation into the extracellular matrix. *The Journal of Biological Chemistry*, *261*(9), 4337–4345.

Ihn, H. (2002). Pathogenesis of fibrosis: Role of TGF-beta and CTGF. *Current Opinion in Rheumatology*, *14*(6), 681–685.

Ikushima, H., Todo, T., Ino, Y., Takahashi, M., Miyazawa, K., & Miyazono, K. (2009). Autocrine TGF-beta signaling maintains tumorigenicity of glioma-initiating cells through Sry-related HMG-box factors. *Cell Stem Cell*, *5*(5), 504–514.

Izzo, F., Mercogliano, F., Venturutti, L., Tkach, M., Inurrigarro, G., Schillaci, R., et al. (2014). Progesterone receptor activation downregulates GATA3 by transcriptional repression and increased protein turnover promoting breast tumor growth. *Breast Cancer Research*, *16*(6), 491.

Jacob, C., Grabner, H., Atanasoski, S., & Suter, U. (2008). Expression and localization of Ski determine cell type-specific TGFβ signaling effects on the cell cycle. *The Journal of Cell Biology*, *182*(3), 519–530.

Jang, S. M., Kim, J. W., Kim, D., Kim, C. H., An, J. H., Choi, K. H., et al. (2013). Sox4-mediated caldesmon expression facilitates differentiation of skeletal myoblasts. *Journal of Cell Science*, *126*(Pt. 22), 5178–5188.

Javelaud, D., van Kempen, L., Alexaki, V. I., Le Scolan, E., Luo, K., & Mauviel, A. (2011). Efficient TGF-β/SMAD signaling in human melanoma cells associated with high c-SKI/SnoN expression. *Molecular Cancer*, *10*(1), 2.

Jenkins, G. (2008). The role of proteases in transforming growth factor-beta activation. *The International Journal of Biochemistry & Cell Biology*, *40*(6-7), 1068–1078.
Jeon, E. S., Moon, H. J., Lee, M. J., Song, H. Y., Kim, Y. M., Bae, Y. C., et al. (2006). Sphingosylphosphorylcholine induces differentiation of human mesenchymal stem cells into smooth-muscle-like cells through a TGF-beta-dependent mechanism. *Journal of Cell Science*, *119*(Pt. 23), 4994–5005.
Jiang, Y., Lin, X., Tang, Z., Lee, C., Tian, G., Du, Y., et al. (2017). Critical role of caveolin-1 in ocular neovascularization and multitargeted antiangiogenic effects of cavtratin via JNK. *Proceedings of the National Academy of Sciences of the United States of America*, *114*(40), 10737–10742.
Joly, M. S., Martin, R. P., Mitra-Kaushik, S., Phillips, L., D'Angona, A., Richards, S. M., et al. (2014). Transient low-dose methotrexate generates B regulatory cells that mediate antigen-specific tolerance to alglucosidase alfa. *Journal of Immunology (Baltimore, Md.: 1950)*, *193*(8), 3947–3958.
Kang, Y., He, W., Tulley, S., Gupta, G. P., Serganova, I., Chen, C. R., et al. (2005). Breast cancer bone metastasis mediated by the Smad tumor suppressor pathway. *Proceedings of the National Academy of Sciences of the United States of America*, *102*(39), 13909–13914.
Kang, S. W., Kim, J. L., Kwon, G. T., Lee, Y. J., Park, J. H., Lim, S. S., et al. (2011). Sensitive fern (Onoclea sensibilis) extract suppresses proliferation and migration of vascular smooth muscle cells inflamed by neighboring macrophages. *Biological & Pharmaceutical Bulletin*, *34*(11), 1717–1723.
Kast, J., Hanecker, P., Beaufort, N., Giese, A., Joutel, A., Dichgans, M., et al. (2014). Sequestration of latent TGF-β binding protein 1 into CADASIL-related Notch3-ECD deposits. *Acta Neuropathologica Communications*, *2*, 96.
Khan, S. A., Joyce, J., & Tsuda, T. (2012). Quantification of active and total transforming growth factor-β levels in serum and solid organ tissues by bioassay. *BMC Research Notes*, *5*, 636.
Kim, S. V., Xiang, W. V., Kwak, C., Yang, Y., Lin, X. W., et al. (2013). GPR15-mediated homing controls immune homeostasis in the large intestine mucosa. *Science*, *340*, 1456–1459.
Kim, D. H., Yoo, T. H., Lee, S. H., Kang, H. Y., Nam, B. Y., Kwak, S. J., et al. (2012). Gamma linolenic acid exerts anti-inflammatory and anti-fibrotic effects in diabetic nephropathy. *Yonsei Medical Journal*, *53*(6), 1165–1175.
Koli, K., Saharinen, J., Hyytiäinen, M., Penttinen, C., & Keski-Oja, J. (2001). Latency, activation, and binding proteins of TGF-beta. *Microscopy Research and Technique*, *52*(4), 354–362.
Kong, J., Du, J., Wang, Y., Yang, M., Gao, J., Wei, X., et al. (2016). Focal adhesion molecule Kindlin-1 mediates activation of TGF-β signaling by interacting with TGF-βRI, SARA and Smad3 in colorectal cancer cells. *Oncotarget*, 7(46), 76224–76237.
Kubiczkova, L., Sedlarikova, L., Hajek, R., & Sevcikova, S. (2012). TGF-beta—An excellent servant but a bad master. *Journal of Translational Medicine*, *10*, 183.
Lampugnani, M. G. (1999). Cell migration into a wounded area in vitro. *Methods in Molecular Biology*, *96*, 177–182.
Le Scolan, E., Zhu, Q., Wang, L., Bandyopadhyay, A., Javelaud, D., Mauviel, A., et al. (2008). Transforming growth factor-beta suppresses the ability of Ski to inhibit tumor metastasis by inducing its degradation. *Cancer Research*, *68*(9), 3277–3285.
Leask, A., & Abraham, D. J. (2004). TGF-beta signaling and the fibrotic response. *FASEB Journal: Official Publication of the Federation of American Societies for Experimental Biology*, *18*(7), 816–827.
Lee, W. K., Chung, K. W., Kim, G. H., & Kim, S. J. (2013). Gallotannin causes differentiation and inflammation via ERK-1/-2 and p38 kinase pathways in rabbit articular chondrocytes. *Molecular Medicine Reports*, 7(2), 701–707.

Lee, S. M., Gao, B., & Fang, D. (2008). FoxP3 maintains Treg unresponsiveness by selectively inhibiting the promoter DNA-binding activity of AP-1. *Blood*, *111*(7), 3599–3606.

Lee, S. H., Jung, Y. D., Choi, Y. S., & Lee, Y. M. (2015). Targeting of RUNX3 by miR-130a and miR-495 cooperatively increases cell proliferation and tumor angiogenesis in gastric cancer cells. *Oncotarget*, *6*(32), 33269–33278.

Li, X. Z., Feng, J. T., Hu, C. P., Chen, Z. Q., Gu, Q. H., & Nie, H. P. (2010). Effects of Arkadia on airway remodeling through enhancing TGF-beta signaling in allergic rats. *Laboratory Investigation*, *90*(7), 997–1003.

Li, Y., Tsun, A., Gao, Z., Han, Z., Gao, Y., Li, Z., et al. (2013). 60-kDa Tat-interactive protein (TIP60) positively regulates Th-inducing POK (ThPOK)-mediated repression of eomesodermin in human CD4+ T cells. *The Journal of Biological Chemistry*, *288*(22), 15537–15546.

Li, C. W., Wang, Q., Li, J., Hu, M., Shi, S. J., Li, Z. W., et al. (2016). Silver nanoparticles/chitosan oligosaccharide/poly(vinyl alcohol) nanofiber promotes wound healing by activating TGFβ1/Smad signaling pathway. *International Journal of Nanomedicine*, *11*, 373–386.

Li, S. S., Wang, L. L., Liu, M., Jiang, S. K., Zhang, M., Tian, Z. L., et al. (2016). Cannabinoid $CB_2$ receptors are involved in the regulation of fibrogenesis during skin wound repair in mice. *Molecular Medicine Reports*, *13*(4), 3441–3450.

Lin, M. C., Lin, J. J., Hsu, C. L., Juan, H. F., Lou, P. J., & Huang, M. C. (2017). GATA3 interacts with and stabilizes HIF-1α to enhance cancer cell invasiveness. *Oncogene*, *36*(30), 4243–4252.

Ling, X., Xu, C., Fan, C., Zhong, K., Li, F., & Wang, X. (2014). FL118 induces p53-dependent senescence in colorectal cancer cells by promoting degradation of MdmX. *Cancer Research*, *74*(24), 7487–7497.

Liu, Y., Meyer, C., Müller, A., Herweck, F., Li, Q., Müllenbach, R., et al. (2011). IL-13 induces connective tissue growth factor in rat hepatic stellate cells via TGF-β-independent Smad signaling. *Journal of Immunology*, *187*(5), 2814–2823.

Liu, T., Wang, P., Cong, M., Xu, Y., Jia, J., & You, H. (2013). The CYP2E1 inhibitor DDC up-regulates MMP-1 expression in hepatic stellate cells via an ERK1/2- and Akt-dependent mechanism. *Bioscience Reports*, *33*(3), pii: e00041.

Liu, T., Wang, P., Cong, M., Zhang, D., Liu, L., Li, H., et al. (2016). Matrix metalloproteinase-1 induction by diethyldithiocarbamate is regulated via Akt and ERK/miR222/ETS-1 pathways in hepatic stellate cells. *Bioscience Reports*, *36*(4), e00371.

Liu, F., Zhang, Y., Men, T., Jiang, X., Yang, C., Li, H., et al. (2017). Quantitative proteomic analysis of gastric cancer tissue reveals novel proteins in platelet-derived growth factor B signaling pathway. *Oncotarget*, *8*(13), 22059–22075.

Luo, K., Stroschein, S. L., Wang, W., Chen, D., Martens, E., Zhou, S., et al. (1999). The Ski oncoprotein interacts with the Smad proteins to repress TGFbeta signaling. *Genes & Development*, *13*(17), 2196–2206.

Macias-Silva, M., Li, W., Leu, J. I., Crissey, M. A., & Taub, R. (2002). Up-regulated transcriptional repressors SnoN and Ski bind Smad proteins to antagonize transforming growth factor-beta signals during liver regeneration. *The Journal of Biological Chemistry*, *277*(32), 28483–28490.

Mancini, D., Monteagudo, J., Suárez-Fariñas, M., Bander, J., Varshney, R., Gonzalez, J., et al. (2018). New methodologies to accurately assess circulating active transforming growth factor-β1 levels: Implications for evaluating heart failure and the impact of left ventricular assist devices. *Translational Research*, *192*, 15–29.

Maneechotesuwan, K., Xin, Y., Ito, K., Jazrawi, E., Lee, K. Y., Usmani, O. S., et al. (2007). Regulation of Th2 cytokine genes by p38 MAPK-mediated phosphorylation of GATA-3. *Journal of Immunology*, *178*(4), 2491–2498.

Manojlovic, Z., Blackmon, J., & Stefanovic, B. (2013). Tacrolimus (FK506) prevents early stages of ethanol induced hepatic fibrosis by targeting LARP6 dependent mechanism of collagen synthesis. *PLoS One*, *8*(6), e65897.

Mariathasan, S., Turley, S. J., Nickles, D., Castiglioni, A., Yuen, K., Wang, Y., et al. (2018). TGFβ attenuates tumour response to PD-L1 blockade by contributing to exclusion of T cells. *Nature*, *554*(7693), 544–548.

Marshall, J. (2011). Transwell(®) invasion assays. *Methods in Molecular Biology*, *769*, 97–110.

Massagué, J. (1992). Receptors for the TGF-beta family. *Cell*, *69*(7), 1067–1070.

Massagué, J. (2012). TGFbeta signaling in context. *Nature Reviews. Molecular Cell Biology*, *13*(10), 616–630.

Maurya, V. K., Jha, R. K., Kumar, V., Joshi, A., Chadchan, S., Mohan, J. J., et al. (2013). Transforming growth factor-beta 1 (TGF-B1) liberation from its latent complex during embryo implantation and its regulation by estradiol in mouse. *Biology of Reproduction*, *89*(4), 84.

Mazzieri, R., Jurukovski, V., Obata, H., Sung, J., Platt, A., Annes, E., et al. (2005). Expression of truncated latent TGF-beta-binding protein modulates TGF-beta signaling. *Journal of Cell Science*, *118*(Pt. 10), 2177–2187.

Meng, X. M., Nikolic-Paterson, D. J., & Lan, H. Y. (2016). TGF-β: The master regulator of fibrosis. *Nature Reviews. Nephrology*, *12*(6), 325–338.

Milde-Langosch, K., Röder, H., Andritzky, B., Aslan, B., Hemminger, G., Brinkmann, A., et al. (2004). The role of the AP-1 transcription factors c-Fos, FosB, Fra-1 and Fra-2 in the invasion process of mammary carcinomas. *Breast Cancer Research and Treatment*, *86*(2), 139–152.

Miyake, T., Alli, N. S., & McDermott, J. C. (2010). Nuclear function of Smad7 promotes myogenesis. *Molecular and Cellular Biology*, *30*(3), 722–735.

Miyazono, K., Katsuno, Y., Koinuma, D., Ehata, S., & Morikawa, M. (2018). Intracellular and extracellular TGF-β signaling in cancer: Some recent topics. *Frontiers of Medicine*, *12*(4), 387–411.

Montoya, J. G., Holmes, T. H., Anderson, J. N., Maecker, H. T., Rosenberg-Hasson, Y., Valencia, I. J., et al. (2017). Cytokine signature associated with disease severity in chronic fatigue syndrome patients. *Proceedings of the National Academy of Sciences of the United States of America*, *114*(34), E7150–E7158.

Mucida, D., Husain, M. M., Muroi, S., van Wijk, F., Shinnakasu, R., et al. (2013). Transcriptional reprogramming of mature CD4+ helper T cells generates distinct MHC class II-restricted cytotoxic T lymphocytes. *Nature Immunology*, *14*, 281–289.

Naber, H. P., Wiercinska, E., Ten Dijke, P., & van Laar, T. (2011). Spheroid assay to measure TGF-β induced invasion. *Journal of Visualized Experiments*, *57*, e3337.

Nakamura, K., Kitani, A., & Strober, W. (2001). Cell contact-dependent immunosuppression by CD4(+)CD25(+) regulatory T cells is mediated by cell surface-bound transforming growth factor beta. *The Journal of Experimental Medicine*, *194*(5), 629–644.

Nalvarte, I., Damdimopoulos, A. E., Rüegg, J., & Spyrou, G. (2015). The expression and activity of thioredoxin reductase 1 splice variants v1 and v2 regulate the expression of genes associated with differentiation and adhesion. *Bioscience Reports*, *35*(6), e00269.

Nomura, N., Sasamoto, S., Ishii, S., Date, T., Matsui, M., & Ishizaki, R. (1989). Isolation of human cDNA clones of ski and the ski-related gene, sno. *Nucleic Acids Research*, *17*(14), 5489–5500.

Nyati, S., Schinske, K., Ray, D., Nyati, M. K., Ross, B. D., & Rehemtulla, A. (2011). Molecular imaging of TGFβ-induced Smad2/3 phosphorylation reveals a role for receptor tyrosine kinases in modulating TGFβ signaling. *Clinical Cancer Research*, *17*(23), 7424–7439.

Ohji, M., SundarRaj, N., & Thoft, R. A. (1993). Transforming growth factor-beta stimulates collagen and fibronectin synthesis by human corneal stromal fibroblasts in vitro. *Current Eye Research*, *12*(8), 703–709.

Oida, T., & Weiner, H. L. (2011). Murine CD4 T cells produce a new form of TGF-β as measured by a newly developed TGF-β bioassay. *PLoS One*, *6*(4), e18365.
Pan, B., Wang, X., Kojima, S., Nishioka, C., Yokoyama, A., Honda, G., et al. (2017). The fifth epidermal growth factor-like region of thrombomodulin alleviates murine graft-versus-host disease in a G-protein coupled receptor 15 dependent manner. *Biology of Blood and Marrow Transplantation*, *23*(5), 746–756.
Pan, B., Wang, X., Nishioka, C., Honda, G., Yokoyama, A., Zeng, L., et al. (2017). G-protein coupled receptor 15 mediates angiogenesis and cytoprotective function of thrombomodulin. *Scientific Reports*, 7(1), 692.
Pang, K., Ryan, J. F., Baxevanis, A. D., & Martindale, M. Q. (2011). Evolution of the TGF-β signaling pathway and its potential role in the ctenophore, Mnemiopsis leidyi. *PLoS One*, *6*(9), e24152.
Panis, C., Herrera, A. C., Victorino, V. J., Aranome, A. M., & Cecchini, R. (2013). Screening of circulating TGF-β levels and its clinicopathological significance in human breast cancer. *Anticancer Research*, *33*(2), 737–742.
Pannu, J., Nakerakanti, S., Smith, E., ten Dijke, P., & Trojanowska, M. (2007). Transforming growth factor-beta receptor type I-dependent fibrogenic gene program is mediated via activation of Smad1 and ERK1/2 pathways. *The Journal of Biological Chemistry*, *282*(14), 10405–10413.
Penuelas, S., Anido, J., Prieto-Sanchez, R. M., Folch, G., Barba, I., Cuartas, I., et al. (2009). TGF-beta increases glioma-initiating cell self-renewal through the induction of LIF in human glioblastoma. *Cancer Cell*, *15*(4), 315–327.
Pierreux, C. E., Nicolás, F. J., & Hill, C. S. (2000). Transforming growth factor beta-independent shuttling of Smad4 between the cytoplasm and nucleus. *Molecular and Cellular Biology*, *20*(23), 9041–9054.
Pohlers, D., Brenmoehl, J., Löffler, I., Müller, C. K., Leipner, C., Schultze-Mosgau, S., et al. (2009). TGF-beta and fibrosis in different organs—Molecular pathway imprints. *Biochimica et Biophysica Acta*, *1792*(8), 746–756.
Principe, D. R., Doll, J. A., Bauer, J., Jung, B., Munshi, H. G., Bartholin, L., et al. (2014). TGF-beta: Duality of function between tumor prevention and carcinogenesis. *Journal of the National Cancer Institute*, *106*(2), djt369.
Ramos, G. C., Dalbó, S., Leite, D. P., Goldfeder, E., Carvalho, C. R., Vaz, N. M., et al. (2012). The autoimmune nature of post-infarct myocardial healing: Oral tolerance to cardiac antigens as a novel strategy to improve cardiac healing. *Autoimmunity*, *45*(3), 233–244.
Randall, L. A., Wadhwa, M., Thorpe, R., & Mire-Sluis, A. R. (1993). A novel, sensitive bioassay for transforming growth factor beta. *Journal of Immunological Methods*, *164*(1), 61–67.
Rapoza, M. L., Fu, D., & Sendak, R. A. (2006). Development of an in vitro potency assay for therapeutic TGFbeta antagonists: The A549 cell bioassay. *Journal of Immunological Methods*, *316*(1-2), 18–26.
Reis, B. S., Rogoz, A., Costa-Pinto, F. A., Taniuchi, I., & Mucida, D. (2013). Mutual expression of the transcription factors Runx3 and ThPOK regulates intestinal CD4+ T cell immunity. *Nature Immunology*, *14*, 271–280.
Riquelme-Guzmán, C., Contreras, O., & Brandan, E. (2018). Expression of CTGF/CCN2 in response to LPA is stimulated by fibrotic extracellular matrix via the integrin/FAK axis. *American Journal of Physiology. Cell Physiology*, *314*(4), C415–C427.
Roberts, A. B., Lamb, L. C., Newton, D. L., Sporn, M. B., De Larco, J. E., & Todaro, G. J. (1980). Transforming growth factors: Isolation of polypeptides from virally and chemically transformed cells by acid/ethanol extraction. *Proceedings of the National Academy of Sciences of the United States of America*, 77, 3494–3498.

Ruan, H. C., Zhu, X. M., Luo, Q., Liu, A. X., Qian, Y. L., Zhou, C. Y., et al. (2006). Ovarian stimulation with GnRH agonist, but not GnRH antagonist, partially restores the expression of endometrial integrin beta3 and leukaemia-inhibitory factor and improves uterine receptivity in mice. *Human Reproduction*, *21*(10), 2521–2529.

Ruegemer, J. J., Ho, S. N., Augustine, J. A., Schlager, J. W., Bell, M. P., McKean, D. J., et al. (1990). Regulatory effects of transforming growth factor-beta on IL-2- and IL-4-dependent T cell-cycle progression. *Journal of Immunology*, *144*(5), 1767–1776.

Runyan, C. E., Hayashida, T., Hubchak, S., Curley, J. F., & Schnaper, H. W. (2009). Role of SARA (SMAD anchor for receptor activation) in maintenance of epithelial cell phenotype. *The Journal of Biological Chemistry*, *284*(37), 25181–25189.

Sakaki-Yumoto, M., Katsuno, Y., & Derynck, R. (2013). TGF-beta family signaling in stem cells. *Biochimica et Biophysica Acta*, *1830*(2), 2280–2296.

Schnoor, M., Cullen, P., Lorkowski, J., Stolle, K., Robenek, H., Troyer, D., et al. (2008). Production of type VI collagen by human macrophages: A new dimension in macrophage functional heterogeneity. *Journal of Immunology*, *180*(8), 5707–5719.

Semlali, A., Jacques, E., Plante, S., Biardel, S., Milot, J., Laviolette, M., et al. (2008). TGF-beta suppresses EGF-induced MAPK signaling and proliferation in asthmatic epithelial cells. *American Journal of Respiratory Cell and Molecular Biology*, *38*(2), 202–208.

Seoane, J. (2006). Escaping from the TGFbeta anti-proliferative control. *Carcinogenesis*, *27*(11), 2148–2156.

Stolfi, C., Marafini, I., De Simone, V., Pallone, F., & Monteleone, G. (2013). The dual role of Smad7 in the control of cancer growth and metastasis. *International Journal of Molecular Sciences*, *14*(12), 23774–23790.

Stover, D. G., Bierie, B., & Moses, H. L. (2007). A delicate balance: TGF-beta and the tumor microenvironment. *Journal of Cellular Biochemistry*, *101*(4), 851–861.

Tanaka, A., & Sakaguchi, S. (2017). Regulatory T cells in cancer immunotherapy. *Cell Research*, *27*(1), 109–118.

Taylor, S. C., Berkelman, T., Yadav, G., & Hammond, M. (2013). A defined methodology for reliable quantification of Western blot data. *Molecular Biotechnology*, *55*, 217–226.

Tesseur, I., Zou, K., Berber, E., Zhang, H., & Wyss-Coray, T. (2006). Highly sensitive and specific bioassay for measuring bioactive TGF-beta. *BMC Cell Biology*, *7*, 15.

Tian, A., Ma, H., Zhang, R., Tan, W., Wang, X., Wu, B., et al. (2015). Interleukin17A promotes postoperative cognitive dysfunction by triggering β-amyloid accumulation via the transforming growth factor-β (TGFβ)/Smad signaling pathway. *PLoS One*, *10*(10), e0141596.

Tiwari, A., Kumar, R., Ram, J., Sharma, M., & Luthra-Guptasarma, M. (2016). Control of fibrotic changes through the synergistic effects of anti-fibronectin antibody and an RGDS-tagged form of the same antibody. *Scientific Reports*, *6*, 30872.

Travis, M. A., & Sheppard, D. (2014). TGF-β activation and function in immunity. *Annual Review of Immunology*, *32*, 51–82.

Verrecchia, F., & Mauviel, A. (2007). Transforming growth factor-beta and fibrosis. *World Journal of Gastroenterology*, *13*(22), 3056–3062.

Walther, M., Jallow, I. K., Jeffries, D., & Walther, B. (2009). Bioactive TGF-beta levels can be preserved in plasma samples collected into heparin but not EDTA. *Cytokine*, *48*(3), 267–272.

Wang, P., Liu, X., Xu, P., Lu, J., Wang, R., & Mu, W. (2016). Decorin reduces hypertrophic scarring through inhibition of the TGF-β1/Smad signaling pathway in a rat osteomyelitis model. *Experimental and Therapeutic Medicine*, *12*(4), 2102–2108.

Weiss, A., & Attisano, L. (2013). The TGFbeta superfamily signaling pathway. *Wiley Interdisciplinary Reviews: Developmental Biology*, *2*(1), 47–63.

Wickert, L., Chatain, N., Kruschinsky, K., & Gressner, A. M. (2007). Glucocorticoids activate TGF-beta induced PAI-1 and CTGF expression in rat hepatocytes. *Comparative Hepatology*, *6*, 5.

Wipff, P. J., & Hinz, B. (2008). Integrins and the activation of latent transforming growth factor beta1—An intimate relationship. *European Journal of Cell Biology*, *87*(8–9), 601–615.

Wrighton, K. H., Lin, X., & Feng, X. H. (2008). Critical regulation of TGFbeta signaling by Hsp90. *Proceedings of the National Academy of Sciences of the United States of America*, *105*(27), 9244–9249.

Wu, C., Kan, H., Hu, M., Liu, X., Boye, A., Jiang, Y., et al. (2018). Compound Astragalus and Salvia miltiorrhiza extract inhibits hepatocarcinogenesis via modulating TGF-β/TβR and Imp7/8. *Experimental and Therapeutic Medicine*, *16*(2), 1052–1060.

Xie, S. R., An, J. Y., Zheng, L. B., Huo, X. X., Guo, J., Shih, D., et al. (2017). Effects and mechanism of adenovirus-mediated phosphatase and tension homologue deleted on chromosome ten gene on collagen deposition in rat liver fibrosis. *World Journal of Gastroenterology*, *23*(32), 5904–5912.

Yang, J., Ahmed, A., Poon, E., Perusinghe, N., de Haven, B. A., Box, G., et al. (2009). Small-molecule activation of p53 blocks hypoxia-inducible factor 1alpha and vascular endothelial growth factor expression in vivo and leads to tumor cell apoptosis in normoxia and hypoxia. *Molecular and Cellular Biology*, *29*(8), 2243–2253.

Yang, J., Zhang, X., Li, Y., & Liu, Y. (2003). Downregulation of Smad transcriptional corepressors SnoN and Ski in the fibrotic kidney: An amplification mechanism for TGF-beta1 signaling. *Journal of the American Society of Nephrology*, *14*(12), 3167–3177.

Ying, M., Tilghman, J., Wei, Y., Guerrero-Cazares, H., Quinones-Hinojosa, A., Ji, H., et al. (2014). Kruppel-like factor-9 (KLF9) inhibits glioblastoma stemness through global transcription repression and integrin alpha6 inhibition. *The Journal of Biological Chemistry*, *289*(47), 32742–32756.

Yu, N., Li, X., Song, W., Li, D., Yu, D., Zeng, X., et al. (2012). CD4(+)CD25 (+)CD127 (low/−) T cells: A more specific Treg population in human peripheral blood. *Inflammation*, *35*(6), 1773–1780.

Yuan, S. M., & Lin, H. (2018). Expressions of transforming growth factor β1 signaling cytokines in aortic dissection. *Brazilian Journal of Cardiovascular Surgery*, *33*(6), 597–602.

Zhang, Y. E. (2017). Non-Smad signaling pathways of the TGF-β family. *Cold Spring Harbor Perspectives in Biology*, *9*(2), a022129.

Zhang, S. (2018). The role of transforming growth factor β in T helper 17 differentiation. *Immunology*, *155*(1), 24–35.

Zheng, S. G., Wang, J., Wang, P., Gray, J. D., & Horwitz, D. A. (2007). IL-2 is essential for TGF-β to convert naive CD4+CD25− cells to CD25+Foxp3+ regulatory T cells and for expansion of these cells. *Journal of Immunology*, *178*, 2018–2027.

Zhou, T., Wang, Z., Fan, J., Chen, S., Tan, Z., Yang, H., et al. (2015). Angiotensin-converting enzyme-2 overexpression improves atrial remodeling and function in a canine model of atrial fibrillation. *Journal of the American Heart Association*, *4*(3), e001530.

Zhu, Q., Le Scolan, E., Jahchan, N., Ji, X., Xu, A., & Luo, K. (2016). SnoN antagonizes the hippo kinase complex to promote TAZ signaling during breast carcinogenesis. *Developmental Cell*, *37*(5), 399–412.

Zhu, Y., Wang, J., Meng, X., Xie, H., Tan, J., Guo, X., et al. (2017). A positive feedback loop promotes HIF-1α stability through miR-210-mediated suppression of RUNX3 in paraquat-induced EMT. *Journal of Cellular and Molecular Medicine*, *21*(12), 3529–3539.

Zilberberg, L., Todorovic, V., Dabovic, B., Horiguchi, M., Couroussé, T., Sakai, L. Y., et al. (2012). Specificity of latent TGF-β binding protein (LTBP) incorporation into matrix: Role of fibrillins and fibronectin. *Journal of Cellular Physiology*, *227*(12), 3828–3836.

CHAPTER SEVEN

# Isolation of DNA from exosomes

**Sheila Spada*, Nils-Petter Rudqvist, Erik Wennerberg**
Department of Radiation Oncology, Weill Cornell Medicine, New York, NY, United States
*Corresponding author: e-mail address: shs2066@med.cornell.edu

## Contents



## Abstract

Exosomes are small extracellular vesicles released by prokaryotic and eukaryotic cells with a crucial role in cell-to-cell communication in both physiological and pathological conditions. Exosomes contain and transfer active biomolecules, including nucleic acids, proteins and lipids to target recipient cells. In the last decade, many methodologies have been developed for isolating specific exosomal components. In this chapter, we will detail methods to isolate exosomal DNA, considering the crucial role of exosomal DNA in regulating the behavior of recipient cells in multiple settings, including the response of malignant cells to chemo-, radio- and immunotherapy.

## 1. Introduction

Over the past decade, the extracellular vesicle field has emerged as an exponentially growing and explosive research area (Roy, Hochberg, & Jones, 2018).

Exosomes are a subtype of extracellular vesicles of 50–100 nm secreted by all cell types both in eukaryotic and prokaryotic organisms in physiological

*Methods in Enzymology*, Volume 636
ISSN 0076-6879
https://doi.org/10.1016/bs.mie.2020.01.012

and pathological conditions (Thery et al., 2018). Exosomes, the best characterized extracellular vesicles, are released in culture media in in vitro cells culture or in in vivo different biological fluids (Thery, Amigorena, Raposo, & Clayton, 2006), including blood (Caby, Lankar, Vincendeau-Scherrer, Raposo, & Bonnerot, 2005; Theodoraki et al., 2019), lymph (Garcia-Silva et al., 2019), saliva (Iwai, Yamamoto, Yoshida, & Shiba, 2017), urine (Liu et al., 2018), milk (Zonneveld et al., 2014), amniotic liquid (Dixon et al., 2018), semen (Vojtech et al., 2014), cerebrospinal fluid (Tominaga et al., 2014), ascites (Andre et al., 2002), bronchoalveolar (Admyre et al., 2003) and gastrointestinal lavage fluids (van Niel et al., 2001). In addition to the size, exosomes are distinguished by the other subgroups of extracellular vesicles based on their biogenesis. In eukaryotic cells, exosomes come from endosome compartment. Invagination of late endosomal membranes results in intraluminal vesicle (ILVs) formation within the multivesicles body. Intraluminal vesicles are generated by two mechanisms: the endosomal sorting complex required for transport (ESCRT)-dependent or independent. The ESCRT complex includes cytosolic proteins that determine the curvature of the endosomal membrane to form ILVs. In absence of ESCRT, the tetraspanin CD63 and lipids, such as ceramide, with cone-shaped bending properties that may induce the formation of ILVs (Edgar, 2016). ILVs as exosomes are released into the extracellular space upon the fusion between the multivesicles body and the cellular plasmatic membrane (Sahu et al., 2011), a process mediated by the soluble *N*-ethylmaleimide-sensitive factor attachment protein receptors (SNAREs) (Fader, Sanchez, Mestre, & Colombo, 2009). During biogenesis process, exosomes uptake a plethora of functional molecules, including nucleic acid, such as DNA, RNA, microRNA and long-noncoding RNA, proteins, lipids and other metabolites (Jella et al., 2018). By carrying their specific cargo to targeted recipient cells, exosomes are crucial players in intercellular communication affecting the recipient cell behavior (Diamond et al., 2018) resulting in an important role in signal transduction (Galluzzi, Yamazaki, & Kroemer, 2018), angiogenesis, neuronal development and differentiation (Sharma et al., 2019), viral and bacterial infection (Ludwig et al., 2019; Rodrigues, Fan, Lyon, Wan, & Hu, 2018), immune response (Diamond et al., 2018; Kowal & Tkach, 2019; Thery, Duban, et al., 2002), cancer progression (Xu et al., 2018) and treatment response (Sansone et al., 2017).

In addition, in pathological conditions circulating exosomes in body fluids may represent a promising source of minimally invasive biomarkers for easily diagnosis, disease stage and treatment outcomes considering that they possess the biological fingerprints of the producing cells (Bracht, Mayo-de-Las-Casas, Berenguer, Karachaliou, & Rosell, 2018; Thery, Zitvogel, & Amigorena, 2002).

Taking advantage of the biological nature of exosomes, including their stability and their specificity of the delivery system, they are an ideal therapeutic drug or a promising nano-carrier for drug delivery *per se* or engineered (Maso, Grigoletto, Vicent, & Pasut, 2019; Zhang, Buller, & Chopp, 2019). For all these reasons, exosomes may represent an extremely interesting tool in the new era of nanomedicine alone or in combination with classical treatments as well as with immunotherapy for several human disorders.

In this chapter we will focus on the techniques for extraction of DNA, one of the exosomal cargo component, important in disease monitoring and exosome-based therapies (Kalluri & LeBleu, 2016; Kanuma et al., 2017). In this presented protocol exosomes derived from in vitro cell culture media isolated by ultracentrifugation have been used. However, the protocols for DNA extraction can totally be used with exosomes of different origin and using any isolation method. The following DNA extraction performed by different described methods provides high DNA quality compatible with any down-stream applications of molecular biology and high throughput analysis (Fig. 1).

## 2. Lab instruments, equipment and reagents

- T75 flasks for cell culture
- Beckman Optima XPN-100 ultracentrifuge
- Polypropylene centrifuge tubes or centrifuge polycarbonate bottle
- Heater
- Microcentrifuge tube
- Cell culture media
- Exosome-depleted fetal calf serum (SBI System Biosciences, EXO-FBS-250A-1)
- PBS 1 ×
- Table small centrifuge
- Weight scale
- Sucrose
- HEPES
- DNAse I
- Ethylene diamine tetraacetic acid (EDTA)
- Sodium dodecyl sulfate (SDS)

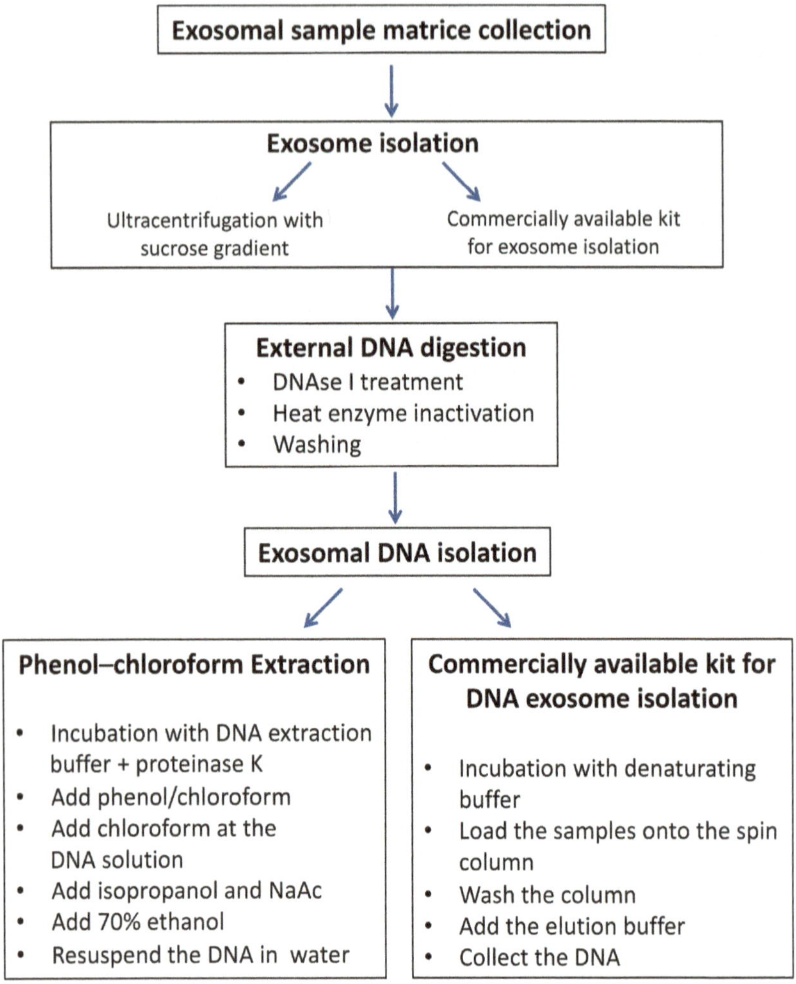


**Fig. 1** Schematic workflow of experimental procedures for DNA extraction from exosomes.

- Tris hydrochloride (Tris–HCl)
- Proteinase K
- Phenol/chloroform
- Isopropanol
- Chloroform
- Sodium acetate (NaAC)
- Ethanol
- Pyrocarbonate (DEPC) treated water

## 3. Exosome isolation

The classical method to isolate exosomes from a wide range of sample matrices is the ultracentrifugation. Recently, the increased attention on exosomes has required rapid and simple techniques to isolate exosomes. Many commercially available kits have been developed to rapidly and easily isolate exosomes and to increase to yield starting from small quantity of the matrix. The isolation of exosomes using kits is based on size exclusion chromatography, flow field-flow fractionation, the immunoaffinity capturing and the precipitation.

Choose the appropriate exosome isolation method considering the quantity and quality of the exosome source and the exosome downstream analysis.

Here, we describe the exosome isolation by ultracentrifugation (Diamond et al., 2018).

To isolate exosomes from in vitro cell culture media:

- Plate the cells in T75 flask. Calculate the number of cells to have 80–90% of confluence the day of the harvesting. Cells are maintained in standard culture conditions at 37 °C and 5% of $CO_2$ (see Note 1)
- The day after proceed with the treatment required by the experiment
- Before starting the collection of the exosomes, replace the classical media with media supplemented with exosomes-depleted fetal calf serum

*Important:* gently wash twice the cells with prewarmed PBS 1 × before adding the media with exosome-depleted serum to eliminate the media supplemented with regular serum (see Note 2)

- Culture the cells in media supplemented with exosome-depleted fetal calf serum for 48 h
- Collect the culture supernatants

(Optional: Count the live cells to estimate how many cells have secreted the exosomes being extracted)

- Spin down the supernatants for 20 min at 2000 × *g* to remove dead cells
- After the centrifugation, move the supernatants in the specific tubes used for the ultracentrifuge (see Note 3)
- To use the ultracentrifuge, identically balanced pair of tubes based on their weight have to be used. Use a scale to adjust the weight
- Once balanced, move the tubes into metal holders and load the rotor
- Ultracentrifuge at 10,000 × *g* at 4 °C for 30 min. The pellet at the bottom of the tubes is cellular debris

- Extract the supernatant which contains the exosomes and place it in new ultracentrifuge tube
- Balance the tubes and ultracentrifuge again at 100,000 × *g* at 4 °C for 70 min
- The pellet obtained is dirty exosome pellet formed by exosomes and contaminant soluble proteins (see Note 4)
- Take off the supernatants for all tubes leaving only 1 mL of supernatants. With a micropipette extract any remaining supernatant
- Add 1 mL of PBS 1 × to one tube and using the micropipette break up the pellet and suspend it in the 1 mL of PBS 1 ×
- Then use that 1 mL of PBS to resuspend and pull all the pellets (see Note 5)
- Add PBS 1 × to fill up the tube
- Wash the dirty exosomes resuspended in PBS at 100,000 × *g* at 4 °C for 70 min
- After the ultracentrifugation, remove the supernatant as much as you can (see Note 6). To increase the purity of the exosome preparation, please follow with the sucrose gradient
- Prepare the sucrose in 20 mM HEPES at different concentrations: 2, 1.65, 1.3, 0.95, 0.6, 0.25 M (Perez-Gonzalez et al., 2017)
- Add a layer of 1.5 mL sucrose concentration sequentially to a new ultracentrifuge tube, ensuring to pipette slowly to create a very distinct separation at each density beginning from the highest concentration (from 2 to 0.25 M) (see Note 7)
- For the exosome layer, add 750 μL of the 0.95 M sucrose containing the exosome at the gradient
- Add the left 750 μL of 0.95 M sucrose
- Finish the gradient loading the less concentrated sucrose fractions
- Balance the tubes
- Ultracentrifuge at 200,000 × *g* at 4 °C for 16 h
- After the ultracentrifugation, remove the first 750 μL as waste
- Collect the next 1.5 mL from each tube. The rest of the liquid in the tube is waste
- Add PBS to fill up the tube and using the 10 mL pipette mix the entire sample up and down
- Balance the tubes
- Ultracentrifuge at 100,000 × *g* at 4 °C for 70 min
- Remove the supernatant with the pipette then using the micropipette remove until only a little bit is left (see Note 8)

- Collect the exosomes and adjust the volume with PBS 1 ×
- Store the isolated exosomes at −80 °C until further analysis

To isolate exosomes using the commercially available kits, please follow the respective protocols provided by the manufacturer.

## 4. External DNA digestion

In order to eliminate contaminating DNA present on the external exosomal membrane, DNA digestion is recommended. Exosomes were treated using DNase I (0.15 units/μL) at 30 °C for 30 min and the reaction was heat inactivated at 70 °C for 5 min in the presence of EDTA 0.5 M in according to manufacturer's protocol. In parallel, equal amounts of untreated exosomes were used as negative control. After enzyme treatment, exosomes were washed with PBS 1 × and centrifuge at 100,000 × *g* for 70 min. After washing the exosomes were resuspended in PBS 1 × (Diamond et al., 2018).

## 5. Exosomal DNA isolation

### 5.1 Classical method for exosomal DNA isolation

Classical DNA isolation methods requires the phenol/chloroform extraction and the subsequent ethanol precipitation (Fischer et al., 2016; Sansone et al., 2017). Briefly:

- Prepare the DNA extraction buffer: SDS 0.5%, Tris–HCl 50 mM pH 8, EDTA 0.1 M (see Note 9)
- Prepare the exosomes in 25 μL of PBS 1 ×
- Incubate the exosome with 450 μL of DNA extraction buffer and 0.1 mg/mL of proteinase K overnight at 56 °C
- Add 500 μL of phenol/chloroform and centrifuge at 13,000 rpm for 5 min at room temperature
- Move the upper phase containing the DNA to a new tube
- Add 500 μL of chloroform and centrifuge at 13,000 rpm for 5 min at room temperature
- Repeat the previous step twice
- Transfer the upper phase to a new tube
- Add 450 μL of isopropanol and 50 μL of NaAc 3 M and centrifuge at 13,000 rpm for 10 min at 4 °C
- Wash the pellet with 750 μL of 70% ethanol and centrifuge at 13,000 rpm for 5 min at 4 °C (see Note 10)

- Dry the DNA pellet (see Note 11)
- Resuspend the pellet in 20 μL of DEPC treated water
- Incubate the DNA suspension at 37 °C for 30 min

## 5.2 Exosomal DNA isolation using commercially available kits

Many kits have been developed for DNA isolation from already isolated exosomes using any exosome isolation method, representing a promising alternative for rapid and efficient DNA isolation, especially if small quantity of exosomes is available from limited amount of initial sample matrice. The kits provide the method and all the components required for a rapid DNA extraction and they share the principle of the method. A general brief protocol is provided:

- Resuspend the exosomes in protein denaturating buffer (each manufacturer names this buffer differently)
- Load the samples onto the spin column and centrifuge (see Note 12)
- Wash the column using the wash buffer (see Note 13)
- Add elution buffer and incubate at room temperature for some minutes, depending on the kit used, and then centrifuge following the manufacturer's recommendations
- To increase the yield, collect the eluted sample and repeat the previous step

Measure the concentration of the DNA obtained using the Nanodrop or the Qubit with high sensitivity kit for DNA quantification.

# 6. Notes

1. When the cells are thawed, an overnight culture before plating is recommended
2. Accurate washes the plated cells with PBS 1 × before adding the media supplemented with exosomes-depleted fetal calf serum are required to avoid bovine exosome contamination
3. If you cannot perform the entire procedure, you can immediately freeze the supernatant at −80 °C and only when you are ready, thaw the tubes at room temperature and proceed with the exosome isolation
4. Manage the ultracentrifuge tubes very slowly without disturbing the exosome pellets
5. Gently pipet PBS 1 × for each exosome pellet wash to avoid losing of the samples

6. Before performing the sucrose gradient, remove any residual supernatant otherwise the sucrose gradient will be thrown off
7. Pipette slowly to make the sucrose gradient
8. Once finished, it is very hard to see a pellet
9. Check the pH of the DNA extraction buffer
10. Remove the ethanol as much as possible without disturb the DNA pellet
11. Do not dry the DNA pellet for very long time
12. Load the sample and the elution buffer onto the center of the column
13. Wash the column accurately

## 7. Conclusions

The rapid explosion of interest on exosomes required a rapid development of exosome-related techniques. Considering the role of the cargo components in the cell-cell communication, an increasing number of methods have been developed to efficiently screen the biomolecules encapsulated into the exosomes improving the quality and the validity of the results.

The methods presented here provide a reliable approach to easily isolate DNA from exosomes for many downstream applications.

A better characterization of exosomal DNA, mirror of the biological condition of the exosome-producing cells, will allow a deeper knowledge of this bioactive genetic material with potential hallmarks as candidate for biomarker in clinic and their transfer to recipient cells, such as dendritic cells, in cell-free therapy will be used to modulate the therapy response.

## Acknowledgments

S.S. is supported by an American-Italian Cancer Foundation Post-doctoral Research Fellowship (2018–2020). E.W. is supported by Breakthrough Fellowship Award (W81XWH-17-1-0029) from the US Department of the Defense (DoD).

## Conflict of interest statement

The authors declare no conflicts of interest.

## References

Admyre, C., Grunewald, J., Thyberg, J., Gripenback, S., Tornling, G., Eklund, A., et al. (2003). Exosomes with major histocompatibility complex class II and co-stimulatory molecules are present in human BAL fluid. *The European Respiratory Journal*, *22*(4), 578–583.

Andre, F., Schartz, N. E., Movassagh, M., Flament, C., Pautier, P., Morice, P., et al. (2002). Malignant effusions and immunogenic tumour-derived exosomes. *Lancet*, *360*(9329), 295–305.

Bracht, J. W. P., Mayo-de-Las-Casas, C., Berenguer, J., Karachaliou, N., & Rosell, R. (2018). The present and future of liquid biopsies in non-small cell lung cancer: Combining four biosources for diagnosis, prognosis, prediction, and disease monitoring. *Current Oncology Reports*, *20*(9), 70.

Caby, M. P., Lankar, D., Vincendeau-Scherrer, C., Raposo, G., & Bonnerot, C. (2005). Exosomal-like vesicles are present in human blood plasma. *International Immunology*, *17*(7), 879–887.

Diamond, J. M., Vanpouille-Box, C., Spada, S., Rudqvist, N. P., Chapman, J. R., Ueberheide, B. M., et al. (2018). Exosomes shuttle TREX1-sensitive IFN-stimulatory dsDNA from irradiated cancer cells to DCs. *Cancer Immunology Research*, *6*(8), 910–920.

Dixon, C. L., Sheller-Miller, S., Saade, G. R., Fortunato, S. J., Lai, A., Palma, C., et al. (2018). Amniotic fluid exosome proteomic profile exhibits unique pathways of term and preterm labor. *Endocrinology*, *159*(5), 2229–2240.

Edgar, J. R. (2016). Q&A: What are exosomes, exactly? *BMC Biology*, *14*, 46.

Fader, C. M., Sanchez, D. G., Mestre, M. B., & Colombo, M. I. (2009). TI-VAMP/VAMP7 and VAMP3/cellubrevin: Two v-SNARE proteins involved in specific steps of the autophagy/multivesicular body pathways. *Biochimica et Biophysica Acta*, *1793*(12), 1901–1916.

Fischer, S., Cornils, K., Speiseder, T., Badbaran, A., Reimer, R., Indenbirken, D., et al. (2016). Indication of horizontal DNA gene transfer by extracellular vesicles. *PLoS One*, *11*(9), e0163665.

Galluzzi, L., Yamazaki, T., & Kroemer, G. (2018). Linking cellular stress responses to systemic homeostasis. *Nature Reviews. Molecular Cell Biology*, *19*(11), 731–745.

Garcia-Silva, S., Benito-Martin, A., Sanchez-Redondo, S., Hernandez-Barranco, A., Ximenez-Embun, P., Nogues, L., et al. (2019). Correction: Use of extracellular vesicles from lymphatic drainage as surrogate markers of melanoma progression and BRAF (V600E) mutation. *The Journal of Experimental Medicine*, *216*(5), 1230.

Iwai, K., Yamamoto, S., Yoshida, M., & Shiba, K. (2017). Isolation of extracellular vesicles in saliva using density gradient ultracentrifugation. *Methods in Molecular Biology*, *1660*, 343–350.

Jella, K. K., Nasti, T. H., Li, Z., Malla, S. R., Buchwald, Z. S., & Khan, M. K. (2018). Exosomes, their biogenesis and role in inter-cellular communication, tumor microenvironment and cancer immunotherapy. *Vaccines (Basel)*, *6*(4), 69.

Kalluri, R., & LeBleu, V. S. (2016). Discovery of double-stranded genomic DNA in circulating exosomes. *Cold Spring Harbor Symposia on Quantitative Biology*, *81*, 275–280.

Kanuma, T., Yamamoto, T., Kobiyama, K., Moriishi, E., Masuta, Y., Kusakabe, T., et al. (2017). CD63-mediated antigen delivery into extracellular vesicles via DNA vaccination results in robust CD8(+) T cell responses. *Journal of Immunology*, *198*(12), 4707–4715.

Kowal, J., & Tkach, M. (2019). Dendritic cell extracellular vesicles. *International Review of Cell and Molecular Biology*, *349*, 213–249.

Liu, Z., Cauvi, D. M., Bernardino, E. M. A., Lara, B., Lizardo, R. E., Hawisher, D., et al. (2018). Isolation and characterization of human urine extracellular vesicles. *Cell Stress & Chaperones*, *23*(5), 943–953.

Ludwig, S., Marczak, L., Sharma, P., Abramowicz, A., Gawin, M., Widlak, P., et al. (2019). Proteomes of exosomes from HPV(+) or HPV(−) head and neck cancer cells: Differential enrichment in immunoregulatory proteins. *Oncoimmunology*, *8*(7), 1593808.

Maso, K., Grigoletto, A., Vicent, M. J., & Pasut, G. (2019). Molecular platforms for targeted drug delivery. *International Review of Cell and Molecular Biology*, *346*, 1–50.

Perez-Gonzalez, R., Gauthier, S. A., Kumar, A., Saito, M., Saito, M., & Levy, E. (2017). A method for isolation of extracellular vesicles and characterization of exosomes from brain extracellular space. *Methods in Molecular Biology*, *1545*, 139–151.

Rodrigues, M., Fan, J., Lyon, C., Wan, M., & Hu, Y. (2018). Role of extracellular vesicles in viral and bacterial infections: Pathogenesis, diagnostics, and therapeutics. *Theranostics*, *8*(10), 2709–2721.

Roy, S., Hochberg, F. H., & Jones, P. S. (2018). Extracellular vesicles: The growth as diagnostics and therapeutics; a survey. *Journal of Extracellular Vesicles*, 7(1), 1438720.

Sahu, R., Kaushik, S., Clement, C. C., Cannizzo, E. S., Scharf, B., Follenzi, A., et al. (2011). Microautophagy of cytosolic proteins by late endosomes. *Developmental Cell*, *20*(1), 131–139.

Sansone, P., Savini, C., Kurelac, I., Chang, Q., Amato, L. B., Strillacci, A., et al. (2017). Packaging and transfer of mitochondrial DNA via exosomes regulate escape from dormancy in hormonal therapy-resistant breast cancer. *Proceedings of the National Academy of Sciences of the United States of America*, *114*(43), E9066–E9075.

Sharma, P., Mesci, P., Carromeu, C., McClatchy, D. R., Schiapparelli, L., Yates, J. R., 3rd, et al. (2019). Exosomes regulate neurogenesis and circuit assembly. *Proceedings of the National Academy of Sciences of the United States of America*, *116*(32), 16086–16094.

Theodoraki, M. N., Yerneni, S., Gooding, W. E., Ohr, J., Clump, D. A., Bauman, J. E., et al. (2019). Circulating exosomes measure responses to therapy in head and neck cancer patients treated with cetuximab, ipilimumab, and IMRT. *Oncoimmunology*, *8*(7), 1593805.

Thery, C., Amigorena, S., Raposo, G., & Clayton, A. (2006). Isolation and characterization of exosomes from cell culture supernatants and biological fluids. *Current Protocols in Cell Biology*, *30*(1), 3.22.1–3.22.29.

Thery, C., Duban, L., Segura, E., Veron, P., Lantz, O., & Amigorena, S. (2002). Indirect activation of naive CD4+ T cells by dendritic cell-derived exosomes. *Nature Immunology*, *3*(12), 1156–1162.

Thery, C., Witwer, K. W., Aikawa, E., Alcaraz, M. J., Anderson, J. D., Andriantsitohaina, R., et al. (2018). Minimal information for studies of extracellular vesicles 2018 (MISEV2018): A position statement of the International society for extracellular vesicles and update of the MISEV2014 guidelines. *Journal of Extracellular Vesicles*, 7(1), 1535750.

Thery, C., Zitvogel, L., & Amigorena, S. (2002). Exosomes: Composition, biogenesis and function. *Nature Reviews. Immunology*, *2*(8), 569–579.

Tominaga, N., Hagiwara, K., Kosaka, N., Honma, K., Nakagama, H., & Ochiya, T. (2014). RPN2-mediated glycosylation of tetraspanin CD63 regulates breast cancer cell malignancy. *Molecular Cancer*, *13*, 134.

van Niel, G., Raposo, G., Candalh, C., Boussac, M., Hershberg, R., Cerf-Bensussan, N., et al. (2001). Intestinal epithelial cells secrete exosome-like vesicles. *Gastroenterology*, *121*(2), 337–349.

Vojtech, L., Woo, S., Hughes, S., Levy, C., Ballweber, L., Sauteraud, R. P., et al. (2014). Exosomes in human semen carry a distinctive repertoire of small non-coding RNAs with potential regulatory functions. *Nucleic Acids Research*, *42*(11), 7290–7304.

Xu, R., Rai, A., Chen, M., Suwakulsiri, W., Greening, D. W., & Simpson, R. J. (2018). Extracellular vesicles in cancer—Implications for future improvements in cancer care. *Nature Reviews. Clinical Oncology*, *15*(10), 617–638.

Zhang, Z. G., Buller, B., & Chopp, M. (2019). Exosomes—Beyond stem cells for restorative therapy in stroke and neurological injury. *Nature Reviews. Neurology*, *15*(4), 193–203.

Zonneveld, M. I., Brisson, A. R., van Herwijnen, M. J., Tan, S., van de Lest, C. H., Redegeld, F. A., et al. (2014). Recovery of extracellular vesicles from human breast milk is influenced by sample collection and vesicle isolation procedures. *Journal of Extracellular Vesicles*, *3*. https://doi.org/10.3402/jev.v3.24215. 25206958. PMCID: PMC4139932.

CHAPTER EIGHT

# Decoding cancer cell death-driven immune cell recruitment: An *in vivo* method for site-of-vaccination analyses

**Shentong Fang[a], Patrizia Agostinis[b,c], Petri Salven[d], Abhishek D. Garg[b,]***

[a]Wihuri Research Institute and Translational Cancer Medicine, University of Helsinki, Helsinki, Finland
[b]Department for Cellular and Molecular Medicine, Cell Death Research & Therapy (CDRT) Unit, KU Leuven, Leuven, Belgium
[c]Center for Cancer Biology (CCB), VIB, Leuven, Belgium
[d]Department of Pathology, Haartman Institute, University of Helsinki, Helsinki, Finland
*Corresponding author: e-mail address: abhishek.garg@kuleuven.be

## Contents



## Abstract

Anticancer vaccines have recently received renewed attention for immunotherapy of at least a subset of cancer-types. Such vaccines mostly involve either killed cancer or tumor cells alone, or combinations thereof with specific (co-incubated) innate immune cells. In recent years, the immunogenic characteristics of the dead or dying cancer cells have emerged as decisive factors behind the success of anticancer vaccines. This has amplified the importance of accounting for immunology of cell death while preparing anticancer vaccines. This, in turn, has increased the emphasis on the immune reactions at

*Methods in Enzymology*, Volume 636
ISSN 0076-6879
https://doi.org/10.1016/bs.mie.2019.04.013

the site-of-vaccination since the therapeutic efficacy of the killed cancer/tumor cell vaccines is contingent upon the nature and characteristics of these reactions at the site-of-injection. In this article, we present a systematic methodology that exploits the murine ear pinna model to study differential immune cell recruitment by dead/dying cancer cells injected *in vivo*, thereby modeling the site-of-injection relevant for anticancer vaccines.

## 1. Introduction

Immunotherapy has irreversibly modified the current landscape of cancer therapy as well as cancer patient management (Chen & Mellman, 2017; Colli et al., 2017; Sharma, Wagner, Wolchok, & Allison, 2011). The speed and consistency with which cancer immunotherapy integrated into treatment regimens for different types of cancers, over the last few years, has been staggering. In exceptional cases, specific immunotherapies have gained regulatory approval as first-line interventions replacing previous standard-of-care (SOC) regimens (Iacovelli et al., 2018; Rivalland, Scott, & John, 2017). In majority of other cases, application of immunotherapies is either stand-alone at second/third-line intervention levels, or in combinatorial setting (concurrently or sequentially) with existing chemo/radio/targeted SOC therapies (Khalil, Smith, Brentjens, & Wolchok, 2016; Mahoney, Rennert, & Freeman, 2015). The type of cancer and the pathological or molecular characteristics of the tumor are two of many important factors driving the application strategy for immunotherapy (Chen & Mellman, 2017).

Owing to these trends, there is growing motivation to upgrade some existing immunotherapeutic modalities that have previously performed sub-optimally in the clinic against cancer, but still have promising properties that may fulfil specific oncoimmunology challenges (Garg, Vandenberk, Van Woensel, et al, 2017). One such modality is, anticancer vaccines. Anticancer vaccines have received renewed attention as putative combinatorial partners for SOC therapies in at least a subset of patients, suffering from specific cancer-types like glioblastoma, lung cancer and melanoma (among others) (Anguille, Smits, Lion, van Tendeloo, & Berneman, 2014; Garg, Coulie, Van den Eynde, & Agostinis, 2017; Garg, Vara Perez, Schaaf, et al., 2017; Romero et al., 2016). Anticancer vaccines are typically prepared ex vivo with patient-derived (cancer/immune) cells and injected thereafter over a specified period of time into the same patient (Bol, Schreibelt, Gerritsen, de Vries, & Figdor, 2016). These vaccines in general act by stimulating the adaptive immune system via innate immune cells, with either specific or a collection of tumor-associated antigens (TAAs), to nurture the formation of TAAs-specific cytotoxic T lymphocytes (CTLs) that can identify and eradicate malignant cells expressing those TAAs (Sabado, Balan, & Bhardwaj, 2017). Since the early days of anticancer vaccines, it has been a matter of debate, whether it is better to apply vaccines based on purified or recombinant TAAs, or to use whole cancer/tumor cells as an unbiased source of TAAs (Romero et al., 2016). This debate has further intensified recently due to the emergence of neoantigens-based strategies (Bobisse, Foukas, Coukos, & Harari, 2016; Verdegaal et al., 2016). Nevertheless, it is clear that in cases involving cancer-types with lack of unique or broadly applicable TAAs (e.g., glioblastoma), it might be better to operate with whole cancer/tumor cells as a TAAs source (Garg, Coulie, et al., 2017).

In principle, such vaccines either involve killed cancer/tumor cells alone (often termed, "tumor lysate"), or could combine the co-incubation of these with specific innate immune cells like dendritic cells (DCs) ex vivo (termed as, DC vaccines), before injection (Garg, Coulie, et al., 2017). In past, the process of killing the cancer cells in the context of anticancer vaccines was solely a means to derive their TAAs while simultaneously incapacitating the cancer cells—latter being an important requirement for good manufacturing/laboratory practices (GMPs or GLPs) (Garg, Coulie, et al., 2017; Sabado et al., 2017). Due to this, less attention was devoted to the type of cancer cell death, thereby often favoring freeze/thaw (F/T)-based necrosis, mainly due to the practical and regulatory ease of handling it. However, in recent years, the immunogenic characteristics of the dead/dying cancer cells and the specific cell death pathways have emerged as decisive factors behind the success of anticancer vaccines (Garg & Agostinis, 2017). Accidental necrosis, such as the one induced by F/T, has been widely established to be poorly immunogenic for most cancer-types (Galluzzi & Kroemer, 2017). In contrast, specific instances of regulated cell death (RCD), either necroptotic or apoptotic, can be appreciably immunogenic (Aaes et al., 2016; Galluzzi et al., 2017) especially (but not only) when they exhibit the characteristics of immunogenic cell death (ICD, for detailed description, please see: (Bezu et al., 2015; Galluzzi, Buqué, Kepp, Zitvogel, & Kroemer, 2017; Garg, Galluzzi, et al., 2015; Garg, More,

Rufo, et al., 2017). In fact, apoptotic ICD-based "next-generation" anticancer vaccines are particularly potent against notoriously immunosuppressive cancer-types like glioblastoma (Garg et al., 2016) or even malignancies of immune cells themselves (e.g., B-cell lymphoma) (Zappasodi et al., 2010).

The emergence of this new knowledge has increased the emphasis on accounting for the immunological profiles of cancer cell death within the anticancer vaccine preparations (Anguille et al., 2014). Moreover, these trends have also allowed some researchers to conceive ICD-based cancer cells-only vaccines, which are producible more rapidly, and in a more cost-effective manner than DC vaccines (Cicchelero, de Rooster, & Sanders, 2014; De Gruijl, van den Eertwegh, Pinedo, & Scheper, 2008). This, in turn, has increased the emphasis on the immune reactions at the site-of-vaccination (Ohlfest et al., 2013). In the DC-based vaccination setup, immune reactions at the site-of-vaccination are of less relative importance (unless they are detrimental to the injected DCs) since upon injection surviving DCs tend to rapidly leave these sites for the nearest lymph nodes to prime the T cells (Aarntzen et al., 2013; Adema, de Vries, Punt, & Figdor, 2005; Ciabattini et al., 2011). However, the efficacy of the killed cancer/tumor cell vaccines is contingent upon the nature and characteristics of the immune reactions at the site-of-injection/vaccination (Mould et al., 2017; Zhou, Johnson, & Orentas, 2007). Emerging research advocates that the local immunological reactions at the site-of-injection of dead/dying cancer cells have a key effect on the ultimate immune cell priming and vaccination efficacy (Ferguson, Choi, & Green, 2011; Ohlfest et al., 2013; Yoon, 2017). This is because, upon injection of dying/dead cancer cells, in parallel with local phagocytic action by the tissue-resident immune cells, a systematic signaling cascade tends to recruit specific immune cells from the blood circulation in a spatiotemporally-defined order (Blander, 2017; Fond & Ravichandran, 2016; Krysko & Vandenabeele, 2008). These processes tend to influence as well as regulate the antigen procurement, processing and presentation by professional antigen-presenting cells (APCs)—the proper execution of which is a major pre-requisite for the success of anticancer vaccines (Gálvez-Cancino et al., 2018; Goyvaerts & Breckpot, 2015).

Unfortunately, there is an extreme dearth of preclinical studies focusing on the immune reactions elicited by dying/dead cancer cells or anticancer vaccines at the site-of-injection (Ohlfest et al., 2013). This further complicates the efforts of correlating vaccination response rates with immunogenic characteristics of the vaccines (Ohlfest et al., 2013). To this end, in order to study better the immune reactions at the site-of-injection of dead/dying cancer cells, we had recently repurposed the murine ear pinna model (Garg, Vandenberk, Fang, et al., 2017). We were successfully able to use this model to study differential immune cell recruitment as well as differential immune cell activation at the site-of-injection, in response to cancer cells dying via disparate cell death pathways (Garg, Vandenberk, Fang, et al., 2017). In fact, this model allowed us to delineate neutrophils as the first innate immune responders (in terms of recruitment from circulation) to cancer cells undergoing ICD, but not F/T-based necrosis or tunicamycin-based tolerogenic apoptosis (Garg, Vandenberk, Fang, et al., 2017) thereby placing these cells at the core of ICD-relevant immune reactions at the site-of-injection. In this article, we provide a systematic methodology to use the murine ear pinna model to study differential immune cell recruitment by dead/dying cancer cells injected *in vivo*, thereby modeling the site-of-injection relevant for anticancer vaccination studies. We also compare this model, with other models used by independent investigators in recent past to study differential immune recruitment by dead/dying cancer cells injected *in vivo*. Of note, intra-dermal inoculation in murine ear pinna is one of the classical immunology models in the field of virology (Lin, Smith, & Tscharke, 2012). It's main use has been for modeling the site of viral infections, especially adenovirus, vaccinia virus and herpes simplex virus, since it is easier to monitor virus-induced lesions on the ear surface (Lin et al., 2012). The adult murine ear pinna is also very appropriate for such analyses since it is highly vascularized, which allows for rapid mobilization of immune cells (García-Caballero et al., 2017). In this sense, our re-purposing of this model for studying immune cell recruitment by dead/dying cancer cells-based vaccines, fits very well with its historical use for immunology analyses.

## 2. Inducing different cancer cell death pathways

The current murine ear pinna inoculation protocol constitutes three main phases (Fig. 1), wherein the first phase involves induction of specific cell death pathways in suitable murine cancer cells. Thereafter, the second phase involves injection of these in the ear pinna of suitable syngeneic murine host and the third phase involves ear pinna processing and analyses at specific recovery time-points (Fig. 1).

### 2.1 Suitable murine cancer cells

For the first phase (Fig. 1), a number of different murine cancer cell lines are utilizable depending upon the preferred cell death stimuli and the specific

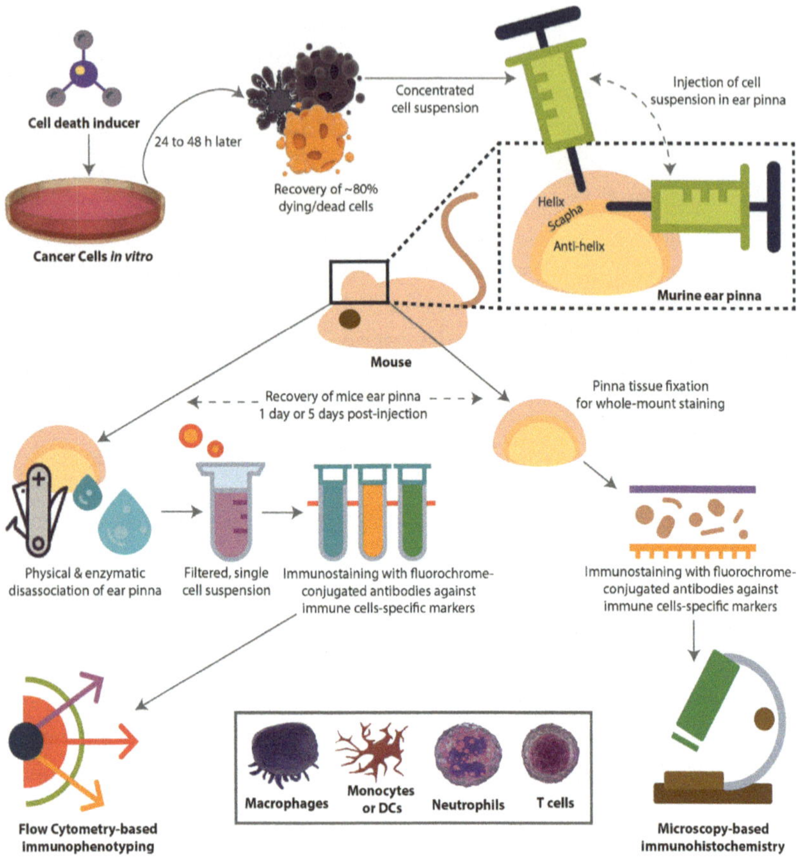


**Fig. 1** A schematic representation of the murine ear pinna model repurposed for studying immune cell infiltrates attracted by dying/dead cancer cells at the site-of-injection.

cell death pathway (Galluzzi et al., 2018; Kepp et al., 2014). This is because, different cancer cells can exhibit different susceptibilities to the cell death inducers of choice, such that in extreme cases, some cells can be completely resistant to a given cell death inducer or its immunogenic effects (Garg, Elsen, et al., 2015; Romano et al., 2018). Moreover, while most cancer cells tend to undergo apoptosis, yet susceptibility to other cell death pathways like necroptosis or ferroptosis can be highly cell line-dependent (due to genetic variability or lack of expression of crucial pro-death proteins) (Dillon & Green, 2016; Fearnhead, Vandenabeele, & Vanden Berghe, 2017). Hence, initial screening analyses for selecting a suitable murine cancer cell line is necessary; further complemented by a proper cell death analyses (Galluzzi et al., 2009). We validated the current model through the widely used,

Lewis lung carcinoma (LLC) cells and the B16 melanoma cells, both syngeneic for the C57BL/6 mice background (Garg, Vandenberk, Fang, et al., 2017). In principle, this protocol is adaptable to most cancer cell lines or cell death stimuli of choice. However, cell lines that have tendency to form "sticky" clusters or clumps when in suspension can pose a formidable challenge for this protocol considering the small injection volume and a relatively small injection area with limited tissue elasticity (i.e., ear pinna). Hence, it is advisable to prefer cell lines that are injectable as per the parameters relevant for ear pinna inoculations (see below). Upon delineation of a suitable cancer cell line, one can proceed with the induction of different cell death subroutines (see below).

## 2.2 Cell death induction

1. Cancer cells should be cultured in suitable cell culture dishes, preferably at least 24 h before commencement of treatment with the suitable cell death stimuli. The 100 mm × 20 mm (round) polystyrene cell culture dishes suitable for cell culture should be fine for most settings. The cells should be plated such that, the cell density should not be more than ∼75% on the day of cell death treatment. Be mindful to disrupt any cell clusters or clumps while processing the cells in order to have a homogenously single cell suspension. The choice of suitable cell culture medium and number of cells to plate in order to reach the correct confluence is cell line-dependent; hence should be tested and optimized beforehand.
2. On the day of treatment (24–48 h after the above plating step, depending upon the cells' growth rate), start with preparing the proper dilutions of the cell death inducing agents of choice, in the suitable cell culture medium. There are large varieties of cell death inducers available that are capable of inducing different cell death pathways depending upon the cancer cell line. For instance, tolerogenic apoptosis is reliably achievable via tunicamycin, thapsigargin, cisplatin or brefeldin A (Martins et al., 2011; Obeid et al., 2007). Similarly, immunogenic apoptosis or apoptotic ICD is readily inducible via mitoxantrone, doxorubicin, oxaliplatin or hypericin-based photodynamic therapy (PDT) (Garg, Galluzzi, et al., 2015). Likewise, well-established necroptosis-inducing cocktails based on tumor necrosis factor (TNF) or TNF-related apoptosis-inducing ligand (TRAIL); or ferroptosis-inducers like erastin or RSL3, are utilizable in a cancer cell-line dependent manner (Galluzzi et al., 2018; Ting, 2018). In our experience, while inducers of

tolerogenic apoptosis tend to work consistently across various cell lines yet necroptotic/apoptotic/ICD inducers may be prone to failure depending upon the cell line (Romano et al., 2018). Therefore, the optimization for these distinct cell death profiles should be performed in advance, using already established protocols and guidelines (Galluzzi et al., 2009; Kepp et al., 2014; Ting, 2018). Of note, the correct concentrations or conditions to use for each cell death stimulus should be optimized in such a manner that after 24 or 48 h of treatment, there should be no more than 70% ± 10% of total dying/dead cells. Please refer to existing protocols and guidelines for cell death analyses (Galluzzi et al., 2009; Kepp et al., 2014; Klionsky et al., 2016; Ting, 2018).

3. Working dilutions of the cell death agent(s) of choice should be prepared in advance, in fresh culture medium. Subsequently, the existing culture medium on the cells should be aspirated and discarded. Thereafter, the above medium with the diluted cell death agent(s) should be introduced onto the cells. Of note, this step is only applicable to adherent cell cultures; for cells, growing in suspension an interim centrifugation step would be required.
4. Let these treatments continue for 24 or 48 h as desired, in order to reach 70% ± 10% of total dying/dead cells. As soon as this cell death proportion is reached scrap off all the cells with a cell scraper, along with the media (in order to also recover the floating cells). Collect all these cells with the media and centrifuge in a suitable tube at 1600 rpm for 5 min. Post-centrifugation, discard the media and re-suspend the cells in 2–5 mL phosphate buffered saline (PBS).
5. In case you would like to have a condition with primary necrosis, then for a relevant cell pellet after the above centrifugation step, discard the media, add 3 mL of PBS and treat the cell pellet with at least three cycles of freezing (−50 to −70 °C) and thawing (37 °C). Normally three cycles should be sufficient to achieve 70% ± 10% of total dead cells. If not, then please adjust the number of F/T cycles accordingly.
6. Subsequently, count the number of cells via regular Trypan Blue staining procedure combined with hemocytometer-based counting. Please count both live as well as dead cells.
7. Following counting, centrifuge and pellet the cells again at 1600 rpm for 5 min.
8. Resuspend each pellet in PBS at a "density" of approximately 3 million cells/200 μL. The "vaccine" or cellular suspension meant for injection is now ready.

# 3. Murine ear pinna model

## 3.1 Mice, materials, equipment and reagents

- Adult syngeneic mice (8–12 weeks old)
- Cellular suspension of interest for injection (see Section 2.2)
- Insulin syringe
- 70% ethanol (EtOH) solution
- Suitable hair removal cream (e.g., Veet)
- PBS
- 70 μm nylon mesh filters
- Dissection microscope
- Flow cytometer (e.g., BD FACSCanto system)
- Small-animal butterfly catheter infusion set (BD Vacutainer®)
- 10 mL Syringe
- Small forceps, 0.2 mm
- Dissecting forceps
- Scissors
- Collagenase H (Roche)
- Dispase II (Roche)
- 1 × Hank's balanced salt solution (HBSS) (Invitrogen)
- Ethylenediaminetetraacetic acid (EDTA) (Sigma)
- 1 × HBSS without $Ca^{2+}$ and $Mg^{2+}$ (Invitrogen)
- 10 × Red blood cells (RBC) lysis buffer (BioLegend)
- Heat-inactivate fetal bovine serum (FBS) (Promega)
- Fc block antibody, e.g., TruStain FcX (anti-mouse CD16/32) antibody (BioLegend)
- 5 mL polystyrene round-bottom tubes (BD)
- 15 mL falcon tubes
- 1.5 mL Eppendorf tubes
- Sodium azide
- Paraformaldehyde (PFA)
- Bovine serum albumin (BSA)
- Normal donkey serum
- Triton-X 100
- 10-cm culture dishes
- 10-cm silicon culture dish plates
- Insect pins
- Coverslips

- Aluminum foil
- Orbital shaker
- Parafilm

## 3.2 Reagent setup

*Digestion solution*: Freshly made HBSS containing 1 mg/mL Collagenase H and 0.8 U/mL Dispase II

*Digestion "Stop" Solution*: 2% heat-inactivated FBS in 1 × HBSS containing 2 mM EDTA (Store at 4 °C)

*RBC lysis buffer solution*: 1 × working concentration in deionised water (prepared from 10 × stock)

*Staining buffer*: 2% heat-inactivated FBS in 1 × HBSS without $Ca^{2+}$ and $Mg^{2+}$ (Store at 4 °C)

*Blocking buffer*: 0.3% Triton-X 100, 0.2% BSA, 0.05% NaN3, 5% normal donkey serum in PBS

*Washing buffer*: 0.3% Triton-X 100 in PBS

## 3.3 Ear pinna injections and recovery

Anesthetize the animals as approved in the animal (ethical) experimentation protocol. Thereafter, inject a total of approximately 3 million cells/200 μL PBS, slowly between the skin and cartilage on the dorsal side of the ear. In principle, one can utilize both ear pinna of the same mice for injections.

*Critical step*: Injection site is preferably in the center of the ear and not too close to the ear base. Inject slowly and avoid big vessels. Target, as much as possible, the helix and/or scapha areas of the ear pinna while avoiding the anti-helix area (Fig. 1).

During the injection, if it is successful, you will observe an inflation of the intra-dermal space. After 1–5 days, you would be able to see discolorations at the site-of-injection.

## 3.4 Ear pinna processing for flow cytometry-based immunophenotyping

The following steps should be carried out at a suitable recovery time-point. In our experience, the earliest recovery time-point for the ear pinna should be no sooner than 1 day post-injection and the final time-point should be no later than 5–6 days post-injection (in order to stably as well as reproducibly capture the respective immune cell infiltrates).

1. Prepare the digestion solution for the tissue.

2. Carryout deep anaesthesia of the animals as approved in the animal ethics experimentation protocol. Wait until there is no response to the foot-pinch test.
3. Next, carryout the *trans-cardiac perfusion surgical procedure* (to replace systemic blood with PBS) in order to ensure the (almost complete) removal of interference from blood-associated circulatory/peripheral immune cells. Briefly, spray the mouse with 70% EtOH and make a lateral incision in the skin just below the rib cage. Make a small incision in the diaphragm and cut in the dorsal direction of the (entire length of the) ribs, on both lateral and contra-lateral side to expose the pleural cavity. Lift up the sternum to have a clear vision of the heart. Insert the butterfly needle, through a small incision, into the (posterior end of the) left ventricle (Fig. 2). Next, make an incision in the right atrium with scissors (Fig. 2). The heart is ready for perfusion.

   *Important*: Trans-cardiac perfusion is a standard surgical procedure that is widely utilized for various research purposes, however, it does require prior training. So it is recommended to consult a trained

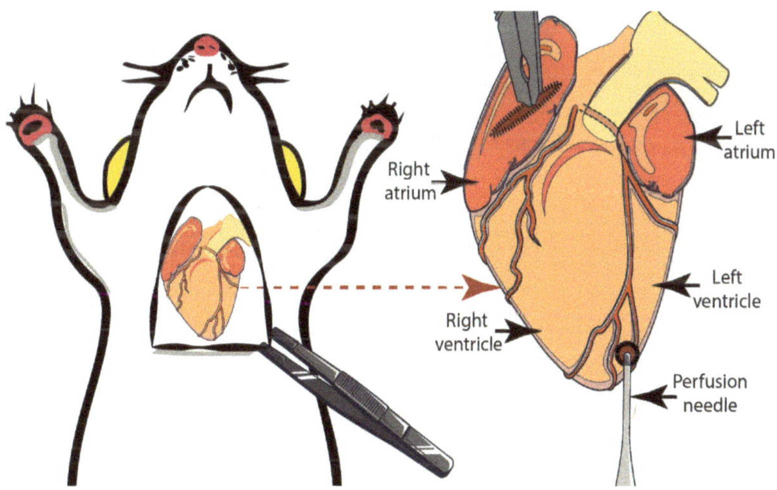


**Fig. 2** A schematic representation of trans-cardiac surgical perfusion in mouse. Under deep anaesthesia, carefully make incisions on the skin close to diaphragm, and around the rib cage, with the final aim of exposing the pleural cavity in order to gain a clear view of the heart. Insert the butterfly perfusion needle through a very small incision in the posterior end of the left ventricle (directed toward the ascending aorta). Subsequently, make a relatively long incision in the right atrium without damaging the descending aorta. With this, the heart is ready for (slow) perfusion with PBS (either via a perfusion apparatus or 20 mL syringe).

personnel for support. Nevertheless we would also like to refer to some publications that describe this procedure in great details, accompanied by an illustrative video-based demonstration (Gage, Kipke, & Shain, 2012; Tremblay, Riad, & Majewska, 2010).

4. Start pumping 1 × PBS, slowly and carefully, through the perfusion apparatus or a 20 mL syringe. Allow 1 × PBS perfusion to continue for around 2 min (approximately 20 mL total volume of PBS). An almost immediate sign of a successful perfusion is a change in the color, toward pale, of the liver and extremities.

   *Note*: Instead of perfusion apparatus, perfusion with 10 mL syringe is also a viable option.

5. Dissect the ear from mice and remove the hair, skin folds at the pinna base and unnecessary surrounding tissues followed by manual mincing (with scalpel and scissors) of the pinna tissue in a 10 mm plate

   *Important*: When you remove unnecessary surrounding tissues, it is important to keep at least 1–2 mm distance from your injection area.

6. Add digestion solution to the 10 mm plate and transfer the solution plus minced tissue to a 15-mL falcon tube using 5 mL pipette, which should be then transferred to a 37 °C incubator for 30 min. Please re-suspend the suspension every 5–10 min during this incubation period.

7. Add digestion "stop" medium to the falcon tube, pipette up and down and filter the content through a 70-μm nylon mesh filter.

8. Centrifuge the filtrate at 400 × *g* at 4 °C for 5 min and discard the supernatant.

9. Re-suspend the pellet in 1 × RBC lysis solution to remove any residual RBCs. Incubate on ice for 4–5 min with intermittent shaking. The reaction can be stopped by diluting the solution with 20–30 mL of 1 × PBS. Thereafter centrifuge the cells (400 × *g*) and discard the supernatant.

10. Re-suspend the resulting pellet in 100 μL of staining buffer and incubate on ice with murine Fc block antibody for 5 min. These cell suspensions can now be transferred to 1.5-mL microcentrifuge tubes, especially if multiple samples are to be handled.

11. For antibody-based immunostaining, prepare 10 μL of staining buffer containing the appropriate dilution of immune cells-labeling (fluorochrome-conjugated) primary antibodies, intended for immunophenotyping. Add and mix well in the cell suspensions.

    *Important*: Depending upon the immune cells (or subsets thereof) that one needs to detect/quantify, a suitable panel of antibodies for immunophenotyping should be designed. Since the choice of immune cells to be detected and suitability of fluorochromes varies from project-to-project or user-to-user, we cannot recommend a panel of antibodies for prioritization. However, we refer the readers to the recent guidelines for flow cytometry-based cell sorting in immunological studies that can help in this regard (Cossarizza et al., 2017).

*Critical step*: Remember to set up control samples with no primary antibodies, every antibody alone and fluorescence minus one (FMOs) controls.

12. Incubate the sample in dark on ice for 30 min.
13. Add 1 mL of staining buffer per tube and centrifuge the cell suspension for 5 min at 400 × *g* (at 4 °C).
14. Discard the supernatant and re-suspend the pellets in 0.2–0.3 mL of staining buffer.
15. Transfer the cell suspension to 5-mL polystyrene round-bottom tubes with mesh cap, and score via a suitable flow cytometry machine.

    *Optional:* Sample can be fixed in 1% PFA, washed with PBS and stored at 4 °C for later analysis.

## 3.5 Ear pinna processing for immunohistochemistry-based analyses

1. Anesthetize the animals as approved in the animal ethics experimentation protocol. Wait until there is no response to the foot-pinch test.
2. Transfer the mouse to a fume hood. Spray the mouse with 70% EtOH and make an incision in the skin to expose the thorax. Cut in a dorsal direction of the ribs to have a clear vision of the heart. Insert the butterfly needle into the left ventricle and start pumping 1 × PBS through a perfusion apparatus. Cut the right atrium with scissors (i.e., perform the trans-cardiac surgical perfusion procedure, as detailed earlier and in Fig. 2).
3. Allow ice-cold 1 × PBS perfusion to continue for around 2 min (approximately a total of 20 mL PBS) and then continue with ice cold 1% PFA for around 2–3 min (approximately a total of 20 mL PBS).
4. Dissect the ear from mice on a culture dish plate. Clean the hair from the ear with Veet cream using a cotton swab. Wash away the Veet cream with PBS with a cotton swab. Remove the tissue folds in the base of the pinna.
5. Split the ear into two coronal halves with forceps. Collect both dorsal and ventral coronal sides. Pin to a silicone plate with the "inside" of the

pinna facing-up before further cleaning. Thereafter re-fix (1% PFA), wash and immunostain using standard whole mount immunostaining method (Fang, Pentinmikko, Ilmonen, & Salven, 2012).

*Critical step*: Carefully clean the cartilage away from the tissue.

6. Refix the ear on the silicone plate with 4% PFA at room temperature for 1 h on an orbital shaker.
7. Wash with PBS for 5 min. Repeat two more times.

   *Optional:* Sample can be stored in PBS overnight. For long time storage before the staining, samples can be submerged in PBS $+NaN_3$ and stored at 4 °C.
8. Permeabilize tissue with 0.3% PBS-Triton-X for 15 min.
9. Add blocking buffer to the samples for 1 h at room temperature on an orbital shaker (1 mL per well for a 12-well plate or 3 mL for a 10 cm dish).
10. Antibody staining: Prepare blocking buffer containing the appropriate dilution of intended primary antibodies.
11. Remove the blocking buffer from the sample and add antibody mix to the sample silicone plate. Incubate the samples on an orbital shaker for 2 days at 4 °C. Use the parafilm to seal the silicone plate with the lid.
12. Remove the primary antibody mix and add washing buffer. Wash the samples on an orbital shaker at room temperature for 0.5–1 h. Repeat this step four more times.
13. Prepare the appropriate secondary antibody mix (if applicable) in blocking buffer. Replace washing buffer with secondary antibody mix. Incubate the samples on an orbital shaker overnight at 4 °C.
14. Remove the secondary antibody mix and add washing buffer. Wash the samples on an orbital shaker at room temperature for 0.5–1 h. Repeat this step at least five more times.

    *Optional:* Sample can be left in washing buffer overnight at 4 °C, at this juncture.
15. Remove the washing buffer and add 4% PFA for 10 min on an orbital shaker to post-fix the samples.
16. Remove 4% PFA and add PBS. Wash the samples on an orbital shaker three times for 30 min each.
17. Label microscopy slides with relevant sample IDs and antibodies used in the staining procedure.
18. Transfer the plate to the dissection microscope. Use small forceps to move the pins away. Transfer the ear to the slides and make it flat under the microscope (Cut and clean the unnecessary tissues away if you still need to, use a fluorescent dissection microscope, if possible).

19. Use a tissue paper to absorb excess PBS by placing it near the ear sample. Add three drops of mounting medium to the ear sample. Tilt the slide to help in the spreading of the mounting medium and add a coverslip on the top.
20. Remove excess mounting medium and seal the coverslip with Cytoseal. Keep the samples in dark.

## 4. Data analysis

We have successfully used this ear pinna model for flow cytometry-based immunophenotyping of immune cell recruitment in response to dying/dead cancer cells (Garg, Vandenberk, Fang, et al., 2017; Fig. 3). This ear pinna model can be used for both flow cytometry-based immunophenotyping and immunohistochemistry-based analyses on different ear pinna derived from the same mouse. Proper tissue preparation, clearing, immunostaining and mounting can allow full visualization of both lymphatic and blood vessels as well as immune cells during the recruitment process in response to dying/dead cancer cells.

For flow cytometry-based immunophenotyping of infiltrating immune cells (Garg, Vandenberk, Fang, et al., 2017), the first step in flow cytometry analysis should be to exclude debris based on the cell size and to only include singlet hits in further analyses (based on Forward scatter/FSC-area/width or

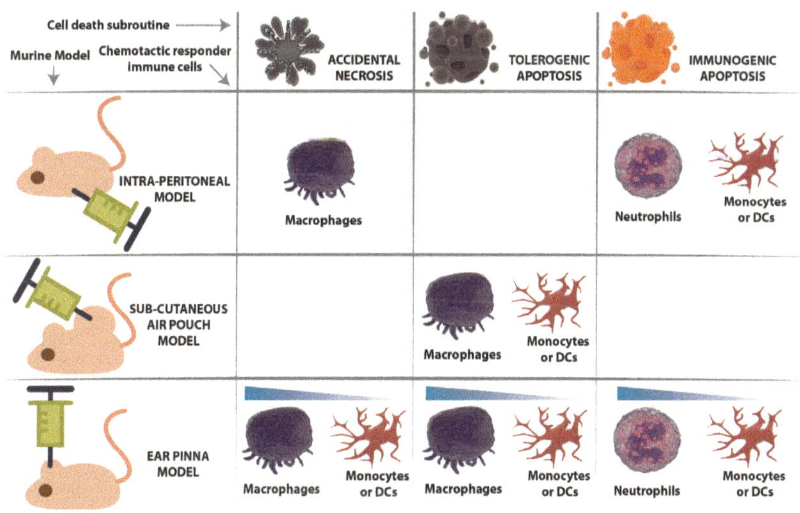


**Fig. 3** A broad comparison of immune cell attraction patterns in response to dying/dead cells exhibiting different immunological profiles, reported across three different murine models.

FSC-area/height), followed by live/dead cell gating, based on staining with an exclusion dye, to exclude dead cells (especially if samples are not fixed). After selecting all live singlet hits, one can select the white blood cells (WBCs) populations based on CD45 expression. Within the WBC population, one can use CD3 and B220/CD20 staining to broadly score T cells and B cells. On the other hand, for the myeloid cells within the WBC populations, F4/80, Ly6G/Ly6C, CD11b, NK1.1 and CD11c can be used in different combinations to score for neutrophils ($CD11b^{+}Ly6G^{+}$), macrophages ($CD11b^{+}F4/80^{+}$), monocytes ($CD11b^{+}$ $F4/80^{-}CD11c^{-}$), NK cells ($NK1.1^{+}$), dendritic cells ($CD11c^{+}$) or myeloid DCs ($CD11c^{+}Ly6C^{+}$) (Garg, Vandenberk, Fang, et al., 2017).

For immunohistochemistry-based immunophenotyping, one ought to be focussing on the interaction between infiltrating immune cells and the blood vasculature (in order to further compliment the above flow cytometry analyses). Herein we advise to start with PECAM1/CD45 staining. PECAM1 can mark both blood vascular endothelial cells and lymphatic endothelial cells in the ear, which can be distinguished based on their morphology (more specific markers can also be involved if multi-channel fluorescence microscopy is feasible). Infiltrating immune cells can be generally observed around the pinna stromal area "loosened" by the injection of dying/dead cancer cells. For specific immune cell-types, one should first test the relevant antibodies.

## 5. Ear pinna versus air-pouch and intra-peritoneal murine models

To the best of our knowledge, our recent study was the first one to repurpose the murine ear pinna model for studying differential immune cell recruitment by cancer cells undergoing different cell death subroutines (Garg, Vandenberk, Fang, et al., 2017). Thus, it is reasonable to expect that some researchers wanting to exploit our model would want to first understand whether results obtained herein are unique (i.e., applicable to only ear pinna or intra-dermal milieu), or do they reconcile with results obtained in other murine models exploiting much different locales.

In recent past, at least two different murine models have been repurposed to study dying cells-driven immune cell recruitment, i.e., murine subcutaneous (*s.c.*) air-pouch model (Duarte, Vasko, & Fehrenbacher, 2012; Gaspar, Sakai, & Gaspari, 2014; Saribas, Ergun, Mamuk, Köseoglu-Eser, & Melli, 2012) and murine intra-peritoneal (*i.p.*) model (Wu et al., 2013;

Yuan et al., 2017). The air-pouch model requires artificial creation of a sterile air-filled "sac" by *s.c.* injecting a particular amount of sterile air on the back of the mice (Duarte et al., 2012; Gaspar et al., 2014). Once such an air-pouch has been created, dying/dead cancer cells can be injected within this pouch (Elliott et al., 2009; Sandilos et al., 2012). Subsequently, at desired recovery time-points post-injection, lavage fluid collected from this pouch can be immunophenotyped (Elliott et al., 2009; Sandilos et al., 2012). In contrast, the *i.p.* model operates with direct *i.p.* injection of dying/dead cancer cells (Ayna et al., 2012; Krysko et al., 2011). Subsequently, at desired recovery time-points post-injection, lavage fluid is collected from the peritoneum for immunophenotyping (Ayna et al., 2012; Krysko et al., 2011). A relative comparison of results obtained in *s.c.* air-pouch model or *i.p.* model with murine ear pinna model shows broad consistency in results across models (Fig. 3) such that, accidental necrosis and tolerogenic apoptosis tend to attract macrophages and/or monocytes/DCs; whereas, immunogenic apoptosis tends to attract largely neutrophils and monocytes/DCs (Fig. 3). Although there are finer quantitative and qualitative differences between results obtained with these models (notwithstanding differences in terms of cell death stimuli and cell line models utilized in these different studies), this comparison at least vouches for the broad applicability of the results obtained with our murine ear pinna model.

Finally, all three murine models discussed above have unique merits as well as specific drawbacks. Hence, to understand dying/dead cancer cells-driven immune cell recruitment, it might be desirable to study at least two out of these three models in parallel. However, due to various practical constrains, researchers may sometimes need to prioritize one model. To this end, we believe that our murine ear pinna model has some crucial advantages over the *s.c.* air-pouch and *i.p.* models. For instance, the *s.c.* air-pouch model is much more artificial (since this *s.c.* air-pouch is effectively a new, hitherto non-existing, tissue milieu) and relatively time-consuming (since the air-pouch tissue lining takes time to establish) (Duarte et al., 2012; Saribas et al., 2012; Wang, Wu, Mou, Chen, & Deng, 2013). Our model involves an already existing tissue milieu which can be immediately exploited (Li et al., 2012; Marino-Marmolejo et al., 2016). Similarly, the *i.p.* model involves a very large surface area (i.e., the murine peritoneum), wherein the dying/dead cells and signals derived thereof (i.e., danger signals, cytokine or chemokines) are much more likely to get diffused and hence require high quantities to form effective chemotactic gradients (Romano et al., 2018; Skupsky, McCann, Nossal, & Losert, 2007; Thomas, Kleist, & Volkman, 2018). Thus immune cell recruitment by dying/dead cancer cells with "mild" immunological profiles is susceptible to under appreciation. Moreover, the need to derive the peritoneal lavage fluid may be a rate-limiting factor for recovering immune cells exhibiting sub-optimal yet biologically-meaningful chemotactic recruitment (Romano et al., 2018). In comparison, our model "traps" the dying/dead cancer cells in a relatively small space within the ear pinna thereby limiting their diffusion and thereby allowing formation of a better chemotactic gradient. Moreover, the entire injection area (i.e., the ear pinna) is analysed thereby decreasing chances of missing threshold amounts of recruited immune cells.

Last but not least, from the perspective of clinical cancer immunotherapy, the anticancer vaccines are often injected in the patients mostly via the intra-dermal or sub-cutaneous route (Bol et al., 2016; Garg, 2017; Garg, Vara Perez, Schaaf, et al., 2017; Melero et al., 2014; Sabado et al., 2017). Therefore, our model has a high (but not absolute) probability of capturing the immune-contexture of a clinically relevant site-of-vaccination. Of note, the clinical validation of our results is still pending and hence, it would be interesting to see whether our model indeed captures the clinical reality for anticancer vaccination.

## Acknowledgments

This work got support from grants of Academy of Finland and Sigrid Juselius Foundation to P.S., Research Foundation Flanders or FWO (G0584.12N, K202313N, GA01111N) and KU Leuven (C16/15/073) to P.A., and KU Leuven (POR/16/040) to A.D.G. We would also like to acknowledge the support of FWO/FNRS Excellence of Science (EOS) grant (30837538) to P.A. and A.D.G. A.D.G. is a recipient of the European Molecular Biology Organization (EMBO) Short-term Fellowship (2013); and the FWO Postdoctoral Fellowship (2013–2016 and 2016–2019). We acknowledge Servier Medical Art (https://smart.servier.com/) and https://www.swifticons.com for providing open access to their figure/icon artworks.

## References

Aaes, T. L., Kaczmarek, A., Delvaeye, T., De Craene, B., De Koker, S., Heyndrickx, L., et al. (2016). Vaccination with necroptotic cancer cells induces efficient anti-tumor immunity. *Cell Reports*, *15*(2), 274–287. https://doi.org/10.1016/j.celrep.2016.03.037.

Aarntzen, E. H., Srinivas, M., Schreibelt, G., Heerschap, A., Punt, C. J., Figdor, C. G., et al. (2013). Reducing cell number improves the homing of dendritic cells to lymph nodes upon intradermal vaccination. *Oncoimmunology*, *2*(7), e24661. https://doi.org/10.4161/onci.24661.

Adema, G. J., de Vries, I. J. M., Punt, C. J. A., & Figdor, C. G. (2005). Migration of dendritic cell based cancer vaccines: In vivo veritas? *Current Opinion in Immunology*, *17*(2), 170–174. https://doi.org/10.1016/j.coi.2005.01.004.

Anguille, S., Smits, E. L., Lion, E., van Tendeloo, V. F., & Berneman, Z. N. (2014). Clinical use of dendritic cells for cancer therapy. *The Lancet Oncology*, *15*(7), e257–e267. https://doi.org/10.1016/S1470-2045(13)70585-0.

Ayna, G., Krysko, D. V., Kaczmarek, A., Petrovski, G., Vandenabeele, P., & Fésüs, L. (2012). ATP release from dying autophagic cells and their phagocytosis are crucial for inflammasome activation in macrophages. *PLoS One, 7*(6), e40069. https://doi.org/10.1371/journal.pone.0040069.

Bezu, L., Gomes-de-Silva, L. C., Dewitte, H., Breckpot, K., Fucikova, J., Spisek, R., et al. (2015). Combinatorial strategies for the induction of immunogenic cell death. *Frontiers in Immunology, 6*, 187. https://doi.org/10.3389/fimmu.2015.00187.

Blander, J. M. (2017). The many ways tissue phagocytes respond to dying cells. *Immunological Reviews, 277*(1), 158–173. https://doi.org/10.1111/imr.12537.

Bobisse, S., Foukas, P. G., Coukos, G., & Harari, A. (2016). Neoantigen-based cancer immunotherapy. *Annals of Translational Medicine, 4*(14), 262. https://doi.org/10.21037/atm.2016.06.17.

Bol, K. F., Schreibelt, G., Gerritsen, W. R., de Vries, I. J. M., & Figdor, C. G. (2016). Dendritic cell-based immunotherapy: State of the art and beyond. *Clinical Cancer Research, 22*(8), 1897–1906. https://doi.org/10.1158/1078-0432.CCR-15-1399.

Chen, D. S., & Mellman, I. (2017). Elements of cancer immunity and the cancer-immune set point. *Nature, 541*(7637), 321–330. https://doi.org/10.1038/nature21349.

Ciabattini, A., Pettini, E., Fiorino, F., Prota, G., Pozzi, G., & Medaglini, D. (2011). Distribution of primed T cells and antigen-loaded antigen presenting cells following intranasal immunization in mice. *PLoS One, 6*(4), e19346. https://doi.org/10.1371/journal.pone.0019346.

Cicchelero, L., de Rooster, H., & Sanders, N. N. (2014). Various ways to improve whole cancer cell vaccines. *Expert Review of Vaccines, 13*(6), 721–735. https://doi.org/10.1586/14760584.2014.911093.

Colli, L. M., Machiela, M. J., Zhang, H., Myers, T. A., Jessop, L., Delattre, O., et al. (2017). Landscape of combination immunotherapy and targeted therapy to improve cancer management. *Cancer Research*, 77(13), 3666–3671. https://doi.org/10.1158/0008-5472.CAN-16-3338.

Cossarizza, A., Chang, H.-D., Radbruch, A., Akdis, M., Andrä, I., Annunziato, F., et al. (2017). Guidelines for the use of flow cytometry and cell sorting in immunological studies. *European Journal of Immunology, 47*(10), 1584–1797. https://doi.org/10.1002/eji.201646632.

De Gruijl, T. D., van den Eertwegh, A. J., Pinedo, H. M., & Scheper, R. J. (2008). Whole-cell cancer vaccination: From autologous to allogeneic tumor- and dendritic cell-based vaccines. *Cancer Immunology, Immunotherapy, 57*(10), 1569–1577. https://doi.org/10.1007/s00262-008-0536-z.

Dillon, C. P., & Green, D. R. (2016). Molecular cell biology of apoptosis and necroptosis in cancer. *Advances in Experimental Medicine and Biology, 930*, 1–23. https://doi.org/10.1007/978-3-319-39406-0_1.

Duarte, D. B., Vasko, M. R., & Fehrenbacher, J. C. (2012). Models of inflammation: Carrageenan air pouch. *Current Protocols in Pharmacology, 56*(1), 5.6.1–5.6.8. Editorial Board S.J. Enna (Editor-in-Chief) ... [et al.], Chapter 5, Unit 5.6. https://doi.org/10.1002/0471141755.ph0506s56.

Elliott, M. R., Chekeni, F. B., Trampont, P. C., Lazarowski, E. R., Kadl, A., Walk, S. F., et al. (2009). Nucleotides released by apoptotic cells act as a find-me signal to promote phagocytic clearance. *Nature, 461*(7261), 282–286. https://doi.org/10.1038/nature08296.

Fang, S., Pentinmikko, N., Ilmonen, M., & Salven, P. (2012). Dual action of TGF-β induces vascular growth in vivo through recruitment of angiogenic VEGF-producing hematopoietic effector cells. *Angiogenesis, 15*(3), 511–519. https://doi.org/10.1007/s10456-012-9278-9.

Fearnhead, H. O., Vandenabeele, P., & Vanden Berghe, T. (2017). How do we fit ferroptosis in the family of regulated cell death? *Cell Death and Differentiation, 24*(12), 1991–1998. https://doi.org/10.1038/cdd.2017.149.

Ferguson, T. A., Choi, J., & Green, D. R. (2011). Armed response: How dying cells influence T-cell functions. *Immunological Reviews, 241*(1), 77–88. https://doi.org/10.1111/j.1600-065X.2011.01006.x.

Fond, A. M., & Ravichandran, K. S. (2016). Clearance of dying cells by phagocytes: Mechanisms and implications for disease pathogenesis. *Advances in Experimental Medicine and Biology, 930*, 25–49. https://doi.org/10.1007/978-3-319-39406-0_2.

Gage, G. J., Kipke, D. R., & Shain, W. (2012). Whole animal perfusion fixation for rodents. *Journal of Visualized Experiments, 65*. https://doi.org/10.3791/3564e3564.

Galluzzi, L., Aaronson, S. A., Abrams, J., Alnemri, E. S., Andrews, D. W., Baehrecke, E. H., et al. (2009). Guidelines for the use and interpretation of assays for monitoring cell death in higher eukaryotes. *Cell Death and Differentiation, 16*(8), 1093–1107. https://doi.org/10.1038/cdd.2009.44.

Galluzzi, L., Baehrecke, E. H., Ballabio, A., Boya, P., Bravo-San Pedro, J. M., Cecconi, F., et al. (2017). Molecular definitions of autophagy and related processes. *The EMBO Journal, 36*(13), 1811–1836. https://doi.org/10.15252/embj.201796697.

Galluzzi, L., Buqué, A., Kepp, O., Zitvogel, L., & Kroemer, G. (2017). Immunogenic cell death in cancer and infectious disease. *Nature Reviews. Immunology, 17*(2), 97–111. https://doi.org/10.1038/nri.2016.107.

Galluzzi, L., & Kroemer, G. (2017). Secondary necrosis: Accidental no more. *Trends in Cancer, 3*(1), 1–2. https://doi.org/10.1016/j.trecan.2016.12.001.

Galluzzi, L., Vitale, I., Aaronson, S. A., Abrams, J. M., Adam, D., Agostinis, P., et al. (2018). Molecular mechanisms of cell death: Recommendations of the nomenclature committee on cell death 2018. *Cell Death and Differentiation, 25*(3), 486–541. https://doi.org/10.1038/s41418-017-0012-4.

Gálvez-Cancino, F., López, E., Menares, E., Díaz, X., Flores, C., Cáceres, P., et al. (2018). Vaccination-induced skin-resident memory CD8+ T cells mediate strong protection against cutaneous melanoma. *Oncoimmunology*, 7(7), e1442163. https://doi.org/10.1080/2162402X.2018.1442163.

García-Caballero, M., Van de Velde, M., Blacher, S., Lambert, V., Balsat, C., Erpicum, C., et al. (2017). Modeling pre-metastatic lymphvascular niche in the mouse ear sponge assay. *Scientific Reports,* 7, 41494. https://doi.org/10.1038/srep41494.

Garg, A. D., & Agostinis, P. (2017). Cell death and immunity in cancer: From danger signals to mimicry of pathogen defense responses. *Immunological Reviews, 280*(1), 126–148. https://doi.org/10.1111/imr.12574.

Garg, A. D., Coulie, P. G., Van den Eynde, B. J., & Agostinis, P. (2017). Integrating next-generation dendritic cell vaccines into the current cancer immunotherapy landscape. *Trends in Immunology, 38*(8), 577–593. https://doi.org/10.1016/j.it.2017.05.006.

Garg, A. D., Elsen, S., Krysko, D. V., Vandenabeele, P., de Witte, P., & Agostinis, P. (2015). Resistance to anticancer vaccination effect is controlled by a cancer cell-autonomous phenotype that disrupts immunogenic phagocytic removal. *Oncotarget, 6*(29), 26841–26860. https://doi.org/10.18632/oncotarget.4754.

Garg, A. D., Galluzzi, L., Apetoh, L., Baert, T., Birge, R. B., Bravo-San Pedro, J. M., et al. (2015). Molecular and translational classifications of damps in immunogenic cell death. *Frontiers in Immunology, 6*, 588. https://doi.org/10.3389/fimmu.2015.00588.

Garg, A. D., More, S., Rufo, N., Mece, O., Sassano, M. L., Agostinis, P., et al. (2017). Trial watch: Immunogenic cell death induction by anticancer chemotherapeutics. *Oncoimmunology, 6*(12), e1386829. https://doi.org/10.1080/2162402X.2017.1386829.

Garg, A. D., Vandenberk, L., Fang, S., Fasche, T., Van Eygen, S., Maes, J., et al. (2017). Pathogen response-like recruitment and activation of neutrophils by sterile immunogenic dying cells drives neutrophil-mediated residual cell killing. *Cell Death and Differentiation, 24*(5), 832–843. https://doi.org/10.1038/cdd.2017.15.

Garg, A. D., Vandenberk, L., Koks, C., Verschuere, T., Boon, L., Van Gool, S. W., et al. (2016). Dendritic cell vaccines based on immunogenic cell death elicit danger signals and T cell-driven rejection of high-grade glioma. *Science Translational Medicine*, *8*(328), 328ra27. https://doi.org/10.1126/scitranslmed.aae0105.

Garg, A. D., Vandenberk, L., Van Woensel, M., Belmans, J., Schaaf, M., Boon, L., et al. (2017). Preclinical efficacy of immune-checkpoint monotherapy does not recapitulate corresponding biomarkers-based clinical predictions in glioblastoma. *Oncoimmunology*, *6*(4), e1295903. https://doi.org/10.1080/2162402X.2017.1295903.

Garg, A. D., Vara Perez, M., Schaaf, M., Agostinis, P., Zitvogel, L., Kroemer, G., et al. (2017). Trial watch: Dendritic cell-based anticancer immunotherapy. *Oncoimmunology*, *6*(7), e1328341. https://doi.org/10.1080/2162402X.2017.1328341.

Gaspar, E. B., Sakai, Y. I., & Gaspari, E. D. (2014). A mouse air pouch model for evaluating the immune response to *Taenia crassiceps* infection. *Experimental Parasitology*, *137*, 66–73. https://doi.org/10.1016/j.exppara.2013.12.005.

Goyvaerts, C., & Breckpot, K. (2015). Pros and cons of antigen-presenting cell targeted tumor vaccines. *Journal of Immunology Research 2015*, 785634. https://doi.org/10.1155/2015/785634.

Iacovelli, R., Ciccarese, C., Bria, E., Bimbatti, D., Fantinel, E., Mosillo, C., et al. (2018). Immunotherapy versus standard of care in metastatic renal cell carcinoma. A systematic review and meta-analysis. *Cancer Treatment Reviews*, *70*, 112–117. https://doi.org/10.1016/j.ctrv.2018.08.007.

Kepp, O., Senovilla, L., Vitale, I., Vacchelli, E., Adjemian, S., Agostinis, P., et al. (2014). Consensus guidelines for the detection of immunogenic cell death. *Oncoimmunology*, *3*(9), e955691. https://doi.org/10.4161/21624011.2014.955691.

Khalil, D. N., Smith, E. L., Brentjens, R. J., & Wolchok, J. D. (2016). The future of cancer treatment: Immunomodulation, CARs and combination immunotherapy. *Nature Reviews. Clinical Oncology*, *13*(5), 273–290. https://doi.org/10.1038/nrclinonc.2016.25.

Klionsky, D. J., Abdelmohsen, K., Abe, A., Abedin, M. J., Abeliovich, H., Acevedo Arozena, A., et al. (2016). Guidelines for the use and interpretation of assays for monitoring autophagy (3rd edition). *Autophagy*, *12*(1), 1–222. https://doi.org/10.1080/15548627.2015.1100356.

Krysko, D. V., Kaczmarek, A., Krysko, O., Heyndrickx, L., Woznicki, J., Bogaert, P., et al. (2011). TLR-2 and TLR-9 are sensors of apoptosis in a mouse model of doxorubicin-induced acute inflammation. *Cell Death and Differentiation*, *18*(8), 1316–1325. https://doi.org/10.1038/cdd.2011.4.

Krysko, D. V., & Vandenabeele, P. (2008). From regulation of dying cell engulfment to development of anti-cancer therapy. *Cell Death and Differentiation*, *15*(1), 29–38. https://doi.org/10.1038/sj.cdd.4402271.

Li, J. L., Goh, C. C., Keeble, J. L., Qin, J. S., Roediger, B., Jain, R., et al. (2012). Intravital multiphoton imaging of immune responses in the mouse ear skin. *Nature Protocols*, 7(2), 221–234. https://doi.org/10.1038/nprot.2011.438.

Lin, L. C. W., Smith, S. A., & Tscharke, D. C. (2012). An intradermal model for vaccinia virus pathogenesis in mice. *Methods in Molecular Biology*, *890*, 147–159. https://doi.org/10.1007/978-1-61779-876-4_9.

Mahoney, K. M., Rennert, P. D., & Freeman, G. J. (2015). Combination cancer immunotherapy and new immunomodulatory targets. *Nature Reviews. Drug Discovery*, *14*(8), 561–584. https://doi.org/10.1038/nrd4591.

Marino-Marmolejo, E. N., Flores-Hernández, F. Y., Flores-Valdez, M. A., García-Morales, L. F., González-Villegas, A. C., & Bravo-Madrigal, J. (2016). A quantitative model for dermal infection and oedema in BALB/c mice pinna. *BMC Microbiology*, *16*(1), 290. https://doi.org/10.1186/s12866-016-0907-0.

Martins, I., Kepp, O., Schlemmer, F., Adjemian, S., Tailler, M., Shen, S., et al. (2011). Restoration of the immunogenicity of cisplatin-induced cancer cell death by endoplasmic reticulum stress. *Oncogene*, *30*(10), 1147–1158. https://doi.org/10.1038/onc.2010.500.

Melero, I., Gaudernack, G., Gerritsen, W., Huber, C., Parmiani, G., Scholl, S., et al. (2014). Therapeutic vaccines for cancer: An overview of clinical trials. *Nature Reviews. Clinical Oncology*, *11*(9), 509–524. https://doi.org/10.1038/nrclinonc.2014.111.

Mould, R. C., AuYeung, A. W. K., van Vloten, J. P., Susta, L., Mutsaers, A. J., Petrik, J. J., et al. (2017). Enhancing immune responses to cancer vaccines using multi-site injections. *Scientific Reports*, 7(1), 8322. https://doi.org/10.1038/s41598-017-08665-9.

Obeid, M., Tesniere, A., Ghiringhelli, F., Fimia, G. M., Apetoh, L., Perfettini, J.-L., et al. (2007). Calreticulin exposure dictates the immunogenicity of cancer cell death. *Nature Medicine*, *13*(1), 54–61. https://doi.org/10.1038/nm1523.

Ohlfest, J. R., Andersen, B. M., Litterman, A. J., Xia, J., Pennell, C. A., Swier, L. E., et al. (2013). Vaccine injection site matters: Qualitative and quantitative defects in CD8 T cells primed as a function of proximity to the tumor in a murine glioma model. *Journal of Immunology*, *190*(2), 613–620. https://doi.org/10.4049/jimmunol.1201557.

Rivalland, G., Scott, A. M., & John, T. (2017). Standard of care in immunotherapy trials: Challenges and considerations. *Human Vaccines & Immunotherapeutics*, *13*(9), 2164–2178. https://doi.org/10.1080/21645515.2016.1277845.

Romano, E., Rufo, N., Korf, H., Mathieu, C., Garg, A. D., & Agostinis, P. (2018). BNIP3 modulates the interface between B16-F10 melanoma cells and immune cells. *Oncotarget*, *9*(25), 17631–17644. https://doi.org/10.18632/oncotarget.24815.

Romero, P., Banchereau, J., Bhardwaj, N., Cockett, M., Disis, M. L., Dranoff, G., et al. (2016). The human vaccines project: A roadmap for cancer vaccine development. *Science Translational Medicine*, *8*(334), 334ps9. https://doi.org/10.1126/scitranslmed.aaf0685.

Sabado, R. L., Balan, S., & Bhardwaj, N. (2017). Dendritic cell-based immunotherapy. *Cell Research*, *27*(1), 74–95. https://doi.org/10.1038/cr.2016.157.

Sandilos, J. K., Chiu, Y.-H., Chekeni, F. B., Armstrong, A. J., Walk, S. F., Ravichandran, K. S., et al. (2012). Pannexin 1, an ATP release channel, is activated by caspase cleavage of its pore-associated C-terminal autoinhibitory region. *The Journal of Biological Chemistry*, *287*(14), 11303–11311. https://doi.org/10.1074/jbc.M111.323378.

Saribas, Z., Ergun, H., Mamuk, S., Köseoglu-Eser, O., & Melli, M. (2012). Critical appraisal of air pouch infection model in rats. *Annals of Clinical and Laboratory Science*, *42*(1), 50–56.

Sharma, P., Wagner, K., Wolchok, J. D., & Allison, J. P. (2011). Novel cancer immunotherapy agents with survival benefit: Recent successes and next steps. *Nature Reviews. Cancer*, *11*(11), 805–812. https://doi.org/10.1038/nrc3153.

Skupsky, R., McCann, C., Nossal, R., & Losert, W. (2007). Bias in the gradient-sensing response of chemotactic cells. *Journal of Theoretical Biology*, *247*(2), 242–258. https://doi.org/10.1016/j.jtbi.2007.02.016.

Thomas, M. A., Kleist, A. B., & Volkman, B. F. (2018). Decoding the chemotactic signal. *Journal of Leukocyte Biology*, *104*, 359–374. https://doi.org/10.1002/JLB.1MR0218-044.

Ting, A. T. (2018). Tools in the art of studying necroptosis. *Methods in Molecular Biology*, *1857*, 1–9. https://doi.org/10.1007/978-1-4939-8754-2_1.

Tremblay, M.-E., Riad, M., & Majewska, A. (2010). Preparation of mouse brain tissue for immunoelectron microscopy. *Journal of Visualized Experiments, 41,* e2021. https://doi.org/10.3791/2021.

Verdegaal, E. M. E., de Miranda, N. F. C. C., Visser, M., Harryvan, T., van Buuren, M. M., Andersen, R. S., et al. (2016). Neoantigen landscape dynamics during human melanoma-T cell interactions. *Nature*, *536*(7614), 91–95. https://doi.org/10.1038/nature18945.

Wang, Y., Wu, N.-N., Mou, Y.-Q., Chen, L., & Deng, Z.-L. (2013). Inhibitory effects of recombinant IL-4 and recombinant IL-13 on UHMWPE-induced bone destruction in the murine air pouch model. *The Journal of Surgical Research, 180*(2), e73–e81. https://doi.org/10.1016/j.jss.2012.04.016.

Wu, C.-C., Chuang, Y.-T., Hsu, Y.-T., Huang, J.-T., Wu, T. C., Hung, C.-F., et al. (2013). Intra-peritoneal hyperthermia combining α-galactosylceramide in the treatment of ovarian cancer. *PLoS One, 8*(7), e69336. https://doi.org/10.1371/journal.pone.0069336.

Yoon, K. W. (2017). Dead cell phagocytosis and innate immune checkpoint. *BMB Reports, 50*(10), 496–503. https://doi.org/10.5483/BMBRep.2017.50.10.147.

Yuan, M., Li, D., An, M., Li, Q., Zhang, L., & Wang, G. (2017). Rediscovering peritoneal macrophages in a murine endometriosis model. *Human Reproduction, 32*(1), 94–102. https://doi.org/10.1093/humrep/dew274.

Zappasodi, R., Pupa, S. M., Ghedini, G. C., Bongarzone, I., Magni, M., Cabras, A. D., et al. (2010). Improved clinical outcome in indolent B-cell lymphoma patients vaccinated with autologous tumor cells experiencing immunogenic death. *Cancer Research, 70*(22), 9062–9072. https://doi.org/10.1158/0008-5472.CAN-10-1825.

Zhou, Q., Johnson, B. D., & Orentas, R. J. (2007). Cellular immune response to an engineered cell-based tumor vaccine at the vaccination site. *Cellular Immunology, 245*(2), 91–102. https://doi.org/10.1016/j.cellimm.2007.04.004.

CHAPTER NINE

# Computational methods in tumor immunology

**Bhavneet Bhinder**[a,b], **Olivier Elemento**[a,b,*]
[a]Caryl and Israel Englander Institute for Precision Medicine, Weill Cornell Medicine, New York, NY, United States
[b]Department of Physiology and Biophysics, Weill Cornell Medicine, New York, NY, United States
*Corresponding author: e-mail address: ole2001@med.cornell.edu

## Contents



## Abstract

The remarkable success of cancer immunotherapies, especially the checkpoint blocking antibodies, in a subset of patients has reinvigorated the study of tumor-immune crosstalk and its role in heterogeneity of response. High-throughput sequencing and imaging technologies can help recapitulate various aspects of the tumor ecosystem. Computational approaches provide an arsenal of tools to efficiently analyze, quantify and integrate multiple parameters of tumor immunity mined from these diverse but complementary high-throughput datasets. This chapter describes numerous such computational approaches in tumor immunology that leverage high-throughput data from

*Methods in Enzymology*, Volume 636
ISSN 0076-6879
https://doi.org/10.1016/bs.mie.2020.01.001

diverse sources (genomic, transcriptomics, epigenomics and digitized histopathology images) to systematically interrogate tumor immunity in context of its microenvironment, and to identify mechanisms that confer resistance or sensitivity to cancer therapies, in particular immunotherapy.

## 1. Introduction

Tumor cells interact with a broad range of factors and cell types within their microenvironment that influence tumor dynamics, ability to metastasize as well as clinical responses to therapy (Binnewies et al., 2018; Fouad & Aanei, 2017; Fridman, Zitvogel, Sautès-Fridman, & Kroemer, 2017; Joyce & Fearon, 2015). The unprecedented success of immunotherapies in multiple cancer types including melanomas, lung, head and neck, renal, gastric, and urothelial cancers, has fueled the interest in understanding the concerted role of tumor-immune interactions in regulating tumor progression. Immunotherapies, in particular immune checkpoint blocking (ICB) antibodies, have shown a durable clinical benefit in patients who respond well to it. But, the initial response rates to immunotherapy are low (approx. 20%), associated costs of treatment are very high (approx. $150,000/year), and robust cancer-agnostic biomarkers for patient selection are lacking (Cyriac & Gandhi, 2018). At present, the only FDA-approved companion diagnostic biomarkers to prescribe ICB for advanced cancers are *PD-L1* expression levels in tumor tissue (immunohistochemistry based) and/or tumor microsatellite Instability (MSI)/DNA mismatch repair deficiency (dMMR) (Tray, Weber, & Adams, 2018). Despite their recommended use in clinical decision making, the use of *PD-L1* level or MSI status as independent biomarkers of patient prognosis have led to inconsistent conclusions (Yi et al., 2018). Tumors are in a state of constant crosstalk with their microenvironment to survive and tumors are also very heterogeneous from patient to patient, hence it is highly unlikely that an independent variable would suffice as a biomarker to predict patient outcome (Bhinder & Elemento, 2017; Blank, Haanen, Ribas, & Schumacher, 2016). A multimodal approach, which incorporates various aspects of tumor immunology, is needed to address the unmet clinical need of suitable patients' selection and to maximize the cost-benefit nexus for cancer patient care.

Predicting patient responses to precision cancer therapies, including immunotherapies, demands an understanding of how tumors escape immune surveillance. An efficient antitumor-immune response would require the presence of functional immune effectors and tumor-associated immunogenic antigens (derived from genomic or transcriptomic alterations, also called neoantigens), both in adequate quantities. This equation could be disrupted at multiple levels to trigger immune escape: (1) tumors may not have sufficient immune effector cells (i.e., the tumor microenvironment (TME) is not inflamed), (2) TME is inflamed but the immune effector cells are dysfunctional, or (3) TME is inflamed but the neoantigens are inadequately presented (Fig. 1) (Jiang et al., 2018; Rosenthal et al., 2019). Based on these principles, Rosenthal et al. segregated treatment naïve lung tumors (TRACERx consortium) into those with low immune evasion (immune inflamed TME or no evidence of immune escape) versus those with high immune evasion (immune depleted TME or evidence of immune escape); immune escape was determined by the presence of HLA-LOH (human leukocyte antigens—loss of heterozygosity), disruptions in genes of the cellular antigen processing machinery or loss of predicted neoantigens (Rosenthal et al., 2019). Based on this stratification, the group showed that the patients with low immune evasion associated with a significantly longer disease-free survival (DFS) upon systemic therapies. By quantifying immune

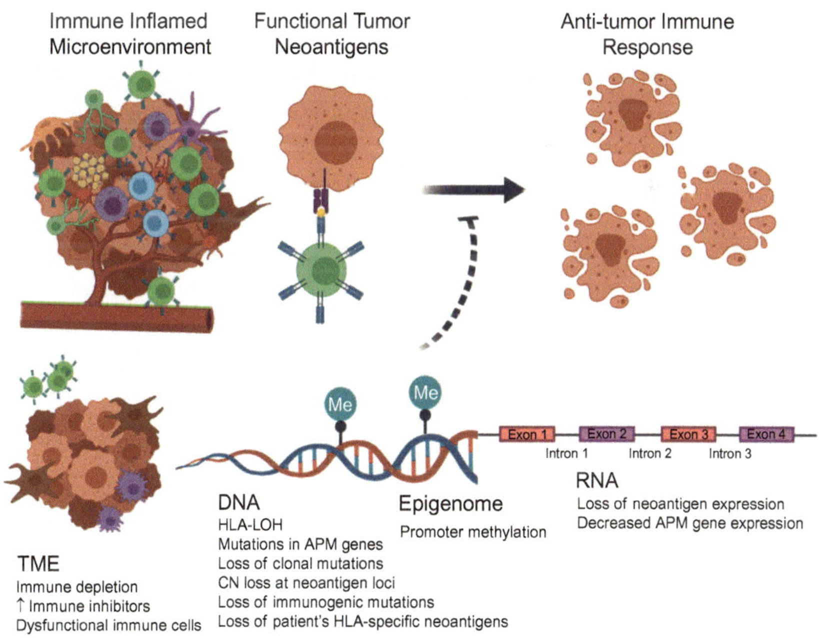


**Fig. 1** Tumor-immune escape mechanisms that thwart antitumor-immune response. CN, copy number; APM, antigen processing machinery; TME, tumor microenvironment; LOH, loss of heterozygosity.

evasion from multiple parameters of tumor immunology and showing its association with DFS, this study emphasized the clinical relevance of studying tumor characteristics in context of its immune microenvironment.

Immunogenomics studies suggest that the tumor and its immune microenvironment continually co-evolve under selection pressures from each other. Rosenthal et al. show that in non-small cell lung cancer (NSCLC), specific genomic profiles correlate with specific immune profiles and that, while the genomically heterogenous tumors associate with sparse immune infiltration, the homogenous tumors associate with denser immune infiltration (Rosenthal et al., 2019). Smyrk et al. show that MSI-H tumors positively associate with higher infiltration of cytotoxic T-cells in colorectal cancers (CRC) (Smyrk, Watson, Kaul, & Lynch, 2001). Schwitalle et al. isolated infiltrating T-cells from tumor tissue and peripheral blood of CRC patients and healthy donors, and performed an ELISpot assay to test T-cell reactivity to neoantigens predicted from frameshift mutations; highly reactive neoantigen-specific T-cells were found not only in MSI-high (MSI-H) CRC patients but also in healthy carriers of MSI-H related mutations, but not in MSI-stable CRC or non-carrier healthy donors (Schwitalle et al., 2008). As neoantigen primed T-cells infiltrate the TME, tumors learn to select against neoantigens that are highly immunogenic (Anagnostou et al., 2017; Riaz et al., 2017; Verdegaal et al., 2016; Vitale et al., 2019). Analyses of sequential lesions from two advanced melanoma patients treated with adoptive T-cell therapy by Verdegaal et al., suggested that the immunogenic neoantigens may be lost over time if neoantigen-specific immune cells are present in the TME (Verdegaal et al., 2016). Predicted immunogenic mutations were reported to be selected against upon treatment with Nivolumab in advanced melanoma patients; similar loss of predicted immunogenic mutations was found in NSCLC tumors that were resistant compared to those that were sensitive to *PD1* or *CTLA4* antibodies (Anagnostou et al., 2017; Riaz et al., 2017). Together this evidence suggests a close relationship between tumor's genomic and immune composition that renders growth advantage to a tumor.

All the examples above demonstrate that a tumor cannot be studied in isolation from its microenvironment and diverse methods to probe the overall tumor immunology must be considered. While experimental approaches in tumor immunology are necessary for true mechanistic insights, they can be laborious and expensive during the initial stages of discovery and require prospective planning (such as tissue collection, storage, reagents, experimental equipment). In comparison, computational methods can provide faster alternatives to interrogate patterns in data and to formulate empirical data driven hypotheses that could be later tested experimentally. In addition, computational methods also allow for retrospective analysis of archived datasets. In this chapter we describe the computational methods to study the main tenants of tumor immunology. We provide strategies to leverage the wealth of data obtained from diverse platforms to perform a comprehensive characterization of the TME. The following sections describe the computational approaches and their importance to quantify mainly the genomic and transcriptomic profiles of a tumor: use of mutation burdens and immunogenic alterations to study tumor immunogenicity, use of gene expression and epigenetic profiles to deconvolve TME into its constituent cell types, use of various estimates that classify immune infiltration and use of deep learning to assess the spatial context of immune localization within the TME directly from digitized histopathology slides (Fig. 2). We recommend adopting an integrative multivariate approach with the methods described in this chapter for a global analysis of tumor immunology in a given cancer cohort.

## 2. Genomic alterations and neoantigens

### 2.1 Tumor mutation burden measurement

Tumor mutation burden (TMB) is a quantitative measure of the total number of non-synonymous somatic mutations (may include insertions and deletions) in a given tumor sample normalized to the number of bases in the sequencing region covered per million (Chan et al., 2019; Meléndez et al., 2018; Samstein et al., 2019). High TMB has been reported to be associated with improved prognosis in patients treated with ICB immunotherapies in a variety of cancers such as metastatic melanomas, NSCLC, urothelial cancer and head and neck cancers (Chan et al., 2019; Hellmann et al., 2018; Hugo et al., 2016; Lauss et al., 2017; Le et al., 2015; Miao et al., 2018; Rizvi et al., 2015; Snyder et al., 2014; Van Allen et al., 2015). In CheckMate 227 study, high TMB of $\geq$10 Mutations/Mb—regardless of tumor's *PD-L1* expression level—was shown to be associated with improved progression free survival (PFS) in first-line treatment of NSCLC patients treated with combination immunotherapy (nivolumab plus ipilimumab) compared to those treated with chemotherapy alone (Meléndez et al., 2018). Such findings have led researchers to propose the use of TMB as a likely companion diagnostic for immunotherapies. However, a survey of the ICB clinical trials revealed that the reported thresholds for high TMB are not

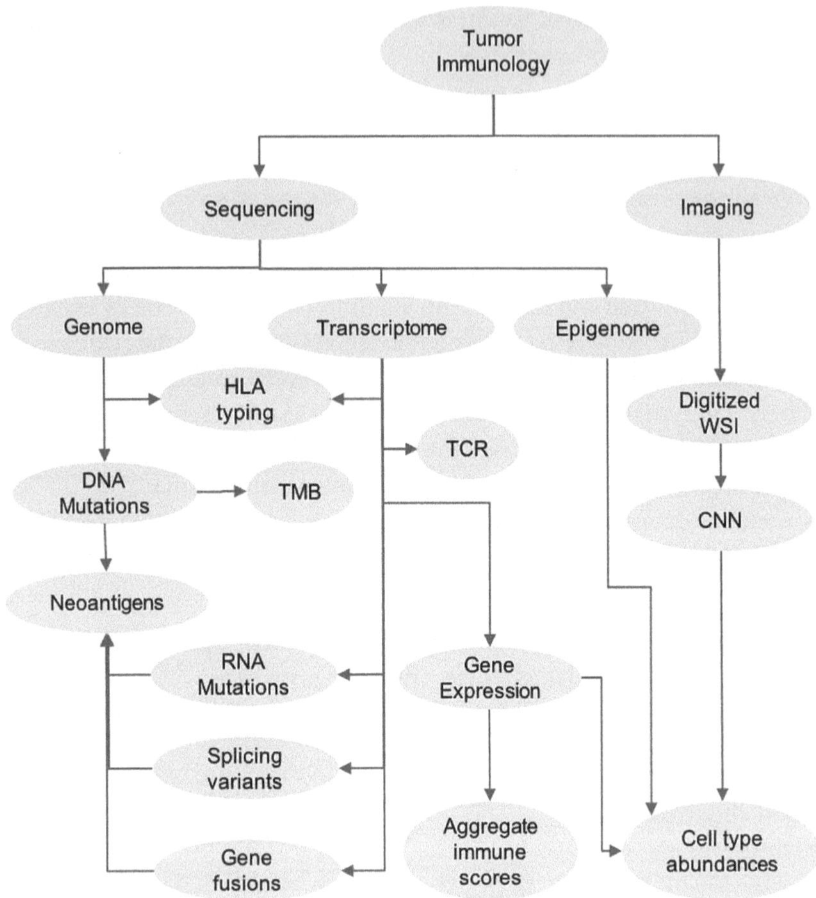


**Fig. 2** Computational avenues in tumor immunology to quantify and visualize the tumor microenvironment. CNN, convolution neural networks; TCR, T-cell receptors; TMB, tumor mutation burden; WSI, whole-slide images.

consistent among different research groups even for the same cancer type (Chan et al., 2019; Meléndez et al., 2018). Unlike is necessary for a metric to be considered a clinical biomarker, there is no universal gold standard for what constitute a high TMB in patients. Key considerations in establishing a clinically relevant threshold for high TMB are discussed below:

1. *Choice of a high-throughput sequencing panel*: While whole-exome sequencing (WES) is the most commonly used platform for TMB quantification, smaller cost-effective targeted gene panels (e.g., Oncomine, Foundation Medicine, MSK-IMPACT) are also being used for TMB estimation (Chan et al., 2019; Meléndez et al., 2018). In addition, whole-genome

sequencing is becoming accessible to identify genomic variants at a much larger scale compared to both WES and targeted panels. Methods that quantify TMB from minimally invasive techniques, such as plasma cell free DNA are also being extensively developed (Volckmar et al., 2018). As the variety of available platforms and panels keep expanding, the magnitude by which variations in sequencing regions covered by different capture kits impact TMB calculation remains to be tested. The minimum recommended size of a targeted panel required to quantify TMB is estimated to be around 300 genes or 1 Mb genomic coverage, but WES panels cover from 39 Mb up to as high as 70 Mb of the genomic region (Meléndez et al., 2018). Even among targeted panels, the genomic regions captured by the sequencing kits vary, e.g., only 50% of the genes from the MSK-IMPACT (targeting 468 genes) are also covered by the Foundation Medicine panel (targeting 315 genes) (Meléndez et al., 2018). Additionally, the mutation categories included for TMB calculation may vary by the choice of sequencing panel, e.g., TMB is commonly calculated using non-synonymous somatic mutations, but inclusion of synonymous mutations in TMB calculation for targeted panels is also reported (Chan et al., 2019). Such factors may affect the specificity of TMB measurements.

2. *Technical factors*: Technical factors include sample collection, library preparation, tissue fixation methods, contamination or degradation (e.g., DNA fragmentation and cytosine deamination are common in Formalin-Fixed Paraffin-Embedded (FFPE) tissues), tumor purities, sequencing depth and sequenced read lengths (Meléndez et al., 2018). Related practices and expertise may vary among different laboratories, and this may impact the sensitivity of mutation detection. As an example, low tumor purity in the sequenced tissue could arise from the choice of tumor region selected for macro-dissection by a pathologist. Also, adoption of protocols for sequencing at high read depths allow for identification of variants occurring at low allelic frequencies, which increases the sensitivity of the assay for mutation detection compared to sequencing at low read depths.
3. *Choice of data processing strategies*: Data analyses variation arise from the choice of a sequencing aligner, variant calling tools and downstream filtering approaches. The pre-processing steps to align the raw sequencing reads to the reference genome are fairly standard (Fig. 3). Once the data has been aligned, a variant calling tool is applied to identify a list of somatic mutations; preference for the tool used has largely remained

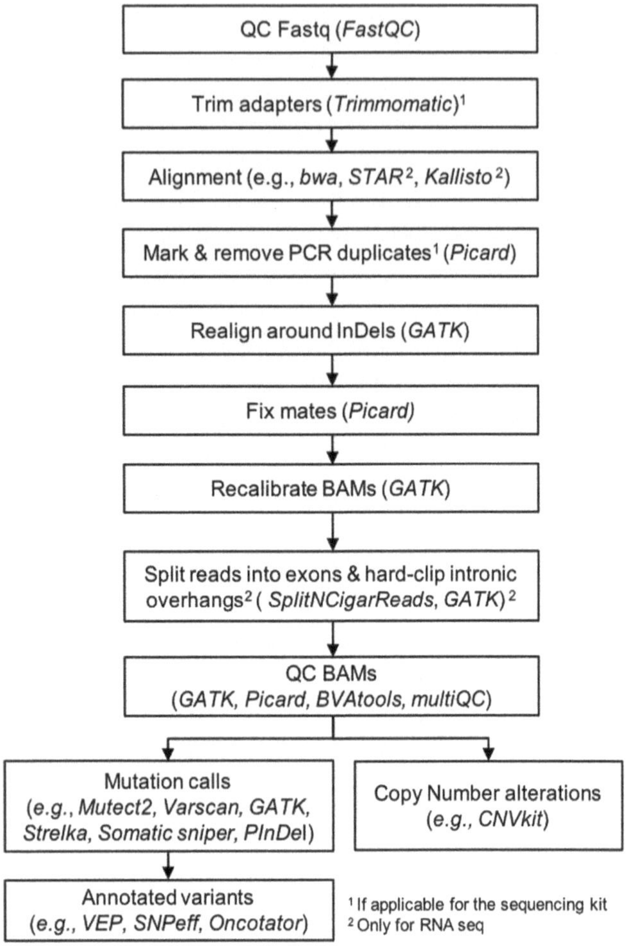


**Fig. 3** Generic workflow for whole-exome and RNA seq data pre-processing and analysis. CNV, copy number variants; GATK, genome analysis toolkit; PCR, polymerase chain reaction; QC, quality control.

subjective. There is poor concordance among mutations identified from different variant callers in general (Krøigård, Thomassen, Lænkholm, Kruse, & Larsen, 2016). Therefore, independent callers are frequently combined for a consensus set of mutations. For example, The Cancer Genome Atlas (TCGA) consortium described a detailed workflow, called MC3, in an effort to harmonize its data sets (Ellrot et al., 2018). MC3 is a combination of seven different mutation and Indel callers (MuTect, MuSE, Somatic Sniper, Radia, VarScan2, Pindel, and

Indelocator) to obtain a consensus list of mutations ($\geq$2 callers) from WES of paired tumor-normal tissues (Ellrot et al., 2018). Suggested steps for downstream filtering in the MC3 pipeline included: (a) set thresholds for minimal allelic read depths to retain a variant, (b) identify and remove predicted artifacts from sample contamination (using a tool called ContEst) or DNA degradation (using a tool called OxoG), (c) exclude false positives identified in the genomic coordinates not covered by the capture kit, and (d) use a Panel of Normals (PoN) to limit platform specific sequencing artifacts. MC3 pipeline is publicly available as a dockerized package for use by the scientific community (Ellrot et al., 2018). Besides, databases such as dbSNP, 1000 Genomes project, and ExAC have been used to filter common germline variants. While mutation analyses for WES are routinely performed using paired tumor-normal tissues, matched normal tissues are mostly not sequenced for targeted panels, hence filtering false positives that are likely germline variants becomes even more important for targeted panels. An added word of caution in the use of publicly available databases is the under-representation of individuals with non-European ancestries which limits the identification of ethnicity specific germline variants and therefore may result in overcalling putative somatic variants in cancer genomes (Garofalo et al., 2016).

4. *Heterogeneity in mutation load*: The total number of mutations per tumor may vary up to 1000 folds across cancers and also within the same cancer type (Chalmers et al., 2017; Chan et al., 2019; Lawrence et al., 2013). Samstein et al. conducted a targeted panel (MSK-IMPACT) sequencing on multiple cancer types ($n=10$) treated with ICB, and found markedly different cutoffs for high TMB (top 20% of the TMB distribution) across cancer types, e.g., TMB cutoffs for CRC, bladder and breast cancers were 52.2, 17.6, and 5.9, respectively (Samstein et al., 2019). High TMB thresholds calculated individually for each cancer type in this study associated with a significantly improved overall survival (OS) in only 4 out of 10 cancer types tested (Samstein et al., 2019).
5. *Hypermutability in tumors*: A subset of tumors, irrespective of the cancer type, can accumulate a very high number of mutations due to a high MSI or dMMR deficiency caused by mutations in DNA repair genes including *MLH1*, *MSH2*, *MSH6* and *PMS2*. In a pan-cancer analysis, Chalmers et al. reported that 16% of the tumors identified as being hypermutated also had a high MSI (Chalmers et al., 2017). Tumors with MSI-H phenotypes also associate with improved clinical response from

ICB (Hellmann et al., 2018; Le et al., 2015; Mandal et al., 2019; Marcus, Lemery, Keegan, & Pazdur, 2019). As an example, MSI-H colorectal cancer patients exhibited such a durable clinical benefit from pembrolizumab, that it led the FDA to approve this drug for any metastatic solid tumor with an MSI-H/dMMR deficient phenotype (Hellmann et al., 2018; Le et al., 2015; Marcus et al., 2019). Since not all hypermutated tumors are MSI-H, a multivariate model that combines MSI status with TMB for improved patient response stratification remains to be tested.

## 2.2 Neoantigens derived from genomic alterations

Neoantigens are tumor-associated mutated peptides that can be recognized as non-self by the patient's immune system. Immunogenic neoantigens bypass the central thymic and peripheral tolerance and can prime the patient's immune cells to mount an antitumor-immune response. Prevalence of certain immunogenic neoantigens in patients have not only driven the unprecedented clinical success of immunotherapies (especially ICB) but are also the basis to develop adoptive T-cell therapy and peptide vaccines (Aurisicchio et al., 2019; Kalaora et al., 2018; Keskin et al., 2019). In cancer, neoantigens arise from an alteration in the patient's genome that gets successfully transcribed and eventually translated and processed into a mutated peptide sequence; this peptide when presented on the surface of a tumor cell gets recognized by cytotoxic T lymphocytes which triggers lysis of the targeted tumor cell. There are two fundamental components that determine the immunogenicity of a tumor-associated neoantigen: (1) ability of the mutated peptide to be processed and bind to the major histocompatibility complex (MHC), also known as human leukocyte antigens (HLA), for presentation and (2) ability of the presented neoantigen to be recognized by the T-cell receptors for a cytotoxic immune response. Out of these, the former component has been most extensively explored and widely studied using computational methods to predict putative neoantigens from high-throughput sequencing datasets.

Complete neoantigen prediction workflows that start from a given list of genomic variants from a user and end in a list of ranked neoantigens from a prediction algorithm are now available as fully automated packages that can be easily applied to whole-exome sequencing datasets. A genomics team at the Washington University School of Medicine was the first to publish an extensive suite of integrated tools for neoantigen discovery called pVAC-Seq. At present similar automated packages from multiple other centers are also readily accessible to the research community and a compiled list of these resources is available (e.g., Antigen. garnish, Epidisco) (Smith et al., 2019). Irrespective of the source, the three key steps for neoantigen discovery have remained consistent among all protocols, as detailed below:

1. *Infer patient's HLA genotype*: There are two broad classes of the highly polymorphic HLA alleles: *Class I* (HLA-A, HLA-B, HLA-C) that bind to peptides of 8–10 amino acids in length, derived from intracellular proteins, that are presented to CD8+ T cells; *Class II* (HLA-DR, HLA-DP, HLA-DQ) that bind to much longer peptides of 13–25 amino acids in length, derived from endocytosed extracellular proteins, that are presented to CD4+ T helper cells (Dimitrov, Garnev, Flower, & Doytchinova, 2010; Gowthaman & Agrewala, 2008). Computational methods to infer HLA Class I alleles directly from the patients sequencing data (both WES and RNAseq) are mature and a complete list of these tools is available elsewhere (Finotello, Rieder, Hackl, & Trajanoski, 2019). Optitype and Polysolver are two of the most commonly applied tools for four-digit HLA Class I typing. Unlike Class I, development of robust HLA class II prediction tools has not been as successful due to two main challenges: (a) since HLA Class II has an open binding groove, the length of peptides that can bind to this complex are highly variable (13–25 amino acids) and dynamic (while only nine amino acids of the peptide actually occupy the binding site, the amino acids flanking the binding core also influence the peptide-HLA binding) and (b) lack of sufficient datasets to train prediction models. Hence, prevalent workflows focus on HLA Class I restricted neoantigen predictions (Dimitrov et al., 2010; Finotello, Rieder, et al., 2019; Gowthaman & Agrewala, 2008).
2. *Extract all possible peptides harboring the mutation*: The nucleotide sequence surrounding the mutation is extracted and translated into its corresponding peptide sequence. With a step-size of one amino acid and a window-size of $n$ amino acids (where $n$ is typically 8–11 amino acids for HLA Class I), all possible $n$-mer peptides harboring the altered amino acid are generated. In general, the total nucleotide length required to extract all possible mutated peptides of length $n$ for a missense mutation can be calculated as $(2n - 1) \times 3$. An important consideration while preparing peptides is to retain the translational open reading frame (ORF) of the mutated gene unless the mutation introduces a frameshift that disrupts the canonical ORF, in which case the peptide with a novel reading frame is translated until the next stop codon is encountered.

Because frameshift mutations alter the ORF, just one frameshift mutation—unlike a point mutation—can produce several novel peptides that are likely to be more dissimilar from self (Turajlic et al., 2017). On another note, different somatic or germline variants may be located in close sequence proximity of a mutation of interest with a haplotype-specific linkage (in-phase). Therefore, the mutated peptides predicted without taking into account such proximal variants would not mimic the peptide actually translated in the tumor, hence affecting the accuracy of neoantigen prediction (Hundal et al., 2019; Wood et al., 2019). Hundal et al. inspected proximal variants co-occurring with missense mutations of interest in 430 tumors from 7 cancer types and estimated that 5.1% of the total missense mutations had in-phase proximal variants (Hundal et al., 2019). They also found that ignoring in-phase proximal variants led to an overall false positive neoantigen prediction rate of 6.9% and a false negative rate of 2.6% (Hundal et al., 2019).

3. *Predict binding affinity of the peptide to the HLA complex*: Once all possible peptides have been collected from step 2, these peptides are tested for their binding affinities to the inferred HLA Class I complex with any of the available off-the-shelf neoantigen prediction tools. A common practice is to use the binding affinity (IC50) threshold of <500nM to select for candidate neoantigens (peptides with IC50 < 50nM are considered strong binders). Existing neoantigen prediction tools are based on one of the following algorithms: (1) artificial neural networks (NetMHC, MHCflurry, EDGE), (2) deep convolution networks (HLA-CNN, DeepSeqPan), (3) position-specific scoring metrices (PSSMHCpan, MixMHCpred), and (4) random forest classifier (ForestMHC) (Finotello, Rieder, et al., 2019). Among these tools, NetMHC (v4, v3 or NetMHCpan) has been the most widely used for neoantigen prediction.

## 2.3 Prioritization of neoantigens

A single immunogenic neoantigen is capable of eliciting a strong clinical response in patients treated with ICB; this was recently shown in a case study of an exceptional responder to *PD1* antibody where just one immunogenic neoantigen from a novel gene fusion led to a remarkable tumor regression (refer to Section 3.1) (Yang et al., 2019). There is a linear positive correlation between the number of tumor-associated neoantigens predicted and the total number of mutations, hence by association hundreds of candidate neoantigens can be predicted for one tumor sample. However, experimental testing has shown that only a tiny fraction of these predicted neoantigens ($\ll$1%) are capable of eliciting an immune response (McGranahan et al., 2016; van Rooij et al., 2013; Yadav et al., 2017; Yang et al., 2019). Hence, in silico identification of singular highly immunogenic neoantigen from a large repertoire of predicted neoantigens is akin to looking for a needle in a haystack, and efficient computational methods to do so are limited. Researchers have identified few characteristics of neoantigens that may be useful parameters in preliminary prioritization of the predicted neoantigens, explained below:

1. *Binding affinity thresholds for immunogenic neoantigens*: As mentioned in Section 2.2, NetMHC is the most commonly used tool for neoantigen prediction in published research; netMHC is trained with artificial neural networks and it predicts the binding affinity (IC50) of the neoantigens with the MHC class I or class II alleles of the patient. IC50 threshold of <500nM has been established as a gold standard to select for candidate neoantigens. This threshold was originally reported in 1994 in context of viral antigens; researchers had tested the immunogenicity of HBV derived viral antigens for HLA-A*0201 in transgenic mice and PBMCs of acute hepatitis patients and had found that the epitopes with binding affinities of <500nM, preferably <50nM, were capable of eliciting a cytotoxic T-cell response (Sette et al., 1994). But the threshold found a widespread use in cancer neoantigen prediction research without an extensive validation in tumor model systems. Tumor derived neoantigens predicted as weak binders, including those with IC50s much above the 500nM cutoff, have been shown to elicit a detectable T-cell response in vivo, which suggest that the threshold need not be adhered to so stringently to exclude candidate peptides (Duan et al., 2014).
2. *Differential agretopicity index for immunogenic neoantigens*: To evaluate how different a predicted neoantigen is from its nonmutated wild-type peptide, Duan et al. introduced differential agretopicity index (DAI). DAI is defined as the difference between the binding affinities of a mutated peptide and its wild-type counterpart (Duan et al., 2014). Duan et al. first predicted MHC class I restricted neoantigens from mutations identified in RNA seq of BALB/c and MethA mouse models. Top ranking neoantigens were selected either based on their predicted binding affinities (from NetMHC v3.0) or based on their DAI; the neoantigens shortlisted with this strategy were synthesized and injected into mouse models to study their effects on tumor growth measured after 1 week of immunization. The group identified that neoantigens selected based

on their DAI showed higher rates of tumor rejection (10/48 tumor protective neoantigens) compared to those selected based on binding affinities alone (0/18 tumor protective neoantigens) (Duan et al., 2014). In addition, Ghorani et al. have showed that higher mean DAI per tumor was associated with improved OS in advanced melanoma and lung cancer cohorts of the TCGA (Ghorani et al., 2018).

3. *Expression of predicted neoantigens*: To ensure that the predicted neoantigens are transcribed and expressed in the tumor cells, tumor-specific expression levels of a neoantigen-associated gene should be incorporated in the prediction workflows, and neoantigens with higher expression levels can be assigned a higher priority. Copy number losses at the genomic loci that correspond to the predicted neoantigens can also be used as an additional filter for neoantigen de-prioritization.
4. *Neoantigen sequence-based considerations*: Neoantigens with sequence similarities to self-antigens would less likely be immunogenic since T-cells reactive to self would be centrally depleted or peripherally tolerated. Hence a predicted neoantigens with high sequence similarities with a known human peptide would most likely be non-immunogenic (Łuksza et al., 2017; Snyder et al., 2014; Wood et al., 2018). In contrast, predicted neoantigens with high sequence similarities with known viral epitopes would most likely be immunogenic; such sequence similarities can be easily found by a protein-BLAST of neoantigen sequences against known viral peptide databases such as the International Epitopes Database (IEDB) (Bubie, Gonzalez-Kozlova, Akers, Villanueva, & Losic, 2019; Łuksza et al., 2017; Snyder et al., 2014; Wood et al., 2018). Location of the mutated amino acid within the neoantigen may also play an important role in its immunogenicity, whether the mutated residue resides at the anchor positions (2 and 9) that influence neoantigen–MHC binding, or at the solvent exposed regions accessible to reactive T-cells (3–7) that influence neoantigen–T-cell binding (Wood et al., 2018; Yadav et al., 2017).
5. *Clonality of the neoantigens*: Mutations identified from sequenced tumors can be either clonal (present in all tumor cells) or subclonal (present in a subset of tumor cells). Based on the parameters of tumor purity and ploidy, computational methods (e.g., ABSOLUTE, PyClone, CHAT, PhyloWGS) can be used to determine the clonality of the mutations identified in a tumor (which also refers to the clonality of the associated neoantigens). Clonal and subclonal populations with distinct genomic profiles within the same tumor constitute intra-tumor heterogeneity (ITH). Tumors with low

ITH (i.e., a more homogenous tumors) have been shown to associate with improved OS, higher immune cytolytic activity and denser levels of CD8 T-cells infiltration (Rosenthal et al., 2019; Wolf et al., 2019). In a retrospective study by McGranahan et al., patient populations segregated by tumors with high clonal neoantigen burden (in the upper quartile) plus low ITH (<1%) associated with improved OS in lung cancer, improved OS in *PD1* antibody treated advanced melanomas, and PFS in *CTLA4* antibody treated advanced melanomas, compared to when the patient populations were segregated by high neoantigen burdens alone (McGranahan et al., 2016). In a limited experimental testing, the group found that only the clonal neoantigens were being recognized by the expanded CD8+ T-cells from the patient PBMC's; although it is also possible that the T-cells reactive to subclonal neoantigens were present but below detectable levels (McGranahan et al., 2016). These findings together with additional reports on correlations between high clonal neoantigen burden and improved ICB responses suggest that prioritization of the neoantigens based on their clonality is a reasonable strategy (Miao et al., 2018; Rosenthal et al., 2019).

6. *Neoantigen fitness models*: Neoantigen fitness models were recently developed as a metric to assess the overall fitness of a tumor clone based on the immunogenicity of its neoantigen repertoire (Balachandran et al., 2017; Łuksza et al., 2017). The models define neoantigen quality as a product of the amplitude (A) and the T-cell recognition potential (R) of a neoantigen. Amplitude A was calculated as the fold change between the predicted binding affinities of the neoantigen and its paired wild-type peptide. Recognition potential R was calculated from the sequence similarities between the neoantigens and known viral epitopes. Higher proportions of high quality neoantigens (>median) were shown to be associated with improved OS in patients with pancreatic cancer from the TCGA (Balachandran et al., 2017). Neoantigen quality was also used to determine tumor fitness as: negative (maximum (A × R)) after iterating over all the neoantigens in a clone. The same group also showed that high tumor fitness associated with poor OS in melanoma and lung cancer patients treated with ICB (Łuksza et al., 2017). However, work by a different research group was not able to replicate the significant associations between more high quality neoantigens and improved OS in patients with melanoma and lung cancer from the TCGA (Balachandran et al., 2017). We postulate that the neoantigen quality and fitness models still need to be independently validated in broader cancer datasets.

# 3. Transcriptome derived neoantigens

Cancer driven chromosomal rearrangements (intra-chromosomal or inter-chromosomal), and aberrant splicing events can also produce tumor-specific neoantigens. The basic steps of neoantigen discovery are same as described for genomic alterations in Sections 2.2 and 2.3. In addition, some studies described below have performed in silico validations of neoantigens peptides with previously published mass spectrometry datasets using tools such as SEQUEST, MSGFPlus or SearchGUI for neoantigen prioritization (Kahles et al., 2018; Smart et al., 2018; Wang, Wang, Alam, Hoeppner, & Yang, 2019). These approaches validate whether a predicted neoantigen is translated and/or binds to a given HLA complex, but the immunostimulatory properties of that neoantigen still need to be tested. Although this section focuses on the transcriptome derived neoantigens, it must be noted that the role of disease specific post-translational modifications (PTMs) in creation of neoantigens is also being studied (Anderton, 2004; Cobbold et al., 2013; Hosen et al., 2017; Kumai et al., 2014). PTMs—such as phosphorylation, acetylation, deamidation, glycosylation—can modify the immunogenicity of a peptide, i.-e., alter a peptide's binding affinity with the MHC complex or the T-cell receptors (TCR), or, may change the conformation of cell surface proteins to expose epitopes previously hidden by the protein's native confirmation (Anderton, 2004; Cobbold et al., 2013; Gan, Burgess, Clayton, & Scott, 2012; Hosen et al., 2017; Kumai et al., 2014).

## 3.1 Neoantigens from gene fusions

Gene fusions are hybrid transcripts that result from chromosomal rearrangements; these hybrid transcripts can form novel non-self-peptides that maybe immunogenic. In a case study reported by Yang et al., one fusion associated neoantigen was so potent that it independently drove an exceptional response to *PD1* antibody in a head and neck cancer patient (Yang et al., 2019). This was particularly interesting because the patient did not have the typical tumor features for ICB responsiveness, i.e., the patient had no driver mutations, no significant copy number alterations or structural variants, low TMB, poor immune infiltration and scored negative for *PD-L1* staining. However, the patient had one driver gene fusion (*DEK–AFF2*) that was identified in the primary tumor using RNA seq and confirmed in the tumor tissue using fluorescence in situ hybridization (FISH). To investigate the immunogenicity of this fusion, all possible nine amino acid long peptides translated from this gene fusion as well as other DNA mutations were extracted and their binding affinities to HLA Class I alleles were predicted using NetMHCpan v.4.0. These peptides were synthesized and tested for peptide-specific T-cell reactivity in an ELISpot assay. Among all peptides tested, only the fusion associated neoantigen (DKESEEEVS) was found to be strongly immunostimulatory. TCR-Vβ deep sequencing performed on this patient's PBMCs confirmed an enrichment of TCR clones specific for this fusion peptide, not found pre-treatment but detected in high levels on-treatment (Yang et al., 2019). This detailed study elegantly demonstrated the role of fusion associated neoantigens in driving response.

Like frameshift mutations, out-of-frame fusions may also produce numerous novel peptides that are likely to be more dissimilar from self. While novel peptides from in-frame fusions would be limited to the junction regions, novel peptides from out-of-frame fusions, due to a shift in the codons at the junctions, can extend until the first downstream premature stop codon is encountered. The first step for fusion neoantigen prediction is to identify gene fusions from the RNA seq and several tools to do so exist. Yang et al. had used FusionCatcher to identify the highly immunogenic gene fusion in their study (Yang et al., 2019). Rathe et al. adopted a more integrated approach to identify fusions in osteosarcoma; they selected a set of fusions that overlapped between fused transcripts identified from deFuse and *denovo* transcripts identified from Trinity (Rathe et al., 2019). Gao et al. developed a strategy applied pan cancer that systematically merged fusion results from three fusion tools (STAR-Fusion, EricScript, and Breakfast) (Gao et al., 2018). A series of rules followed to combine the results including: (a) retain fusions identified by ≥2 tools (except those identified in STAR-fusion due to its high sensitivity), (b) require minimum five split reads supporting the fusion, (c) remove fusions with the same breakpoints in ≥10 samples across multiple cancers, (d) remove fusions also identified in TCGA normal samples, GTEx samples, or in non-cancer cells, and (e) remove fusions associated with uncharacterized genes, immunoglobulin, mitochondrial genes, and fusions in same gene or gene paralogs. For the actual step of fusion neoantigen prediction, automated pipelines are also available and INTEGRATE-neo was one of the first published (Zhang, Mardis, & Maher, 2017). The INTEGRATE-neo workflow begins by identifying gene fusions using an internal tool called INTEGRATE which predicts all novel fusion junctions. The ORF of the 5′ gene fusion partner is retained to translate the fused transcript. Subsequent steps for neoantigen prediction are in principle similar to those described in Section 2.2. Apart from INTEGRATE-neo, NeoepitopePred, pVACfuse and antigen. Garnish also enable fusion neoantigen discovery (Hundal et al., 2016; Wood et al., 2019).

The studies on fusion derived neoantigens have led to some interesting observations. In pan-cancer analyses, fusion neoantigens were found at a higher frequency in tumors with immune depleted microenvironments (i.e., lower leukocyte fractions and lower *CD8A* expression) or in tumors with HLA allelic loss (HLA-LOH) (Rathe et al., 2019; Yang et al., 2019). Additionally, tumors harboring driver fusions also associated with lower mutational burdens (Gao et al., 2018). Overall, immunogenic fusions in tumor are an exciting class of clinically relevant neoantigens.

## 3.2 Neoantigens from alternative splicing events

Aberrant splicing events that include exon or intron creation, intron retention, exon skip, or new terminal start, and end may create novel peptides. Several strategies to identify alternative splicing associated neoantigens from RNA seq have been published. Kahles et al. reported a method to identify alternative splicing events with a focus on novel exon-exon junctions and applied it to all cancer cohorts of the TCGA (Kahles et al., 2018). Briefly, the sequenced reads were aligned to the human genome using the STAR aligner (with parameters that re-created the index for non-annotated junctions), the aligned reads were used to augment the splicing graphs (from SplAdder toolkit) with the number of reads that supported each alternate junction. To identify tumor-specific novel junctions, splicing variants that were also identified in normal tissues (from TCGA and GTEx) were removed. Alternate splicing events were found to occur at a much higher rate in tumor tissues than in normal tissues (Kahles et al., 2018). Next, the nucleotide sequences surrounding these novel exon-exon splice junctions were extracted and translated for neoantigen prediction.

In another work, Smart et al. published a strategy to identify neoantigens derived from intron retention in melanoma patients treated with ICB (Smart et al., 2018). The RNA seq reads were aligned to the augmented hg19 genome built (i.e., containing both exon and intron sequences) using Kallisto and retained intronic loci were identified using the KMA (*k*-mer alignment) algorithm. For each retained intronic loci, nucleotide sequences plus ORF orientations were obtained from the UCSC Table Browser and translated into all possible 9–10 amino acid long peptides. Peptides that overlapped with the known human reference proteome or normal skin tissue proteins were removed. The remaining peptides were translated for neoantigen prediction. The number of intron retention neoantigens identified in this study did not correlate significantly with the number of neoantigens derived from somatic mutations. Additionally, the intron neoantigen load neither associated with clinical benefit from ICB, nor with the expression of cytolytic markers *GZMA* or *PRF1*. These findings suggest that the quality rather than quantity of alternate splicing neoantigens must be investigated for clinical significance.

DNA mutations in the proximity of a known splice site can disrupt the canonical splice site, hence creating a novel alternate splice site. Such splice-site-creating mutations (SCMs) are shaping up a new arena of neoantigen discovery. Jayasinghe et al. introduced a method called MiSplice (available as a suite of Perl programs) which was applied pan cancer to identify novel splice sites in RNA seq transcripts that may have been created by DNA mutations, i.e., SCMs (Jayasinghe et al., 2018). MiSplice first identified all splice junctions within 20 nt of a mutation, filtered out canonical junctions, and then scored the remaining novel junctions for their splice site strength by MaxEntScan. The novel junctions were further filtered by comparing the number of reads supporting a novel junction between cases (i.e., samples with that mutation) versus controls (i.e., samples without that mutation). The mutations that corresponded to the final set of novel splice junctions were identified as SCMs and through this analysis, Jayasinghe et al. reported that 26% and 11% of the SCM mutations in the TCGA cohorts were previously mis-labeled as missense and silent, respectively. The authors, by way of computational neoantigen prediction methods, also report that SCMs can create neoantigens with higher predicted immunogenicity, thereby presenting a new area of neoantigen discovery.

# 4. Estimating immune infiltration from transcriptomics

## 4.1 Deconvolution tools for virtual dissection of the TME

Experimental techniques such as immunohistochemistry (IHC), immunofluorescence, flow cytometry and mass cytometry have been traditionally used to inspect the composition of the TME. With the availability of large amounts of transcriptomics sequencing data from a variety of cancer types virtual dissection of the TME through computational methods is now possible. Deconvolution of the TME constitutes identification and quantification of its distinct resident or infiltrating cell types. Computational methods provide a cost-time-labor effective advantage over the experimental techniques, and these can be applied to new as well as legacy expression profiles from bulk tumors. Roughly 47 different deconvolution methods based on different mathematical models have been published (Table 1), out of which two tools (EPIC and quanTIseq) were developed specifically for RNA seq datasets. These approaches use either gene marker lists (which look for enrichment in marker genes) or expression profiles (which deconstruct bulk tissue profile as a linear weighted sum of its constituent cell type profiles).

While computational deconvolution has revealed disease specific associations between distinct immune cell types and patient responses (Gentles et al., 2015; Şenbabaoğlu et al., 2016), the tools used for deconvolution do have some limitations (Avila, Vandesompele, Mestdagh, & De Preter, 2018; Finotello & Trajanoski, 2018; Mohammadi, Zuckerman, Goldsmith, & Grama, 2017). Some of the factors that may confound the cell type estimations include multicollinearity (i.e., separation of closely related cell types that share the same lineage), bias toward certain cell types due to variable mRNA composition of cell types, limited availability of reference cell types or high tumor purity of the tissues to deconvolve. Quality of the reference signatures determine how robust the deconvolution is. Reference signatures are constructed by applying differential gene expression analyses to purified cell type expression profiles; genes with a minimum intra-group variability and maximum inter-group variability are selected. Some of the newer methods also described iterative selection of the most informative deconvolution markers from the reference profiles (Jew et al., 2019; Li et al., 2016). While reference signatures for most deconvolution methods were initially developed and validated for microarrays, their systematic validation for RNA seq is lacking. To minimize inconsistencies between the bulk tissue and the reference signatures, the tool quanTIseq has been set-up to first re-process the raw sequencing reads from the bulk RNA seq with their standard protocols prior to the actual deconvolution (Finotello et al., 2019). Current trends show a shift toward the use of single-cell profiles to construct reference signatures for deconvolution (Frishberg et al., 2019; Jew et al., 2019; Menden, Marouf, Dalmia, Heutink, & Bonn, 2019; Newman et al., 2019; Wang, Park, Susztak, Zhang, & Li, 2019) (Table 1). Methods that use single-cell expression profiles as references have incorporated normalization steps to reduce cross platform variations; these steps include batch correction between the bulk tissue and reference profiles using ComBat or applying linear transformations learnt from the reference profiles to the bulk tissue (Finotello & Trajanoski, 2018; Jew et al., 2019; Johnson, Li, & Rabinovic, 2007; Li et al., 2016; Newman et al., 2019).

Table 1 List of available deconvolution tools and methods.

| Method/Tool | Description | Year | Base | Algorithm | Reference source | Availability | |
|---|---|---|---|---|---|---|---|
| ABIS | ABsolute Immune Signal | 2019 | GEP | Robust linear modeling | RNAseq, Microarray | https://github.com/giannimonaco/ABIS | R Shiny app |
| BayICE | Bayesian deconvolution model for Intracellular Component Exploration | 2019 | GEP (semi-reference-based) | Bayesian inference (hierarchical) with stochastic search variable selection | RNAseq | https://github.com/AshTai/BayICE | R package |
| Bisque | – | 2019 | GEP, GML | NNLS | scRNA Seq | https://github.com/cozygene/bisque | R package |
| CIBERSORTx | CIBERSORT Extension | 2019 | GEP | SVM (NMF for differentially expressed genes, NNLS for cell type specific GEP inferred) | scRNA Seq | https://cibersortx.stanford.edu | Web tool |
| CPM/scBio | Cell Population Mapping | 2019 | GEP | SVM with multiple iterations | scRNA Seq | https://github.com/amitfrish/scBio | R package |

*Continued*

**Table 1** List of available deconvolution tools and methods.—cont'd

| Method/Tool | Description | Year | Base | Algorithm | Reference source | Availability | |
|---|---|---|---|---|---|---|---|
| DECODER | DE novo COmpartment DEconvolution and weight estimation of tumoR | 2019 | de novo | Weighted NMF | Ref free | https://github.com/laurapeng/decoder | Matlab |
| dtangle | – | 2019 | GEP w/o GML | Linear mixing model | Microarray | https://github.com/gjhunt/dtangle | R package |
| DWLS | Dampened weighted least squares | 2019 | GEP | Weighted ordinary least squares | scRNA Seq | https://bitbucket.org/yuanlab/dwls/src/default/ | R package |
| FARDEEP | Fast And Robust Deconvolution of Expression Profiles | 2019 | GEP | Adaptive least trimmed squares with NNLS | RNAseq, Microarray | https://github.com/YuningHao/FARDEEP | R package |
| ICTD | Inference of cell types and deconvolution | 2019 | GML | Constrained NMF | NA | https://github.com/changwn/ICTD | R package |
| linseed | LINear Subspace identification for gene Expression Deconvolution | 2019 | de novo | Singular value decomposition (SVD) | Ref free | https://github.com/ctlab/LinSeed | R package |

| | | | | | | | |
|---|---|---|---|---|---|---|---|
| MuSiC | MUlti-Subject SIngle Cell deconvolution | 2019 | GEP | Weighted NNLS | scRNA Seq | https://github.com/xuranw/MuSiC | R package |
| NITUMID | Non-negative matrix factorization-based Immune-TUmor MIcroenvironment Deconvolution | 2019 | GML | Semi-supervised NMF | Microarray | https://github.com/tdw1221/NITUMID | R package |
| quanTIseq | Quantify fractions of 10 immune cell types from bulk RNA-sequencing data | 2019 | GEP | Constrained least square regression | RNAseq | https://icbi.i-med.ac.at/quantiseq | bash/Singularity/Docker |
| Scaden | – | 2019 | GEP | Deep neural network | scRNA Seq | https://scaden.readthedocs.io/en/latest/ | Python |
| SCDC | Single-Cell RNA-seq reference sets for bulk gene expression DeConvolution | 2019 | GEP | ENSEMBLE method (weighted NNLS, Grid search, least absolute deviation regression) | scRNA Seq | https://github.com/meichendong/SCDC | R package |

*Continued*

**Table 1** List of available deconvolution tools and methods.—cont'd

| Method/Tool | Description | Year | Base | Algorithm | Reference source | Availability | |
|---|---|---|---|---|---|---|---|
| DeMixT | Deconvolution for Mixed Transcriptomes | 2018 | de novo *w/o GML* | Linear additive models with 3 component deconvolution (iterated conditional modes algorithm) | Ref free | https://bioconductor.org/packages/release/bioc/html/DeMixT.html | R package |
| singscore | Single sample Score | 2018 | GML | Rank (ordered by expression level) | NA | https://bioconductor.org/packages/singscore | R package |
| EPIC | Estimating the Proportion of Immune and Cancer cells | 2017 | GEP (semi-reference-based) | Constrained least square regression | RNAseq | https://gfellerlab.shinyapps.io/EPIC_1-1 | R Shiny app |
| Infino | Infer infiltrate expression phenotypes | 2017 | GEP | Bayesian inference | RNAseq | https://github.com/hammerlab/infino | Python |
| TIminer | – | 2017 | GML | Pre-ranked GSEA | NA | http://icbi.i-med.ac.at/software/timiner/timiner.shtml | Python |

| | | | | | | | |
|---|---|---|---|---|---|---|---|
| WSCUnmix | Weighted Simplicial Complex Unmixing | 2017 | de novo | Simplicial complex inference | Ref free | https://github.com/tedroman/WSCUnmix | Matlab |
| xCell | – | 2017 | GML | ssGSEA with linear transformation of ES scores | RNAseq, Microarray | http://xcell.ucsf.edu | R package, Web tool |
| Bseq-SC | – | 2016 | GEP | CIBERSORT model (see below) | scRNA Seq | http://github.com/shenorrlab/bseq-sc | R package |
| CAM | Convex Analysis of Mixtures | 2016 | de novo | Linear latent variable model & theory of convex sets | Ref free | https://bioconductor.org/packages/devel/bioc/html/CAMTHC.html | R package |
| ImmQuant | – | 2016 | GEP, GML | Elastic net | Microarray | http://csgi.tau.ac.il/ImmQuant/ | Web tool |
| MCPcounter | Microenvironment Cell Populations-counter | 2016 | GML | log2 geometric mean of markers | Microarray | https://github.com/ebecht/MCPcounter | R package |
| TIMER | Tumor IMmune Estimation Resource | 2016 | GEP, tumor purity | Linear least squares regression | Microarray | https://icbi.i-med.ac.at/software/timiner/doc/index.html | Web app |

Continued

**Table 1** List of available deconvolution tools and methods.—cont'd

| Method/Tool | Description | Year | Base | Algorithm | Reference source | Availability | |
|---|---|---|---|---|---|---|---|
| VoCAL | Variation of Cell-type Abundance Loci | 2016 | GEP, GML | Elastic net, GWAS of immune traits | (ImmuneQuant gp) | https://cran.r-project.org/web/packages/ComICS/index.html | R package |
| CIBERSORT | Cell-type Identification By Estimating Relative Subsets Of RNA Transcripts | 2015 | GEP | SVM | Microarray | http://cibersort.stanford.edu/ | R package, Web tool |
| UNDO | UNsupervised DecOnvolution | 2015 | de novo | Linear latent variable model | Ref free | https://www.bioconductor.org/packages/release/bioc/html/UNDO.html | R package |
| DCQ | Digital Cell Counter | 2014 | GEP, GML | Elastic net | FACS, Microarray | https://cran.r-project.org/web/packages/ComICS/index.html | R package |
| MMAD | Microarray Microdissection with Analysis of Differences | 2014 | GEP, GML | Maximum likelihood over the residual sum of squares | Microarray | http://sourceforge.net/projects/mmad/ | Matlab |

| | | | | | | | |
|---|---|---|---|---|---|---|---|
| DeconRNASeq | – | 2013 | GEP | Quadratic programming | RNAseq | https://www.bioconductor.org/packages/release/bioc/html/DeconRNASeq.html | R package |
| DSA | Digital Sorting Algorithm | 2013 | GML | Quadratic programming | NA | https://github.com/zhandong/DSA | R package |
| ESTIMATE | Estimation of STromal and Immune cells in MAlignant Tumor tissues using Expression data | 2013 | GEP | ssGSEA | RNAseq | https://bioinformatics.mdanderson.org/estimate/rpackage.html | R package |
| ISOpure | – | 2013 | GEP | Bayesian inference (2 compartment Tumor/unmatched Normal) | Microarray | https://cran.r-project.org/web/packages/ISOpureR/index.html | R package |
| CTen | Cell Type ENrichment | 2012 | GML | Fisher's Exact test (−log10 of BH adjusted *P*-values as enrichment score) | Microarray | http://www.influenza-x.org/~jshoemaker/cten/ | Web tool |

*Continued*

**Table 1** List of available deconvolution tools and methods.—cont'd

| Method/Tool | Description | Year | Base | Algorithm | Reference source | Availability | |
|---|---|---|---|---|---|---|---|
| PERT | Perturbation model | 2012 | GEP | NNML with multiplicative perturbation | Microarray | https://github.com/gquon/PERT | Octave |
| ssNMF | Semi-Supervised Non-negative matrix factorization | 2012 | GML | Semi-supervised NMF | NA | https://github.com/rforge/cellmix/tree/master/pkg/CellMix | R package |
| qprog | Quadratic programming | 2011 | GEP | Quadratic programming (Constrained least square regression) | NA | https://github.com/rforge/cellmix/tree/master/pkg/CellMix | R package |
| SPEC | Subset Prediction from Enrichment Correlation | 2011 | GML | Weighted GSEA | Microarray | http://clip.med.yale.edu/SPEC/ | R package |
| Deconf | Deconfounding | 2010 | de novo | Least squares NMF | Ref free | https://github.com/rforge/cellmix/tree/master/pkg/CellMix | R package |
| lsfit | – | 2009 | GEP | Linear least squares regression | Microarray | https://github.com/rforge/cellmix/tree/master/pkg/CellMix | R package |

| | | | | | | | |
|---|---|---|---|---|---|---|---|
| ssGSEA | Single sample gene set enrichment analysis | 2009 | GML | Empirical cumulative distribution of ranked marker gene | NA | https://www.bioconductor.org/packages/release/bioc/html/GSVA.html | R package |
| ssKL | – | 2004 | GML | Semi-supervised NMF algorithm for KL divergence | NA | https://github.com/rforge/cellmix/tree/master/pkg/CellMix | R package |
| ssFrobenius | – | 2001 | GML | Semi-supervised NMF for Euclidean distance | NA | https://github.com/rforge/cellmix/tree/master/pkg/CellMix | R package |
| CellMix | Tool Suites | 2013 | cs-lsfit, cs-nnls, cs-qprog, csSAM, deconf, DSA, DSection, lsfit, meanProfile, nnls, qprog, ssFrobenius, ssKL | | | https://github.com/rforge/cellmix/tree/master/pkg/CellMix | R package |
| immunedeconv | Tool Suites | 2019 | CIBERSORT, EPIC, MCP counter, quanTIseq, TIMER, xCell | | | https://github.com/grst/immunedeconv | Python |

ES, enrichment score; GEP, gene expression profile; GML, gene marker list; GSEA, gene set enrichment analysis; NMF, non-negative matrix factorization; NNLS, non-negative least squares regression; NNML, non-negative maximum likelihood; SVM, support vector regression; scRNA, single-cell RNA seq; GWAS, genome-wide association study.

## 4.2 Classification of tumors into broad immune clusters

Gene expression profiles have been used to classify tumors into broad immune categories that can be leveraged to understand the underlying mechanisms of immune evasion and may explain resistance to immunotherapies. Numerous published studies have used gene signatures-based clustering approaches that identify immune and/or stromal cells clusters in cancer and study their associations with clinical outcomes (Charoentong et al., 2017; Danaher et al., 2017; Iglesia et al., 2016; Jiang et al., 2018; Prat et al., 2017; Şenbabaoğlu et al., 2016). Unsupervised NMF has also been used to identify de novo immune and stromal components from expression profiles in pancreatic and liver cancers (Moffitt et al., 2015; Sia et al., 2017). Selected methodologies are described below:

1. *T-cell inflamed versus depleted classification*: Spranger et al. used a clustering approach to classify metastatic melanoma tumors into either hot (T-cell inflamed) or cold (T-cell depleted) groups (Spranger, Bao, & Gajewski, 2015). Unsupervised hierarchal agglomerative clustering ($K=12$ and Euclidean distance) was applied to the melanoma expression profiles, and the sub-cluster with known T-cell-marker genes was selected for consensus clustering ("ConsensusClusterPlus" R package, 2000 iterations with 80% melanoma tumor resampling, and Euclidean distance) which resulted in the identification of stable hot and cold tumor clusters. Pathway analysis of the differentially expressed genes between the hot and cold tumors revealed upregulation of the β-catenin signaling pathway in the cold tumors. Using mouse melanoma models with active β-catenin signaling, the researchers validated that this pathway was indeed associated with poor T-cell infiltration, driven mechanistically via suppression of dendritic cells; reconstitution of the dendritic cells in the mouse models reversed the resistance to *PD1* and *CTLA4* antibodies (Spranger et al., 2015). Similar clustering approach has been also used to classify and study hot/cold tumors in colorectal and urothelial cancers (Sweis et al., 2016; Xue et al., 2019).
2. *Identification of six immune subtypes in cancer*: In a pan-cancer TCGA study, Thorsson et al. applied a supervised clustering approach to classify 9126 tumors into 6 immune subtypes (Thorsson et al., 2018). First, 83 published expression signatures known to be associated with immune activity in cancer were collected and used as references to calculate signatures specific ssGSEA scores across all TCGA tumors. ssGSEA scores were then converted into a matrix of their spearman correlations and a

weighted gene correlation network analysis (WGCNA) was applied to this matrix. Most representative signatures from each WGCNA module were then evaluated for their predictive strength using a model-based clustering approach called mclust, followed by an expert manual inspection; this resulted in a final set of five robust signatures. Second, to select the optimal number of immune clusters (*K*) for the TCGA cohort, mclust was applied iteratively to the normalized ssGSEA scores from 21 random tumor sample subsets (with 50% resampling). An optimal *K* that maximized the Bayesian Information Criteria (BIC) in each clustering model was selected and an average *K* was determined at a 6-cluster solution. Finally, with $K=6$, mclust models were fit to each of the 256 randomly resampled TCGA tumor subsets (with 50% resampling); consensus clustering ("clue" R package) was applied to all clustering results to obtain the final cluster assignments. This resulted in the classification of the entire TCGA cohort into any of the six immune subtypes which they called: wound healing, IFN-γ dominant, inflammatory, lymphocyte depleted, immunologically quiet, and TGF-β-dominant. Each immune subtype associated with distinct genomic profiles, regulatory networks and survival benefits; the inflammatory cluster associated with significantly improved OS relative to the wound healing cluster (Thorsson et al., 2018).

3. *Immune-phenotypes*: In a pan-cancer study by Tamborero et al., GSVA enrichment scores for 16 immune cells were calculated for 9174 solid tumors from 29 different cancer types (Tamborero et al., 2018). Hierarchal agglomerative clustering (Euclidean distance, ward linkage) was applied to these scores with higher weights assigned to the cytotoxic cells. This led to the identification of six broad immune clusters with variable levels of cytotoxic content; these clusters were called immune-phenotypes. Each immune-phenotype associated with distinctive mutational profiles and cellular pathways; increase in cytotoxicity of the immune-phenotypes associated with significantly improved OS in 3/29 cancer types tested (endometrial, adrenal and breast cancers) (Tamborero et al., 2018).

## 4.3 Aggregate immune scores and predictors of immunotherapy responses

Measures that quantify immune content from tumor's bulk gene expression profiles and provide aggregated scores are compiled below:

1. *Cytolytic score (CYT)*: CYT was first defined as the geometric mean of expression of genes that mediate cytolysis (*GZMA* and *PRF1*) to quantify the tumor-associated cytolytic activity in 18 cancer types by Rooney, Shukla, Wu, Getz, and Hacohen (2015); positive associations were identified between CYT and neoantigen load as well as mutations in antigen presenting machinery genes, which could be the likely tumor intrinsic mechanism of immune evasion (Rooney et al., 2015). In advanced melanomas, pre-therapy CYT levels were found to be significantly higher in responders versus non-responders to *CTLA4* antibody, but similar pattern was not found in tumors responsive to *PD1* antibody (Hugo et al., 2016; Van Allen et al., 2015). A recent immune landscape study for 36 hematological malignancies further expanded CYT as the geometric mean of 5 genes (*GZMA*, *GZMH*, *GZMM*, *GNLY* and *PRF1*) that captured the abundance of CD8+ cytotoxic T lymphocytes as well as natural killer cells (Dufva et al., 2019).
2. *Innate anti-PD-1 resistance (IPRES)*: Hugo et al. described a transcriptional signature called IPRES to stratify *PD1* antibody responsive melanoma patients (Hugo et al., 2016). First, they sampled a variety of existing gene sets to select for gene sets that significantly correlate with response. To do this, GSVA enrichment scores were calculated for each gene set in each tumor sample; gene sets with absolute median GSVA score difference of $\geq 10\%$ (FDR $\leq 25\%$) between responders versus non-responders were selected as the core gene sets for the IPRES signatures. Second, GSVA scores from IPRES gene sets were standardized (using $z$-score) across all samples for each gene set. All $z$-scores for a given tumor were averaged to give one IPRES enrichment score per tumor. The tumors depleted in IPRES showed improved survival compared to those enriched in IPRES.
3. *Immunophenoscore*: Charoentong et al. developed a predictive score for response to ICB called the immunophenoscore (Charoentong et al., 2017). Using the TCGA expression cohorts, the authors trained a random forest classifier for high and low cytolytic activity in each cancer type using various immune related features. Out-of-bag cross-validation approach was used to rank and select the most predictive features; finalized features included representative genes from four immune categories: MHC genes (e.g., *HLA-A*), immunomodulators (e.g., *PD1*), immune effectors (e.g., CD8+ T-cells) and immune suppressors (e.g., Tregs). Sample-wise normalized expression values ($z$-scores) were obtained for each predictive

feature (for cell types average $z$-score of metagenes that define the cell type was calculated). A weighted average of the $z$-scores corresponding to each of the four immune categories was calculated, these four aggregated $z$-scores were then summed and scaled between 0 and 10 to give an immunophenoscore per sample. The immunophenoscore was found to be associated with significantly improved OS in one-third of the cancer types of the TCGA. The immunophenoscore was also reported to be a better predictor of ICB response (AUC = 1) compared to CYT, mutation load and expression of checkpoint genes (*PD1*, *PD-L1*, *CTLA4*), all tested independently (Charoentong et al., 2017).

4. *Immunopredictive score (IMPRES)*: Auslander et al. developed IMPRES trained on neuroblastoma to classify tumors with spontaneous regression versus progression; this model was latter applied to 265 published metastatic melanoma samples treated with ICB (Auslander et al., 2018). Pairwise binary logical relationships between quantile-normalized expression levels of 28 selected immune checkpoint genes were obtained (i.e., if expression of gene $i$ < gene $j$, then 1, else 0). Most predictive features of response were identified from these pair-wise values using greedy (hill-climbing) aggregative feature selection (500 iterations, five-fold cross-validation) procedure; 15 features that scored consistently high were chosen. Using these features, an IMPRES score was calculated as the total number of 1s for each tumor sample (range 0–15). High IMPRES scores (>cohort median) were positively associated with improved OS in metastatic melanoma patients treated with ICB. The IMPRES prediction accuracy (AUC) for publicly available ICB datasets was found to be within 0.77–0.96 (Auslander et al., 2018).
5. *Tumor-immune dysfunction and exclusion (TIDE) score*: Tumors can either adopt mechanisms that lead to immune exclusion so that low levels of immune cells infiltrate the TME or inclusion of immune cells in dysfunctional states. Based on this concept, Jiang et al. described the TIDE framework to identify cancer specific gene signatures for (a) T-cell dysfunction and (b) T-cell exclusion (Jiang et al., 2018). For T-cell dysfunction signature, the cytotoxic T-lymphocyte (CTL) levels were first quantified as the average expression levels of *CD8A*, *CD8B*, *GZMA*, *GZMB* and *PRF1*. Multiplicative Cox-PH survival regression models were then applied to identify interactions between individual gene expression levels and the CTL levels so as to select for genes that would significantly reduce the positive association between high CTL and

improved patient survival. The TIDE dysfunction score for each gene was then defined as the interaction coefficient *d* from the survival model divided by its standard error. For T-cell exclusion signature, the authors derived this signature form the gene expression (microarrays) profiles of purified immunosuppressive cell types (i.e., cancer-associated fibroblasts, myeloid-derived suppressor cells and the tumor-associated macrophages M2). To study whether TIDE scores are predictive of ICB response, the authors segregated tumors into high or low CTL levels (threshold = average CTL): (a) for tumors with high CTL, Pearson correlations between tumor gene expression profile and T-cell dysfunction signature were calculated and (b) for tumors with low CTL, Pearson correlations between tumor gene expression profile and T-cell exclusion signature were calculated. Scaling and merging these Pearson correlations provided TIDE prediction scores. The authors claimed that their TIDE prediction scores gave a better prediction accuracy (AUC = ~0.80) for ICB response compared to some of the existing criteria for response prediction (e.g., *PD-L1* expression, mutation load) melanoma patients treated with ICB. However, the response prediction to immunotherapy using TIDE is currently limited to melanoma and NSCLC. The authors also require that the input tumors expression profiles be normalized either to a related normal tissue profiles or to mixed cancer expression profiles.

6. *IFN-γ-related T-cell inflammation score*: Merck tested an 18-gene T-cell inflammation signature as a clinical diagnostic for *PD1* antibody response (Ayers et al., 2017; Rozeman et al., 2017). Elastic net penalized regression model was trained with fivefold cross-validation on pembrolizumab response datasets (KEYNOTE-012 (bladder, gastric, HNSCC, and triple-negative breast cancer) and KEYNOTE-028 (NCT02054806) (anal canal, biliary, colorectal, esophageal, and ovarian cancer)) to identify a signature of 18 predictive immune genes. To calculate the signature score, non-zero regression coefficients from this model were used as penalty weights to calculate the final weighted sum of the normalized expression of the 18 genes in the signature (Ayers et al., 2017). The average AUC for the signature score across the nine cancer types used to fit the model was reported to be 0.75. In an independent validation cohort of 96 HNSCC tumors, signature score showed improved performance (AUC = 0.75) over *PD-L1* levels from IHC (AUC = 0.65) in predicting objective response to pembrolizumab. However, based on the findings in this report, the AUC achieved for the signature may not be ideal for its use as a clinical diagnostic for response and requires rigorous validation.

7. *T-cell clonality*: Treatment specific qualitative and quantitative shifts occur in the TCR repertoires after ICB immunotherapies (Hopkins et al., 2018; Khunger, Rytlewski, Fields, Yusko, & Tarhini, 2019; Page et al., 2016; Riaz et al., 2017). High clonal expansion of T-cells upon treatment with *CTLA4* antibody has been reported to associate with improved clinical response in melanoma, breast and pancreatic cancers (Hopkins et al., 2018; Khunger et al., 2019; Page et al., 2016). TCR abundance and diversity is quantified using measures such as TCR clonality (using Shannon's diversity index or Shannon's entropy), which includes the TCR richness (number of unique TCR clonotypes) and evenness (distribution of TCR clonotypes). While bulk RNA seq data can act as a surrogate to profile TCR repertoires in the TME with tools such as MiXCR (Bolotin et al., 2015; Finotello, Rieder, et al., 2019), high-throughput TCR sequencing technologies offer an improved resolution and efficiency in characterizing the T-cell clonotypes compared to the bulk RNA seq.

## 4.4 Single-cell transcriptomics for enhanced resolution analysis of the TME

Single-cell RNA-sequencing (scRNAseq) techniques provide unprecedented resolution at the level of single-cell entities. In the last few years, the field has evolved from low-throughput technologies (e.g., CEL-Seq) to high-throughput ones that capture up to 80,000 cells per run in $<10$ min (e.g., $10\times$ Chromium). Consequently, the high dimensionality datasets obtained from scRNAseq technologies have grown exponentially, fueling biological research as well as contributing to global initiatives such as The Human Cell Atlas for comprehensive molecular profiling of individual cell types in the human body (Regev et al., 2017). Plethora of tools are being developed to process raw data from scRNAseq, and a detailed list of such tools is available (Davis et al., 2018); the most popular tool is Seurat available as an R package (Stuart et al., 2019). scRNAseq has enabled detailed assessments of tumor heterogeneity and sub-population variability at a very high resolution and has facilitated clinically relevant findings (Clarke et al., 2019; Ren, Kang, & Zhang, 2018). For example, scRNAseq analyses performed on tumors from lung cancer patients treated with *PD1* antibody revealed that the patients who responded well to therapy had elevated levels of tissue-resident memory ($T_{RM}$) cells after treatment compared to pre-treatment (Clarke et al., 2019). scRNAseq has created an opportunity to study disease associated dynamic cell states which include developmental transitions in cells and branching cell lineages. This is a big advantage of scRNAseq over bulk RNAseq as it allows for inference of cell trajectories. Cell trajectories can be represented as cyclic, linear, bifurcating or tree-like topologies that capture the TME on a continuum of cell states; this level of granularity is unattainable through bulk RNAseq alone. Over 50 different methods for trajectory inference have been developed (Saelens, Cannoodt, Todorov, & Saeys, 2019). Comparative studies that surveyed numerous existing methods for trajectory inference reported that Monocle2, Slingshot, and TSCAN generally outperform other methods evaluated (Saelens et al., 2019; Soneson & Robinson, 2018). scRNAseq expression profiles are also being increasingly used to compile reference signatures for methods that deconvolve TME from its bulk expression profiles (Frishberg et al., 2019; Jew et al., 2019; Menden et al., 2019; Newman et al., 2019; Wang, Park, Susztak, et al., 2019). Rapid innovations in the scRNAseq technologies and the associated data analyses methods will continue to improve strategies that handle high noise associated with scRNAseq outputs and will enable automated annotation of deconvolved cell maps; these added capabilities will expand the application of scRNAseq to make discoveries in tumor immunology.

## 5. Epigenetics based immune profiling using DNA methylation

DNA methylation refers to the hereditary epigenetic modifications of the DNA, where a methyl group is covalently transferred to the cytosine in the CG dinucleotide (CpG) context (Jin, Li, & Robertson, 2011). Since DNA methylation patterns are known to be lineage and cell type specific (Jin et al., 2011; Teschendorff & Relton, 2018), they can be used to deconvolve complex tissues into its constituent cell types. In this respect, DNA methylation profile of a given tissue can be viewed as a linear weighted sum of its constituent cell type specific methylation profiles. Advancements in technological capabilities to quantify the human methylome using platforms ranging from HumanMethylation27 (270,000 CpG sites) to HumanMethylation450 (485,000 CpG sites) and now HumanMethylationEPIC array (860,000 CpG sites) and whole-genome bisulfite sequencing has provided large genome-scale data repositories that could be leveraged for deconvolution purposes.

Deconvolution methods for methylation data were originally developed to adjust for confounding effects of cell type heterogeneity on phenotypes of interest in epigenome-wide studies (Teschendorff & Zheng, 2017; Jaffe & Irizarry, 2014). These methods gradually found applications in deconvolution of the TME into its resident cell types. Computational tools available to deconvolve bulk DNA methylation profiles can be broadly categorized into two main types: reference based or reference free (Teschendorff & Relton, 2018; Teschendorff & Zheng, 2017); these tools are readily available for implementation as R or python packages (Teschendorff & Relton, 2018). Comparative analysis of these methods had revealed that no single algorithm gives a universally optimum performance (Teschendorff & Zheng, 2017). Prior to deconvolution, raw DNA methylation profiles are first pre-processed and normalized (Koestler et al., 2016; Wang, Wu, & Wang, 2018). Principal component analysis (PCA) is applied to the beta values to identify technical batch effects within the sample cohort; identified batch effects could be corrected by applying ComBat which is a routinely used empirical Bayes based method (Johnson et al., 2007). The cleaned-up data is used for deconvolution.

## 5.1 Reference-based deconvolution

Reference-based deconvolution is a supervised approach that requires cell type specific differentially methylated CpG sites as reference for deconvolution. The most popular reference-based method is the Houseman method which is based on quadratic programming (implemented using "estimateCellCounts" function in the minfi Bioconductor package) that gives normalized coefficients representative of cell type proportions (Houseman et al., 2012). For the purpose of immune cell type deconvolution, Houseman method has been extensively used and has led to clinically relevant findings, e.g., high neutrophil to lymphocyte ratios inferred from DNA methylation profiles were found to associate with an increased risk of cancer or poor survival upon systemic therapies in various cancer types including gliomas and lung cancer (Grieshober et al., 2018; Koestler et al., 2017; Wiencke et al., 2017).

Chakravarthy et al. adapted CIBERSORT, which is a popular expression-based deconvolution tool, to estimate cell type abundances from methylation profiles of HPV negative head and neck tumors from the TCGA and called this implementation methylCIBERSORT (Chakravarthy et al., 2018). The group compiled reference signatures from methylation arrays of purified fibroblasts and seven immune cell types as follows: (1) beta values were processed and normalized using the *minfi* Bioconductor package (single sample Noob normalization); (2) pair-wise linear models were applied to these beta values using the *limma* package to identify differentially methylated sites where the median difference in beta values of cell types is $\geq 0.25$ (FDR $< 0.01$); and (3) beta values for each differentially methylated site were averaged over all samples of each cell type and the resulting matrix was referred to as the methylation variability profiles (MVP). These MVPs were input as a reference signatures for deconvolution in CIBERSORT. Consensus clustering (clusterCons package with 100 iterations) was applied to the deconvolved cell proportions from the head and neck tumors and two immune clusters were identified: hot (high proportion of immune cells) and cold (low proportion of immune cells). This data was then used to train an elastic net model (with five-fold cross-validation) that allows classification of tumors into either immune hot or cold based on their deconvolved cell proportions.

Reference-based approaches have also been used to estimate global leukocyte fractions in a tumor (Carter et al., 2012; Frobel et al., 2018; Thorsson et al., 2018). In one such analyses to infer leukocytes in ovarian cancer (Carter et al., 2012), differentially methylated CpG sites were first compiled from pure leukocytes (buffy coat) and normal tissue (fallopian tubes) groups as follows: CpG sites were ordered based on the difference between the averaged beta values from each tissue group; 100 CpG sites each from the top and bottom of the ordered sites were selected as reference signatures. The following mixture model was then applied to each sample for each CpG site from the reference signature: $Tij = LifL + Ti(1 - fL)$ (Carter et al., 2012), where, $fL$ = leucocyte fraction in the sample; $Tij$ = beta value for CpG site i in tumor sample j; $Li$ = average beta value for that CpG site in the peripheral blood leukocytes; $Ti$ = beta value: cohort min if CpG site is from leukocytes | max if CpG site is from normal tissue. The above equation was solved for fL and kernel density was estimated over all differentially methylated signature sites for each sample, the final leukocyte fraction per sample was calculated as the mode of the estimated kernel density (Carter et al., 2012). The leukocyte fractions quantified from this analysis negatively correlated with tumor purity estimates from DNA sequencing data using ABSOLUTE.

While reference-based methods have been successful in estimating TME cell proportions, availability of suitable and reliable pure cell type methylation profiles has been a limiting factor. Besides, methods used to identify robust differentially methylated signatures also need to mature so as better handle collinearity in DNA methylation profiles of closely related cell types (Accomando, Wiencke, Houseman, Nelson, & Kelsey, 2014; Koestler et al., 2016; Teschendorff, Breeze, Zheng, & Beck, 2017; Teschendorff & Relton, 2018). Liang et al. proposed a hybrid strategy to improve identification of cancer specific CpG markers for deconvolution (Liang et al., 2019); in their paper they used cell type estimates from bulk RNAseq (quantified using ssGSEA or TIMER) as response variables to train elastic net regression models with stability selection as follows: starting from a random selection of 10% CpG sites as candidate features, 500 iterations of the model were run to identify a stable signature of CpG markers for that cell type (*glmnet* package in R) (Liang et al., 2019). Because this method used RNAseq based estimates as the ground truth, errors from RNAseq deconvolution will be expected to propagate in such models. Single-cell methylation profiles will enable extraction of better cell type specific reference signatures in future.

### 5.2 Reference-free deconvolution

Reference-free approach is an unsupervised approach where underlying components in a tissue mixture are identified without any prior knowledge of cell type specific methylation profiles. Prominent reference-free methods used for cell type deconvolution are based on variations of non-negative matrix factorization (NMF), e.g., NMF with convex constraints (RefFreeCellMix) (Houseman et al., 2016), regularized NMF (MeDeCom) (Lutsik et al., 2017). The NMF approaches primarily give differentially methylated sites as latent features associated with each of the deconvolved component. The biological interpretation of these components becomes the key challenge. One way to interpret the deconvolved components would be to cluster the components with reference methylation profiles (e.g., hierarchical clustering with person correlation as distance measure and average linkage) and infer cell types based on the reference cell type profiles it most closely clusters with (Lutsik et al., 2017). However, this strategy again relies on the availability of relevant references. Besides, NMF approaches are computationally intensive and require estimation of an optimal *k* (i.e., number of components) for decomposition.

## 6. Deep learning for immune spatial contexture

Histopathology images of tumor specimens provide useful information that may not be adequately captured by the existing high-throughput sequencing techniques. As an example, aside estimating the abundance of numerous immune-stromal cell types within the TME, imaging data also provides information on their spatial density and organization. The spatial context of tumor-immune compartments within the TME, such as immune cells being localized in the tumor center or being trapped at the tumor margins, has major implications in clinical response and patient prognosis (Ali et al., 2016; Fridman, Pagès, Sautès-Fridman, & Galon, 2012; Fridman et al., 2017; Heindl et al., 2018; Pagès et al., 2018; Saltz et al., 2018; Yuan, 2015). As an example, immunoscore—a measure to estimate the tumor-infiltrating lymphocyte (TIL) content in a tumor's core and at its invasive margins—is quantified from the digital pathology images of a tumor specimen; high immunoscore is strongly associated with lowest risk of relapse in stage I–III colon cancer patients who received standard-of-care treatments and is reported as an improved prognostic biomarker compared to other risk factors in colon cancer (Galon et al., 2014; Pagès et al., 2018).

Routine analysis of histopathology images is mostly limited to manual inspection by trained pathologists, a process that is laborious and subjective that may lead to certain inconsistencies due to inter-observer biases. Alternatively, computational techniques for automated analyses of digitized pathology images provide competitive solutions. Deep learning approaches particularly convolutional neural network (CNN) have found applications in classification, object detection or segmentation of biomedical images since the mid-1990s (Litjens et al., 2017). These methods are now emerging as useful tools in oncology to automate digitized histopathology image analyses with high precision and reproducibility. CNNs have been recently applied to IHC or hematoxylin and eosin (H&E) stained whole-slide images to study tissue heterogeneity, cellular phenotypes, and tissue morphology, and to assess cancer or immune or stroma specific biomarkers, to identify cancer subtypes, to classify malignant versus benign skin lesions, as well as to directly predict patient survival outcomes in brain, lung and colorectal cancers (Aprupe, Litjens, Brinker, van der Laak, & Grabe, 2019; Bychkov et al., 2018; Caicedo et al., 2017; Esteva et al., 2017; Hipp & Stumpe, 2018; Kather et al., 2019; Khosravi, Kazemi, Imielinski, Elemento, & Hajirasouliha, 2018; Koelzer, Sirinukunwattana, Rittscher, & Mertz, 2019; Mobadersany et al., 2018; Paeng et al., 2019; Saltz et al., 2018; Wang et al., 2019).

The growing popularity of CNNs in image analyses stems from their ability to learn relevant features directly from the raw pixels of a structured image through series of non-linear transformations. The quality of the images used for training a CNN is critical for prediction accuracies, therefore image preparation steps should be considered prior to training the model which may include: determining adequate thresholds for image quality (i.e., filter out images that are blurred or have tissue-related artifacts, or have high intensity pixels that likely originate from artifacts), and correcting for the uneven illumination of the image due to non-uniform lighting or slide orientation (i.e., apply a correction function built for each batch of images) (Caicedo et al., 2017). Two broad CNN approaches undertaken for molecular profiling of pathology images to improve model's prediction accuracy include: (1) use of an autoencoder to learn unsupervised feature transformations from sampled image patches; in an autoencoder the output is a near identical reconstruction of the input image and the encoded compressed representation of the image is used to initialize the CNN for training or (2) use of an existing database of natural objects (e.g., ImageNet) to pre-train the CNN; this approach, known as transfer learning, eliminates the need to train the CNN from scratch and uses labeled training data to train the last layer of the model or to fine-tune the training parameters using the pre-trained version of the model.

In a pan-cancer work for TIL quantification from digitized histopathology images, Saltz et al. applied CNN (with convolutional autoencoder) to 5455H&E whole-slide images for 13 cancer types obtained from the TCGA (Saltz et al., 2018). To classify regions of lymphocyte-infiltration within a tissue sample, a set of pathology images were first partitioned into $50 \times 50\,\mu m^2$ patches (which was equivalent to $100 \times 100$ square pixels at $20\times$ magnification) and these patches were used to train a semi-supervised CNN initialized by a convolutional autoencoder that learns representation of nuclei and lymphocytes from sampled pathology images. The lymphocyte CNN predicted the probability for TIL infiltration in every patch of the training dataset and if this probability was above a given threshold, the patch was labeled as TIL positive. The training step also involved iterative review of TIL prediction results by expert pathologists followed by fine-tuning and re-training of the model (with cross validation) with updated labels until the desired prediction accuracy for the model was achieved. In addition, images were partitioned into $500 \times 500\,\mu m^2$ patches to train a separate CNN model for necrosis segmentation using DeconvNet; the purpose of this segmentation was to remove falsely predicted TIL positive regions that overlapped with the regions of tissue necrosis. The TIL content in a given tumor sample was quantified as the number of TIL positive patches divided by the total number of patches from that image. TIL fractions estimated from the CNN model showed concordance with the TIL levels manually assigned to 400 patches by 3 independent pathologists (polyserial correlation coefficient = 0.36). Besides estimation of the total TIL content, various indices that define the structural patterns of lymphocyte infiltrations within the TME were also calculated, e.g., number of spatially connected TIL regions (clusters), mean dispersion of a cluster. These indices showed interesting associations with survival in certain cancer types, e.g., higher TIL cluster count associated with improved OS in melanoma, whereas higher spatial dispersion of TIL clusters (quantified as Ball-Hall index) associated with poor OS in breast cancer. TIL maps for the 5455H&E whole-slide images from this study are made publicly available for further research use (Saltz et al., 2018).

Sophisticated multiplexed imaging techniques such as multi-staining panels (e.g., Vectra® PerkinElmer), multiplexed immunofluorescence platforms (e.g., CODEX), spatial imaging coupled with mass spectrometry (e.g., multiplexed ion beam imaging (MiBi)), flow cytometry coupled with mass spectrometry (e.g., CyTOF) are now becoming readily available for multiparametric immune-phenotyping and profiling (Finotello, Rieder, et al., 2019; Koelzer et al., 2019). Simultaneously, rapid machine learning advances toward deeper network architectures (e.g., 19-layer VGG-19, 22-layer GoogleNet/Inception V3, ResNet) are driving improved performance and higher efficiency in image analyses. Taken together, the wealth of imaging datasets that will become available through the state-of-the-art imaging platforms could be adequately mined with standardized deep learning approaches that will not only enrich our understanding of tumor immunology but also facilitate development of cost-effective robust prognostic and diagnostic biomarkers for patient outcomes.

## 7. Summary

High-throughput sequencing data together with the wide array of computational methods enable systematic quantification and study of tumor-immune interactions at a molecular level. This chapter discussed the various computational approaches available to analyze the TME especially in context of its immunogenic potential. Integration of these complementary approaches with clinical variables will allow stratification of patients based on their overall immune status and study mechanisms of immune surveillance and escape operative in these subgroups with implications in cancer precision medicine.

## References

Accomando, W. P., Wiencke, J. K., Houseman, E. A., Nelson, H. H., & Kelsey, K. T. (2014). Quantitative reconstruction of leukocyte subsets using DNA methylation. *Genome Biology*, *15*, R50.

Ali, H. R., Dariush, A., Provenzano, E., Bardwell, H., Abraham, J. E., Iddawela, M., et al. (2016). Computational pathology of pre-treatment biopsies identifies lymphocyte

density as a predictor of response to neoadjuvant chemotherapy in breast cancer. *Breast Cancer Research*, *18*, 21.
Anagnostou, V., Smith, K. N., Forde, P. M., Niknafs, N., Bhattacharya, R., White, J., et al. (2017). Evolution of neoantigen landscape during immune checkpoint blockade in non-small cell lung cancer. *Cancer Discovery*, *7*, 264–276.
Anderton, S. M. (2004). Post-translational modifications of self antigens: Implications for autoimmunity. *Current Opinion in Immunology*, *16*, 753–758.
Aprupe, L., Litjens, G., Brinker, T. J., van der Laak, J., & Grabe, N. (2019). Robust and accurate quantification of biomarkers of immune cells in lung cancer micro-environment using deep convolutional neural networks. *PeerJ*, *7*, e6335.
Aurisicchio, L., Salvatori, E., Lione, L., Bandini, S., Pallocca, M., Maggio, R., et al. (2019). Poly-specific neoantigen-targeted cancer vaccines delay patient derived tumor growth. *Journal of Experimental and Clinical Cancer Research*, *38*, 78.
Auslander, N., Zhang, G., Lee, J. S., Frederick, D. T., Miao, B., Moll, T., et al. (2018). Robust prediction of response to immune checkpoint blockade therapy in metastatic melanoma. *Nature Medicine*, *24*, 1545–1549.
Avila, C. F., Vandesompele, J., Mestdagh, P., & De Preter, K. (2018). Computational deconvolution of transcriptomics data from mixed cell populations. *Bioinformatics*, *34*, 1969–1979.
Ayers, M., Lunceford, J., Nebozhyn, M., Murphy, E., Loboda, A., Kaufman, D. R., et al. (2017). IFN-γ-related mRNA profile predicts clinical response to PD-1 blockade. *Journal of Clinical Investigation*, *127*, 2930–2940.
Balachandran, V. P., Łuksza, M., Zhao, J. N., Makarov, V., Moral, J. A., Remark, R., et al. (2017). Identification of unique neoantigen qualities in long-term survivors of pancreatic cancer. *Nature*, *551*, 512–516.
Bhinder, B., & Elemento, O. (2017). Towards a better cancer precision medicine: Systems biology meets immunotherapy. *Current Opinion in Systems Biology*, *2*, 67–73.
Binnewies, M., Roberts, E. W., Kersten, K., Chan, V., Fearon, D. F., Merad, M., et al. (2018). Understanding the tumor immune microenvironment (TIME) for effective therapy. *Nature Medicine*, *24*, 541–550.
Blank, C. U., Haanen, J. B., Ribas, A., & Schumacher, T. N. (2016). The "cancer immunogram" *Science*, *352*, 658–660.
Bolotin, D. A., Poslavsky, S., Mitrophanov, I., Shugay, M., Mamedov, I. Z., Putintseva, E. V., et al. (2015). MiXCR: Software for comprehensive adaptive immunity profiling. *Nature Methods*, *12*, 380–381.
Bubie, A., Gonzalez-Kozlova, E., Akers, N., Villanueva, A., & Losic, B. (2019). Tumor fitness, immune exhaustion and clinical outcomes: Impact of immune checkpoint inhibitors. *BioRxiv*. https://doi.org/10.1101/679886.
Bychkov, D., Linder, N., Turkki, R., Nordling, S., Kovanen, P. E., Verrill, C., et al. (2018). Deep learning based tissue analysis predicts outcome in colorectal cancer. *Science Reports*, *8*, 3395.
Caicedo, J. C., Cooper, S., Heigwer, F., Warchal, S., Qiu, P., Molnar, C., et al. (2017). Data-analysis strategies for image-based cell profiling. *Nature Methods*, *14*, 849–863.
Carter, S. L., Cibulskis, K., Helman, E., McKenna, A., Shen, H., Zack, T., et al. (2012). Absolute quantification of somatic DNA alterations in human cancer. *Nature Biotechnology*, *30*, 413–421.
Chakravarthy, A., Furness, A., Joshi, K., Ghorani, E., Ford, K., Ward, M. J., et al. (2018). Pan-cancer deconvolution of tumour composition using DNA methylation. *Nature Communications*, *9*, 3220.
Chalmers, Z. R., Connelly, C. F., Fabrizio, D., Gay, L., Ali, S. M., Ennis, R., et al. (2017). Analysis of 100,000 human cancer genomes reveals the landscape of tumor mutational burden. *Genome Medicine*, *9*, 34.

Chan, T. A., Yarchoan, M., Jaffee, E., Swanton, C., Quezada, S. A., Stenzinger, A., et al. (2019). Development of tumor mutation burden as an immunotherapy biomarker: Utility for the oncology clinic. *Annals of Oncology, 30*, 44–56.

Charoentong, P., Finotello, F., Angelova, M., Mayer, C., Efremova, M., Rieder, D., et al. (2017). Pan-cancer immunogenomic analyses reveal genotype-immunophenotype relationships and predictors of response to checkpoint blockade. *Cell Reports, 18*, 248–262.

Clarke, J., Panwar, B., Madrigal, A., Singh, D., Gujar, R., Wood, O., et al. (2019). Single-cell transcriptomic analysis of tissue-resident memory T cells in human lung cancer. *The Journal of Experiential Medicine, 216*, 2128–2149.

Cobbold, M., De La Peña, H., Norris, A., Polefrone, J. M., Qian, J., English, A. M., et al. (2013). MHC class I-associated phosphopeptides are the targets of memory-like immunity in leukemia. *Science Translational Medicine, 5*, 203ra125.

Cyriac, G., & Gandhi, L. (2018). Emerging biomarkers for immune checkpoint inhibition in lung cancer. *Seminars in Cancer Biology, 52*, 269–277.

Danaher, P., Warren, S., Dennis, L., D'Amico, L., White, A., Disis, M. L., et al. (2017). Gene expression markers of tumor infiltrating leukocytes. *Journal for Immunotherapy of Cancer, 5*, 18.

Davis, S., Kutum, R., Zappia, L., Sorenson, J., Kiselev, V., Olivier, P., et al. (2018). *Awesome-single-cell: List of software packages for single-cell data analysis, including RNA-seq, ATAC-seq, etc.* Zenodo. https://doi.org/10.5281/zenodo.1117762.

Dimitrov, I., Garnev, P., Flower, D. R., & Doytchinova, I. (2010). MHC class II binding prediction—A little help from a friend. *Journal of Biomedicine and Biotechnology, 2010*, 705821.

Duan, F., Duitama, J., Al Seesi, S., Ayres, C. M., Corcelli, S. A., Pawashe, A. P., et al. (2014). Genomic and bioinformatic profiling of mutational neoepitopes reveals new rules to predict anticancer immunogenicity. *Journal of Experimental Medicine, 211*, 2231–2248.

Dufva, O., Pölönen, P., Brück, O., Keränen, M., Mehtonen, J., Kumar, A., et al. (2019). Immunogenomic landscape of hematological malignancies. *bioRxiv*. https://doi.org/10.1101/618918.

Ellrot, T. K., Bailey, M. H., Saksena, G., Covington, K. R., Kandoth, C., Stewart, C., et al. (2018). Scalable open science approach for mutation calling of tumor exomes using multiple genomic pipelines. *Cell Systems, 6*, 271–281.

Esteva, A., Kuprel, B., Novoa, R. A., Ko, J., Swetter, S. M., Blau, H. M., et al. (2017). Dermatologist-level classification of skin cancer with deep neural networks. *Nature, 542*, 115–118.

Finotello, F., Mayer, C., Plattner, C., Laschober, G., Rieder, D., Hackl, H., et al. (2019). Molecular and pharmacological modulators of the tumor immune contexture revealed by deconvolution of RNA-seq data. *Genome Medicine, 11*, 34.

Finotello, F., Rieder, D., Hackl, H., & Trajanoski, Z. (2019). Next-generation computational tools for interrogating cancer immunity. *Nature Reviews Genetics, 20*, 724–746.

Finotello, F., & Trajanoski, Z. (2018). Quantifying tumor-infiltrating immune cells from transcriptomics data. *Cancer Immunology Immunotherapy, 67*, 1031–1040.

Fouad, Y. A., & Aanei, C. (2017). Revisiting the hallmarks of cancer. *American Journal of Cancer Research, 7*, 1016–1036.

Fridman, W. H., Pagès, F., Sautès-Fridman, C., & Galon, J. (2012). The immune contexture in human tumours: Impact on clinical outcome. *Nature Reviews Cancer, 12*, 298–306.

Fridman, W. H., Zitvogel, L., Sautès-Fridman, C., & Kroemer, G. (2017). The immune contexture in cancer prognosis and treatment. *Nature Reviews Clinical Oncology, 14*, 717–734.

Frishberg, A., Peshes-Yaloz, N., Cohn, O., Rosentul, D., Steuerman, Y., Valadarsky, L., et al. (2019). Cell composition analysis of bulk genomics using single-cell data. *Nature Methods, 16*, 327–332.

Frobel, J., Božić, T., Lenz, M., Uciechowski, P., Han, Y., Herwartz, R., et al. (2018). Leukocyte counts based on DNA methylation at individual cytosines. *Clinical Chemistry, 64*, 566–575.

Galon, J., Mlecnik, B., Bindea, G., Angell, H. K., Berger, A., Lagorce, C., et al. (2014). Towards the introduction of the 'Immunoscore' in the classification of malignant tumours. *The Journal of Pathology, 232*, 199–209.

Gan, H. K., Burgess, A. W., Clayton, A. H., & Scott, A. M. (2012). Targeting of a conformationally exposed, tumor-specific epitope of EGFR as a strategy for cancer therapy. *Cancer Research, 72*, 2924–2930.

Gao, Q., Liang, W. W., Foltz, S. M., Mutharasu, G., Jayasinghe, R. G., Cao, S., et al. (2018). Driver fusions and their implications in the development and treatment of human cancers. *Cell Reports, 23*, 227–238.

Garofalo, A., Sholl, L., Reardon, B., Taylor-Weiner, A., Amin-Mansour, A., Miao, D., et al. (2016). The impact of tumor profiling approaches and genomic data strategies for cancer precision medicine. *Genome Medicine, 8*, 79.

Gentles, A. J., Newman, A. M., Liu, C. L., Bratman, S. V., Feng, W., Kim, D., et al. (2015). The prognostic landscape of genes and infiltrating immune cells across human cancers. *Nature Medicine, 21*, 938–945.

Ghorani, E., Rosenthal, R., McGranahan, N., Reading, J. L., Lynch, M., Peggs, K. S., et al. (2018). Differential binding affinity of mutated peptides for MHC class I is a predictor of survival in advanced lung cancer and melanoma. *Annals of Oncology, 29*, 271–279.

Gowthaman, U., & Agrewala, J. N. (2008). In silico tools for predicting peptides binding to HLA-class II molecules: More confusion than conclusion. *Journal of Proteome Research, 7*, 154–163.

Grieshober, L., Graw, S., Barnett, M. J., Thornquist, M. D., Goodman, G. E., Chen, C., et al. (2018). Methylation-derived neutrophil-to-lymphocyte ratio and lung cancer risk in heavy smokers. *Cancer Prevention Research (Philadelphia), 11*, 727–734.

Heindl, A., Sestak, I., Naidoo, K., Cuzick, J., Dowsett, M., & Yuan, Y. (2018). Relevance of spatial heterogeneity of immune infiltration for predicting risk of recurrence after endocrine therapy of ER+ breast cancer. *Journal of the National Cancer Institute, 110*, 166–175.

Hellmann, M. D., Ciuleanu, T. E., Pluzanski, A., Lee, J. S., Otterson, G. A., Audigier-Valette, C., et al. (2018). Nivolumab plus Ipilimumab in lung cancer with a high tumor mutational burden. *New England Journal of Medicine, 378*, 2093–2104.

Hipp, J. D., & Stumpe, M. (2018). Abstract PL02-01: Advancing cancer diagnostics with artificial intelligence. *AACR; Cancer Research, 78,* Abstract nr PL02-01.

Hopkins, A. C., Yarchoan, M., Durham, J. N., Yusko, E. C., Rytlewski, J. A., Robins, H. S., et al. (2018). T cell receptor repertoire features associated with survival in immunotherapy-treated pancreatic ductal adenocarcinoma. *JCI Insight, 3,* pii: 122092.

Hosen, N., Matsunaga, Y., Hasegawa, K., Matsuno, H., Nakamura, Y., Makita, M., et al. (2017). The activated conformation of integrin β7 is a novel multiple myeloma-specific target for CAR T cell therapy. *Nature Medicine, 23*, 1436–1443.

Houseman, E. A., Accomando, W. P., Koestler, D. C., Christensen, B. C., Marsit, C. J., Nelson, H. H., et al. (2012). DNA methylation arrays as surrogate measures of cell mixture distribution. *BMC Bioinformatics, 13*, 86.

Houseman, E. A., Kile, M. L., Christiani, D. C., Ince, T. A., Kelsey, K. T., & Marsit, C. J. (2016). Reference-free deconvolution of DNA methylation data and mediation by cell composition effects. *BMC Bioinformatics, 17*, 259.

Hugo, W., Zaretsky, J. M., Sun, L., Song, C., Moreno, B. H., Hu-Lieskovan, S., et al. (2016). Genomic and transcriptomic features of response to anti-PD-1 therapy in metastatic melanoma. *Cell, 165*, 35–44.

Hundal, J., Carreno, B. M., Petti, A. A., Linette, G. P., Griffith, O. L., Mardis, E. R., et al. (2016). pVAC-Seq: A genome-guided in silico approach to identifying tumor neoantigens. *Genome Medicine*, *8*, 11.

Hundal, J., Kiwala, S., Feng, Y. Y., Liu, C. J., Govinda, R., Chapman, W. C., et al. (2019). Accounting for proximal variants improves neoantigen prediction. *Nature Genetics*, *51*, 175–179.

Iglesia, M. D., Parker, J. S., Hoadley, K. A., Serody, J. S., Perou, C. M., & Vincent, B. G. (2016). Genomic analysis of immune cell infiltrates across 11 tumor types. *Journal of the National Cancer Institute*, *108*, djw44. https://doi.org/10.1093/jnci/djw144.

Jaffe, A. E., & Irizarry, R. A. (2014). Accounting for cellular heterogeneity is critical in epigenome-wide association studies. *Genome Biology*, *15*, R31.

Jayasinghe, R. G., Cao, S., Gao, Q., Wendl, M. C., Vo, N. S., Reynolds, S. M., et al. (2018). Systematic analysis of splice-site-creating mutations in cancer. *Cell Reports*, *23*, 270–281.

Jew, B., Alvarez, M., Rahmani, E., Miao, Z., Ko, A., Sul, J. H., et al. (2019). Accurate estimation of cell composition in bulk expression through robust integration of single-cell information. *bioRxiv*. https://doi.org/10.1101/669911.

Jiang, P., Gu, S., Pan, D., Fu, J., Sahu, A., Hu, X., et al. (2018). Signatures of T cell dysfunction and exclusion predict cancer immunotherapy response. *Nature Medicine*, *24*, 1550–1558.

Jin, B., Li, Y., & Robertson, K. D. (2011). DNA methylation: Superior or subordinate in the epigenetic hierarchy? *Genes & Cancer*, *2*, 607–617.

Johnson, W. E., Li, C., & Rabinovic, A. (2007). Adjusting batch effects in microarray expression data using Empirical Bayes methods. *Biostatistics*, *8*, 118–127.

Joyce, J. A., & Fearon, D. T. (2015). T cell exclusion, immune privilege, and the tumor microenvironment. *Science*, *348*, 74–80.

Kahles, A., Lehmann, K. V., Toussaint, N. C., Hüser, M., Stark, S. G., Sachsenberg, T., et al. (2018). Comprehensive analysis of alternative splicing across tumors from 8,705 patients. *Cancer Cell*, *34*, 211–224.

Kalaora, S., Wolf, Y., Feferman, T., Barnea, E., Greenstein, E., Reshef, D., et al. (2018). Combined analysis of antigen presentation and T cell recognition reveals restricted immune responses in melanoma. *Cancer Discovery*, *8*, 1366–1375.

Kather, J. N., Krisam, J., Charoentong, P., Luedde, T., Herpel, E., Weis, C. A., et al. (2019). Predicting survival from colorectal cancer histology slides using deep learning: A retrospective multicenter study. *PLoS Medicine*, *24*, 16.

Keskin, D. B., Anandappa, A. J., Sun, J., Tirosh, I., Mathewson, N. D., Li, S., et al. (2019). Neoantigen vaccine generates intratumoral T cell responses in phase Ib glioblastoma trial. *Nature*, *565*, 234–239.

Khosravi, P., Kazemi, E., Imielinski, M., Elemento, O., & Hajirasouliha, I. (2018). Deep convolutional neural networks enable discrimination of heterogeneous digital pathology images. *eBioMedicine*, *27*, 317–328.

Khunger, A., Rytlewski, J. A., Fields, P., Yusko, E. C., & Tarhini, A. A. (2019). The impact of CTLA-4 blockade and interferon-α on clonality of T-cell repertoire in the tumor microenvironment and peripheral blood of metastatic melanoma patients. *Oncoimmunology*, *8*, e1652538.

Koelzer, V. H., Sirinukunwattana, K., Rittscher, J., & Mertz, K. D. (2019). Precision immunoprofiling by image analysis and artificial intelligence. *Virchows Archiv*, *474*, 511–522.

Koestler, D. C., Jones, M. J., Usset, J., Christensen, B. C., Butler, R. A., Kobor, M. S., et al. (2016). Improving cell mixture deconvolution by identifying optimal DNA methylation libraries (IDOL). *BMC Bioinformatics*, *17*, 120.

Koestler, D. C., Usset, J., Christensen, B. C., Marsit, C. J., Karagas, M. R., Kelsey, K. T., et al. (2017). DNA methylation-derived neutrophil-to-lymphocyte ratio: An epigenetic tool to explore cancer inflammation and outcomes. *Cancer Epidemiology, Biomarkers and Prevention*, *26*, 328–338.

Krøigård, A. B., Thomassen, M., Lænkholm, A. V., Kruse, T. A., & Larsen, M. J. (2016). Evaluation of nine somatic variant callers for detection of somatic mutations in exome and targeted deep sequencing data. *PLoS One, 11,* e0151664.

Kumai, T., Ishibashi, K., Oikawa, K., Matsuda, Y., Aoki, N., Kimura, S., et al. (2014). Induction of tumor-reactive T helper responses by a posttranslational modified epitope from tumor protein p53. *Cancer Immunology, Immunotherapy, 63,* 469–478.

Lauss, M., Donia, M., Harbst, K., Andersen, R., Mitra, S., Rosengren, F., et al. (2017). Mutational and putative neoantigen load predict clinical benefit of adoptive T cell therapy in melanoma. *Nature Communications, 8,* 1738.

Lawrence, M. S., Stojanov, P., Polak, P., Kryukov, G. V., Cibulskis, K., Sivachenko, A., et al. (2013). Mutational heterogeneity in cancer and the search for new cancer-associated genes. *Nature, 499,* 214–218.

Le, D. T., Uram, J. N., Wang, H., Bartlett, B. R., Kemberling, H., Eyring, A. D., et al. (2015). PD-1 blockade in tumors with mismatch-repair deficiency. *New England Journal of Medicine, 372,* 2509–2520.

Li, B., Severson, E., Pignon, J. C., Zhao, H., Li, T., Novak, J., et al. (2016). Comprehensive analyses of tumor immunity: Implications for cancer immunotherapy. *Genome Biology, 17,* 174.

Liang, C., Yu, X., Li, B., Chen, Y. A., Conejo-Garcia, J. R., & Wang, X. (2019). DNA methylation-based immune cell deconvolution in solid tumors. *bioRxiv.* https://doi.org/10.1101/619965.

Litjens, G., Kooi, T., Bejnordi, B. E., Setio, A. A. A., Ciompi, F., Ghafoorian, M., et al. (2017). A survey on deep learning in medical image analysis. *Medical Image Analysis, 42,* 60–88.

Łuksza, M., Riaz, N., Makarov, V., Balachandran, V. P., Hellmann, M. D., Solovyov, A., et al. (2017). A neoantigen fitness model predicts tumour response to checkpoint blockade immunotherapy. *Nature, 551,* 517–520.

Lutsik, P., Slawski, M., Gasparoni, G., Vedeneev, N., Hein, M., & Walter, J. (2017). MeDeCom: Discovery and quantification of latent components of heterogeneous methylomes. *Genome Biology, 18,* 55.

Mandal, R., Samstein, R. M., Lee, K. W., Havel, J. J., Wang, H., Krishna, C., et al. (2019). Genetic diversity of tumors with mismatch repair deficiency influences anti-PD-1 immunotherapy response. *Science, 364,* 485–491.

Marcus, L., Lemery, S. J., Keegan, P., & Pazdur, R. (2019). FDA approval summary: Pembrolizumab for the treatment of microsatellite instability-high solid tumors. *Clinical Cancer Research, 25,* 3753–3758.

McGranahan, N., Furness, A. J., Rosenthal, R., Ramskov, S., Lyngaa, R., Saini, S. K., et al. (2016). Clonal neoantigens elicit T cell immunoreactivity and sensitivity to immune checkpoint blockade. *Science, 351,* 1463–1469.

Meléndez, B., Van Campenhout, C., Rorive, S., Remmelink, M., Salmon, I., & D'Haene, N. (2018). Methods of measurement for tumor mutational burden in tumor tissue. *Translational Lung Cancer Research, 7,* 661–667.

Menden, K., Marouf, M., Dalmia, A., Heutink, P., & Bonn, S. (2019). Deep-learning-based cell composition analysis from tissue expression profiles. *bioRxiv.* https://doi.org/10.1101/659227.

Miao, D., Margolis, C. A., Vokes, N. I., Liu, D., Taylor-Weiner, A., Wankowicz, S. M., et al. (2018). Genomic correlates of response to immune checkpoint blockade in microsatellite-stable solid tumors. *Nature Genetics, 50,* 1271–1281.

Mobadersany, P., Yousefi, S., Amgad, M., Gutman, D. A., Barnholtz-Sloan, J. S., Velázquez Vega, J. E., et al. (2018). Predicting cancer outcomes from histology and genomics using convolutional networks. *Proceedings of the National Academy of Sciences of the United States of America, 115,* E2970–E2979.

Moffitt, R. A., Marayati, R., Flate, E. L., Volmar, K. E., Loeza, S. G., Hoadley, K. A., et al. (2015). Virtual microdissection identifies distinct tumor- and stroma-specific subtypes of pancreatic ductal adenocarcinoma. *Nature Genetics*, *47*, 1168–1178.

Mohammadi, S., Zuckerman, N., Goldsmith, A., & Grama, A. (2017). A critical survey of deconvolution methods for separating cell types in complex tissues. *Institute of Electrical and Electronics Engineers*, *105*, 340–366.

Newman, A. M., Steen, C. B., Liu, C. L., Gentles, A. J., Chaudhuri, A. A., Scherer, F., et al. (2019). Determining cell type abundance and expression from bulk tissues with digital cytometry. *Nature Biotechnology*, *37*, 773–782.

Paeng, K., Jung, G., Lee, S., Cho, S. Y., Cho, E. Y., & Song, S. Y. (2019). Pan-cancer analysis of tumor microenvironment using deep learning-based cancer stroma and immune profiling in H&E images [abstract]. *AACR; Cancer Research*, *79,* Abstract nr 2445.

Page, D. B., Yuan, J., Redmond, D., Wen, Y. H., Durack, J. C., Emerson, R., et al. (2016). Deep sequencing of T-cell receptor DNA as a biomarker of clonally expanded TILs in breast cancer after immunotherapy. *Cancer Immunology*, *4*, 835–844.

Pagès, F., Mlecnik, B., Marliot, F., Bindea, G., Ou, F. S., Bifulco, C., et al. (2018). International validation of the consensus immunoscore for the classification of colon cancer: A prognostic and accuracy study. *Lancet*, *391*, 128–2139.

Prat, A., Navarro, A., Paré, L., Reguart, N., Galván, P., Pascual, T., et al. (2017). Immune-related gene expression profiling after PD-1 blockade in non-small cell lung carcinoma, head and neck squamous cell carcinoma, and melanoma. *Cancer Research*, 77, 3540–3550.

Rathe, S. K., Popescu, F. E., Johnson, J. E., Watson, A. L., Marko, T. A., Moriarity, B. S., et al. (2019). Identification of candidate neoantigens produced by fusion transcripts in human osteosarcomas. *Scientific Reports*, *9*, 358.

Regev, A., Teichmann, S. A., Lander, E. S., Amit, I., Benoist, C., Birney, E., et al. (2017). The human cell atlas. *eLife*, *6,* pii: e27041.

Ren, X., Kang, B., & Zhang, Z. (2018). Understanding tumor ecosystems by single-cell sequencing: Promises and limitations. *Genome Biology*, *19*, 211.

Riaz, N., Havel, J. J., Makarov, V., Desrichard, A., Urba, W. J., Sims, J. S., et al. (2017). Tumor and microenvironment evolution during immunotherapy with nivolumab. *Cell*, *171*, 934–949.

Rizvi, N. A., Hellmann, M. D., Snyder, A., Kvistborg, P., Makarov, V., Havel, J. J., et al. (2015). Mutational landscape determines sensitivity to PD-1 blockade in non-small cell lung cancer. *Science*, *348*, 124–128.

Rooney, M. S., Shukla, S. A., Wu, C. J., Getz, G., & Hacohen, N. (2015). Molecular and genetic properties of tumors associated with local immune cytolytic activity. *Cell*, *160*, 48–61.

Rosenthal, R., Cadieux, E. L., Salgado, R., Bakir, M. A., Moore, D. A., Hiley, C. T., et al. (2019). Neoantigen-directed immune escape in lung cancer evolution. *Nature*, *567*, 479–485.

Rozeman, E. A., Franchi, L., Kuilman, T., Krijgsman, O., van Akkooi, A. C. J., Kvistborg, P., et al. (2017). Biomarker analysis from the OpACIN trial (neo-/adjuvant ipilimumab + nivolumab (IPI + NIVO) in palpable stage 3 melanoma). *Journal for Immunotherapy of Cancer*, *5*(Suppl. 2), 99.

Saelens, W., Cannoodt, R., Todorov, H., & Saeys, Y. (2019). A comparison of single-cell trajectory inference methods: Towards more accurate and robust tools. *Nature Biotechnology*, *37*, 547–554.

Saltz, J., Gupta, R., Hou, L., Kurc, T., Singh, P., Nguyen, V., et al. (2018). Spatial organization and molecular correlation of tumor-infiltrating lymphocytes using deep learning on pathology images. *Cell Reports*, *23*, 181–193.

Samstein, R. M., Lee, C. H., Shoushtari, A. N., Hellmann, M. D., Shen, R., Janjigian, Y. Y., et al. (2019). Tumor mutational load predicts survival after immunotherapy across multiple cancer types. *Nature Genetics*, *51*, 202–206.

Schwitalle, Y., Kloor, M., Eiermann, S., Linnebacher, M., Kienle, P., Knaebel, H. P., et al. (2008). Immune response against frameshift-induced neopeptides in HNPCC patients and healthy HNPCC mutation carriers. *Gastroenterology*, *134*, 988–997.

Şenbabaoğlu, Y., Gejman, R. S., Winer, A. G., Liu, M., Van Allen, E. M., de Velasco, G., et al. (2016). Tumor immune microenvironment characterization in clear cell renal cell carcinoma identifies prognostic and immunotherapeutically relevant messenger RNA signatures. *Genome Biology*, *17*, 231.

Sette, A., Vitiello, A., Reherman, B., Fowler, P., Nayersina, R., Kast, W. M., et al. (1994). The relationship between class I binding affinity and immunogenicity of potential cytotoxic T cell epitopes. *Journal of Immunology*, *153*, 5586–5592.

Sia, D., Jiao, Y., Martinez-Quetglas, I., Kuchuk, O., Villacorta-Martin, C., Castro de Moura, M., et al. (2017). Identification of an immune-specific class of hepatocellular carcinoma, based on molecular features. *Gastroenterology*, *153*, 812–826.

Smart, A. C., Margolis, C. A., Pimentel, H., He, M. X., Miao, D., Adeegbe, D., et al. (2018). Intron retention is a source of neoepitopes in cancer. *Nature Biotechnology*, *36*, 1056–1058.

Smith, C. C., Selitsky, S. R., Chai, S., Armistead, P. M., Vincent, B. G., & Serody, J. S. (2019). Alternative tumour-specific antigens. *Nature Reviews Cancer*, *19*, 465–478.

Smyrk, T. C., Watson, P., Kaul, K., & Lynch, H. T. (2001). Tumour-infiltrating lymphocytes are a marker for microsatellite instability in colorectal carcinoma. *Cancer*, *91*, 2417–2422.

Snyder, A., Makarov, V., Merghoub, T., Yuan, J., Zaretsky, J. M., Desrichard, A., et al. (2014). Genetic basis for clinical response to CTLA-4 blockade in melanoma. *New England Journal of Medicine*, *371*, 2189–2199.

Soneson, C., & Robinson, M. D. (2018). Bias, robustness and scalability in single-cell differential expression analysis. *Nature Methods*, *15*, 255–261.

Spranger, S., Bao, R., & Gajewski, T. F. (2015). Melanoma-intrinsic β-catenin signalling prevents anti-tumour immunity. *Nature*, *523*, 231–235.

Stuart, T., Butler, A., Hoffman, P., Hafemeister, C., Papalexi, E., Mauck, W. M., et al. (2019). Comprehensive integration of single-cell data. *Cell*, *177*, 1888–1902.

Sweis, R. F., Spranger, S., Bao, R., Paner, G. P., Stadler, W. M., Steinberg, G., et al. (2016). Molecular drivers of the non-T-cell-inflamed tumor microenvironment in urothelial bladder cancer. *Cancer Immunology Research*, *4*, 563–568.

Tamborero, D., Rubio-Perez, C., Muiños, F., Sabarinathan, R., Piulats, J. M., Muntasell, A. A., et al. (2018). Pan-cancer landscape of interactions between solid tumors and infiltrating immune cell populations. *Clinical Cancer Research*, *24*, 3717–3728.

Teschendorff, A. E., Breeze, C. E., Zheng, S. C., & Beck, S. (2017). A comparison of reference-based algorithms for correcting cell-type heterogeneity in Epigenome-Wide Association Studies. *BMC Bioinformatics*, *18*, 105.

Teschendorff, A. E., & Relton, C. L. (2018). Statistical and integrative system-level analysis of DNA methylation data. *Nature Reviews Genetics*, *19*, 129–147.

Teschendorff, A. E., & Zheng, S. C. (2017). Cell-type deconvolution in epigenome-wide association studies: A review and recommendations. *Epigenomics*, *9*, 757–768.

Thorsson, V., Gibbs, D. L., Brown, S. D., Wolf, D., Bortone, D. S., Ou Yang, T. H., et al. (2018). The immune landscape of cancer. *Immunity*, *48*, 812–830.

Tray, N., Weber, J. S., & Adams, S. (2018). Predictive biomarkers for checkpoint immunotherapy: Current status and challenges for clinical application. *Cancer Immunology Research*, *6*, 1122–1128.

Turajlic, S., Litchfield, K., Xu, H., Rosenthal, R., McGranahan, N., Reading, J. L., et al. (2017). Insertion-and-deletion-derived tumour-specific neoantigens and the immunogenic phenotype: A pan-cancer analysis. *The Lancet Oncology*, *18*, 1009–1021.

Van Allen, E. M., Miao, D., Schilling, B., Shukla, S. A., Blank, C., Zimmer, L., et al. (2015). Genomic correlates of response to CTLA-4 blockade in metastatic melanoma. *Science*, *350*, 207–211.

van Rooij, N., van Buuren, M. M., Philips, D., Velds, A., Toebes, M., Heemskerk, B., et al. (2013). Tumor exome analysis reveals neoantigen-specific T-cell reactivity in an ipilimumab-responsive melanoma. *Journal of Clinical Oncology*, *31*, e439–e442.

Verdegaal, E. M., de Miranda, N. F., Visser, M., Harryvan, T., van Buuren, M. M., Andersen, R. S., et al. (2016). Neoantigen landscape dynamics during human melanoma-T cell interactions. *Nature*, *536*, 91–95.

Vitale, I., Sistigu, A., Manic, G., Rudqvist, N. P., Trajanoski, Z., & Galluzzi, L. (2019). Mutational and antigenic landscape in tumor progression and cancer immunotherapy. *Trends in Cell Biology*, *29*, 396–416.

Volckmar, A. L., Sültmann, H., Riediger, A., Fioretos, T., Schirmacher, P., Endris, V., et al. (2018). A field guide for cancer diagnostics using cell-free DNA: From principles to practice and clinical applications. *Genes Chromosomes & Cancer*, *57*, 123–139.

Wang, X., Park, J., Susztak, K., Zhang, N. R., & Li, M. (2019). Bulk tissue cell type deconvolution with multi-subject single-cell expression reference. *Nature Communications*, *10*, 380.

Wang, T. Y., Wang, L., Alam, S. K., Hoeppner, L. H., & Yang, R. (2019). ScanNeo: Identifying indel derived neoantigens using RNA-Seq data. *Bioinformatics*, *35*, 4159–4161.

Wang, S., Wang, T., Yang, L., Yang, D. M., Fujimoto, J., Yi, F., et al. (2019). ConvPath: A software tool for lung adenocarcinoma digital pathological image analysis aided by convolutional neural network. *eBioMedicine*, *50*, 103–110.

Wang, Z., Wu, X., & Wang, Y. (2018). A framework for analyzing DNA methylation data from Illumina Infinium HumanMethylation450 BeadChip. *BMC Bioinformatics*, *19*, 115.

Wiencke, J. K., Koestler, D. C., Salas, L. A., Wiemels, J. L., Roy, R. P., Hansen, H. M., et al. (2017). Immunomethylomic approach to explore the blood neutrophil lymphocyte ratio (NLR) in glioma survival. *Clinical Epigenetics*, *2*, 9–10.

Wolf, Y., Bartok, O., Patkar, S., Eli, G. B., Cohen, S., Litchfield, K., et al. (2019). UVB-induced tumor heterogeneity diminishes immune response in melanoma. *Cell*, *179*, 219–235.

Wood, M. A., Nguyen, A., Struck, A., Ellrott, K., Nellore, A., & Thompson, R. F. (2019). Neoepiscope improves neoepitope prediction with multi-variant phasing. *Bioinformatics*, pii: btz653.

Wood, M. A., Paralkar, M., Paralkar, M. P., Nguyen, A., Struck, A. J., Ellrott, K., et al. (2018). Population-level distribution and putative immunogenicity of cancer neoepitopes. *BMC Cancer*, *18*, 414.

Xue, J., Yu, X., Xue, L., Ge, X., Zhao, W., & Peng, W. (2019). Intrinsic β-catenin signaling suppresses CD8+ T-cell infiltration in colorectal cancer. *Biomedicine & Pharmacotherapy*, *115*, 108921.

Yadav, M., Jhunjhunwala, S., Phung, Q. T., Lupardus, P., Tanguay, J., Bumbaca, S., et al. (2017). Predicting immunogenic tumour mutations by combining mass spectrometry and exome sequencing. *Nature*, *515*, 572–576.

Yang, W., Lee, K. W., Srivastava, R. M., Kuo, F., Krishna, C., Chowell, D., et al. (2019). Immunogenic neoantigens derived from gene fusions stimulate T cell responses. *Nature Medicine*, *25*, 767–775.

Yi, M., Jiao, D., Xu, H., Liu, Q., Zhao, W., Han, X., et al. (2018). Biomarkers for predicting efficacy of PD-1/PD-L1 inhibitors. *Molecular Cancer*, *17*, 129.

Yuan, Y. (2015). Modelling the spatial heterogeneity and molecular correlates of lymphocytic infiltration in triple-negative breast cancer. *Journal of the Royal Society Interface, 12,* pii: 20141153.

Zhang, J., Mardis, E. R., & Maher, C. A. (2017). INTEGRATE-neo: A pipeline for personalized gene fusion neoantigen discovery. *Bioinformatics, 33*, 555–557.

## Further reading

Piskol, R., Ramaswami, G., & Li, J. B. (2013). Reliable identification of genomic variants from RNA-seq data. *American Journal of Human Genetics, 93*, 641–651.

Van der Auwera, G. A., Carneiro, M. O., Hartl, C., Poplin, R., Del Angel, G., Levy-Moonshine, A., et al. (2013). From FastQ data to high confidence variant calls: The Genome Analysis Toolkit best practices pipeline. *Current Protocols in Bioinformatics, 43,* 11.10.1–11.10.33.

Yizhak, K., Aguet, F., Kim, J., Hess, J. M., Kübler, K., Grimsby, J., et al. (2019). J RNA sequence analysis reveals macroscopic somatic clonal expansion across normal tissues. *Science, 364,* pii: eaaw0726.

CHAPTER TEN

# Deconvoluting tumor-infiltrating immune cells from RNA-seq data using quanTIseq

**Christina Plattner, Francesca Finotello, Dietmar Rieder***
Division for Bioinformatics, Biocenter, Medical University of Innsbruck, Innsbruck, Austria
*Corresponding author: e-mail address: dietmar.rieder@i-med.ac.at

## Contents



## Abstract

Tumor-infiltrating immune cells comprise various cells of the innate and the adaptive immune system, which influence tumor growth and response to immunotherapy by exerting anti- and protumorigenic functions. Therefore, the quantification of tumor immune infiltrates is of paramount importance for cancer immunology and immunotherapy.

We recently developed quanTIseq, a computational pipeline for the quantification of immune-cell fractions from bulk RNA sequencing (RNA-seq) data from blood or tumor samples. In this chapter, we show the capabilities of quanTIseq by analyzing two publicly available data sets. In the first example, we demonstrate how quanTIseq can be used to quantify circulating immune cells from preprocessed RNA-seq data and how to validate the results using matched flow cytometry data. In the second example, we analyze raw RNA-seq data from bulk tumor samples of melanoma patients collected before and on-treatment with kinase inhibitors to show how quanTIseq can be used to reveal the immunological effects of targeted and conventional drugs.

*Methods in Enzymology*, Volume 636
ISSN 0076-6879
https://doi.org/10.1016/bs.mie.2019.05.056

## 1. Introduction

Tumors do not only consist of malignant cells but also contain immune, stromal, and endothelial cells (Balkwill, Capasso, & Hagemann, 2012). It has been shown that the tumor microenvironment and the tumor immune contexture play a major role in tumor growth and response to tumor therapy (Chen & Mellman, 2017; Fridman, Pagès, Sautès-Fridman, & Galon, 2012; Galon et al., 2006). For instance, alternatively activated (M2) macrophages and regulatory $CD4^+$ T ($T_{reg}$) cells are known to contribute to tumor progression, whereas $CD8^+$ T cells can specifically recognize and kill tumor cells (Fridman et al., 2012).

Traditional methods to quantify cell fractions, like flow cytometry, immune fluorescence (IF) and immunohistochemistry (IHC) still suffer from important limitations, e.g., IHC/IF can only profile tumor-tissue slides that may not be representative of the whole tumor mass. Cutting-edge technologies like single-cell RNA sequencing or mass cytometry are currently too expensive to be used routinely and cannot be applied to archived samples. Moreover, cell-specific differences in the efficiency of single-cell dissociation can bias cell proportions (Finotello & Eduati, 2018). Therefore, there is need for computational methods to quantify the number of immune cells from bulk RNA-seq data.

We recently developed and extensively validated quanTIseq, a computational pipeline for the quantification of immune-cell fractions from blood or tumor samples through deconvolution of transcriptomic data (Finotello et al., 2019). As opposed to microarray-based methods, quanTIseq signature matrix was built from RNA-seq datasets from 10 different immune cell types (Table 1). Moreover, quanTIseq estimates immune-cell fractions referred to the total cellular content of the entire sample, allowing intra- and inter-sample comparison. In contrast to previous deconvolution approaches (e.g., CIBERSORT (Newman et al., 2015), EPIC (Racle, de Jonge, Baumgaertner, Speiser, & Gfeller, 2017); reviewed in Finotello and Trajanoski (2018)). quanTIseq implements a full pipeline for deconvolution of RNA-seq data, from data preprocessing and gene expression quantification, to cell-type deconvolution, simplifying its applicability and ensuring robustness of the full analytical procedure. Given that quanTIseq is available as a containerized software with all required dependencies, it is amenable to a broad user base by requiring only basic computer skills.

**Table 1** Cell types quantified by quanTIseq.

| Cell type | Cell ID in the output file |
|---|---|
| B cells | B.cells |
| Classically activated macrophages (M1) | Macrophages.M1 |
| Alternatively activated macrophages (M2) | Macrophages.M2 |
| Monocytes | Monocytes |
| Neutrophils | Neutrophils |
| Natural killer (NK) cells | NK.cells |
| Nonregulatory $CD4^+$ T cells | T.cells.CD4 |
| $CD8^+$ T cells | T.cells.CD8 |
| Regulatory $CD4^+$ (Treg) cells | Tregs |
| Dendritic cells | Dendritic.cells |
| Other uncharacterized cells | Other |

In this chapter we will describe the quanTIseq pipeline, its implementation and its application using two example RNA-seq data sets from blood and tumor samples. In particular, we will show how to run the pipeline, specify relevant parameters for different sample or data types and, validate quanTIseq results using flow cytometry data.

quanTIseq is freely available at http://icbi.at/quantiseq.

## 2. quanTIseq deconvolution pipeline

quanTIseq is a deconvolution pipeline that quantifies the fractions and densities of 10 different immune cell types relevant for cancer immunology as well as the proportion of uncharacterized cells (e.g., malignant cells in bulk tumors). The cell types estimated by quanTIseq are listed in Table 1 where "T.cells.CD4" corresponds to nonregulatory $CD4^+$ T cells. The total $CD4^+$ T cell fraction can be obtained by summing up "T.cells.CD4" and "Tregs."

The quanTIseq pipeline consists of three main modules (see Fig. 1):

1. *Read preprocessing*: Trimmomatic (Bolger, Lohse, & Usadel, 2014) is used for raw RNA-seq reads (single- or paired-ends) preprocessing, to remove Illumina adapter sequences, trim low-quality read ends, crop long-reads to a maximum length, and discard short reads.

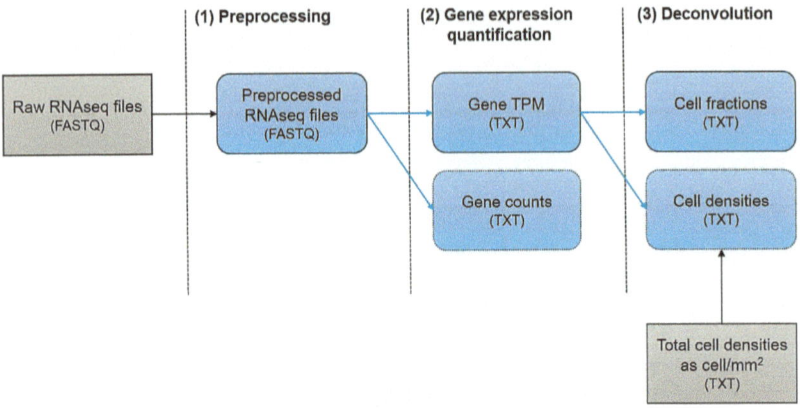


**Fig. 1** The quanTIseq pipeline consists of three main modules. The pipeline can be started from any of these steps by specifying the parameter `--pipelinestart` ("preproc" to start with preprocessing, "expr" to start with quantifying the gene expression or "decon" for the deconvolution of cell fractions). Gene counts and cell densities can be obtained optionally by specifying the parameters accordingly.

To allow for flexibility in raw data preprocessing, different tools, such as Flexbar (Dodt, Roehr, Ahmed, & Dieterich, 2012) or CutAdapt (Martin, 2011), may be applied independently from the quanTIseq pipeline. If preprocessing was performed separately the pipeline can be started at the quantification step (see option `--pipelinestart`).

2. *Gene expression quantification*: Kallisto (Bray, Pimentel, Melsted, & Pachter, 2016) is used to quantify gene expression as transcripts per millions (TPM) and (optionally) as raw counts.
3. *Deconvolution*: this module performs expression normalization, gene reannotation, and deconvolution of cell fractions using constrained least squares regression. In addition, total cell densities per tumor can be used to optionally scale the cell fractions estimated by quanTIseq (see below).

The quanTIseq deconvolution algorithm models the expression of a bulk tumor mixture **M** as a linear combination of the expression of genes in different cell types, whose expression profiles are summarized in a signature matrix **S**, weighted by the relative cell fractions **f** (Fig. 2).

Given an input mixture M and a reference matrix S, quanTIseq estimates the unknown cell fractions **f** using least square regression to solve the following system of linear equations:

$$\boldsymbol{M} = \boldsymbol{S} x f$$

Constrained least square regression is used to force the estimated cell fractions to be nonnegative and their sum to be lower of equal to one. As the

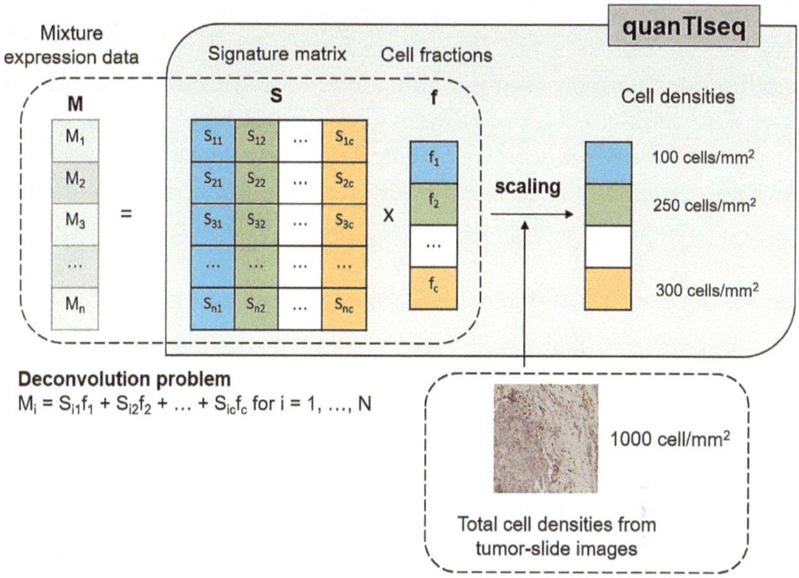


**Fig. 2** Illustration of the deconvolution problem and of quanTIseq deconvolution approach. quanTIseq addresses the deconvolution problem by solving a system of linear equations modeling the expression of the mixture as the linear combination of the expression profiles of single immune cell types, weighted by their (unknown) fractions. Optionally the obtained cell fractions can be scaled to cells/mm$^2$ by using the total cell densities from tumor-slide images.

sum can be lower than one, the residual fraction represent the proportion of "other" uncharacterized cells, namely, cells that are present in a sample but are not represented in the signature matrix (e.g., cancer cells, Table 1). By quantifying the fraction of "other" cells, quanTIseq refers all immune cell fractions to the total cellular content of a sample, thereby allowing intra- and intersample comparison.

quanTIseq uses the "TIL10" signature matrix consisting of 153 genes for RNA-seq data and 170 genes for microarray data sets as they often lack some signature genes (see option `--rmgenes` below).

This signature matrix was built from a compendium of 51 RNA-seq data sets from 10 different immune cell types: B cells, CD4$^+$ and CD8$^+$ T cells, dendritic cells, M1 and M2 macrophages, monocytes, natural killer cells, neutrophils and $T_{reg}$ cells (Finotello et al., 2019).

In case a dataset contains genes with aberrant high expression that would bias the results, it is possible to specify a customized file containing the name of genes which should be removed from the TIL10 signature matrix (see option `--rmgenes` below).

As noted above, quanTIseq can scale the estimated cell fractions to cell densities (e.g., cells/mm$^2$) by considering the density of total cells estimated from histological images. If this information is of interest and enough tissue material is available, a slice from the same tissue block consecutive to the slice that was used for RNA-seq may be used to perform hematoxylin and eosin (H&E) or immunofluorescence (IF) staining and manual or computational determination of total cells per area. For instance, IHCount (https://github.com/mui-icbi/IHCount), a publicly available semiautomated pipeline that was developed in conjunction with quanTIseq, can be used to obtain the overall cell densities as cells/mm$^2$ from H&E or IHC stained tumor tissue slides. IHcount can automatically preprocess whole-slide scans to create smaller image tiles and use a small subset of those to train a pixel classifier for nuclei detection with Ilastik (Sommer, Straehle, Köthe, & Hamprecht, 2011). This classifier is then applied to all tiles of the whole slide and the resulting probability maps together with the original tiles are used as input for a cellprofiler (Carpenter et al., 2006) pipeline. This cellprofiler pipeline, then, automatically segments the nuclei generates the count tables that may be used with quanTIseq.

# 3. Materials

## 3.1 Input files

QuanTIseq processes FASTQ files of raw RNA-seq reads when run from step 1 and 2, or a file containing gene expression levels in TPM units, when run from step 3 (Fig. 1).

For step 1 (`--pipelinestart=preproc`) and step 2 (`--pipelinestart=expr`), the `--inputfile` parameter must indicate the path and name of a tab-delimited text file containing information about the FASTQ files to be analyzed. The file consists of three columns without header (Table 2). For each sample, the first column specifies the sample identifier, the second column the path to the first FASTQ file, and the third column the path to the second FASTQ file, in case of paired-end samples. For single-end sequencing data, the third column must contain the string "None."

When starting the pipeline at step 3 (`--pipelinestart=decon`) the `--inputfile` parameter should specify the path and name of the tab-delimited text file with gene TPM (or microarray expression values) for all samples to be deconvoluted. The first column in this file should contain the gene symbols and the first row the sample identifiers (Table 3). The expression data should be on a nonlogarithmic scale.

**Table 2** Input file format for the quanTIseq pipeline run on FASTQ data.

| | | |
|---|---|---|
| Sample 1 | Input/rnaseq_sample1_1.fastq | Input/rnaseq_sample1_2.fastq |
| Sample 2 | Input/rnaseq_sample2.fastq | None |
| ... | ... | ... |

**Table 3** Input file format for the quanTIseq deconvolution module run on preprocessed expression data.

| Gene | Sample 1 | Sample 2 | ... |
|---|---|---|---|
| UBE2Q2P2 | 0 | 0 | ... |
| GTPBP6 | 75.06235 | 76.78617 | ... |
| ... | ... | ... | ... |

**Table 4** Input file format for the number of total cells from tumor images.

| | |
|---|---|
| Sample 1 | 10767.0926258729 |
| Sample 2 | 6871.40489414807 |
| ... | ... |

To also calculate densities of the deconvoluted cell types, the `--totalcell` option must be used to indicate the path and name of the file containing the number of total cells per $mm^2$ estimated from tumor-slide images of the samples under investigation. This file should contain the sample identifiers in the first column and the number of total cells per $mm^2$ in the second column (Table 4). The sample identifiers should match those of the input file.

Example files and other resources can be downloaded from the quanTIseq website: http://icbi.at/quantiseq.

## 3.2 System requirements

### *3.2.1 Operating system*

The quanTIseq pipeline is available for Mac OS X Sierra or newer and for current Linux systems like CentOS 7, RedHat Linux 7, Fedora 28, Ubuntu 16.04 or later.

### *3.2.2 Software*

To avoid software and library dependency issues and facilitate seamless integration into high performance computing environments, quanTIseq is containerized in Docker for Mac OS X or Singularity for Linux operating systems. Depending on the operating system in use, either Docker or Singularity is required to run quanTIseq. We also recommend installing the R programming language for further analysis of the data as shown in our examples below.

#### *3.2.3 Hardware*

About 4 GB of free disk space is required to download the container image and extra disk space of about twice the data size of the input files should be available for the analyses. Additionally, a minimum of 8 GB RAM is needed to run the pipeline.

Mac hardware must be a 2010 or newer model and you might need to increase the amount of memory and CPUs (in case multithreading is considered to be used) in the MAC OS X Docker settings (Settings > Preferences > Advanced) according to the requirements above depending on the Docker version you are using.

## 4. Methods

### 4.1 Installation

quanTIseq is freely available and embedded in a Singularity image in case of Linux systems or in a Docker image in case of Mac OS X. The two-step installation procedure is explained in the following.

**Mac OS X (based on Docker)**

1. Install Docker (Instructions can be found on the Docker website: https://docs.docker.com/install/).
2. Download the "`quanTIseq_pipeline.sh`" script from the quanTIseq website: http://icbi.at/quantiseq

**Linux (based on Singularity)**

1. Install Singularity (Instructions can be found on the Singularity website: https://www.sylabs.io/docs/).
2. Download the "`quanTIseq_pipeline.sh`" script from the quanTIseq website: http://icbi.at/quantiseq.

The Docker/Singularity image includes all tools and dependencies that are required to run the analysis pipeline and is downloaded automatically the first time that the "`quanTIseq_pipeline.sh`" script is run.

### 4.2 Running quanTIseq

To run the quanTIseq pipeline, execute the following command:

- `bash quanTIseq_pipeline.sh --inputfile=path/to/input_file.txt \`
  `--outputdir=path/to/outputdirectory [options]`

Note that `--inputfile` and `--outputdir` are mandatory parameters.

The quanTIseq pipeline will automatically select Docker or Singularity depending on your operating system. If you are running the quanTIseq pipeline for the first time on Linux, quanTIseq saves the Singularity image file in the current directory ("quantiseq2.img"). All result files are saved in the directory specified by `--outputdir`.

## 4.3 Options

- *help*: prints instructions about how to run quanTIseq.
- *pipelinestart*: step from which the pipeline should be started as shown in Fig. 1: (1) "preproc," (2) "expr," or (3) "decon." Default: "preproc."
- If the option "decon" is set, the input file must be a tab-delimited text file with the gene TPMs (see Section 3).
- *tumor*: specifies whether expression data are from tumor or other samples. If this option is set to TRUE, signature genes with high expression in tumor samples are removed. Default: FALSE.
- We highly recommend setting "`--tumor=TRUE`" when analyzing tumor data. Very low fractions of other cells are estimated in tumor data using the "lsei" method can indicate a wrong setting of this parameter.
- *arrays*: specifies whether expression data are from microarrays (instead of RNA-seq). If TRUE, the "`--rmgenes`" parameter is automatically set to "none." Default: FALSE.
- *method*: deconvolution method to be used: "hampel," "huber," or "bisquare" for robust regression with Huber, Hampel, or Tukey bisquare estimators, respectively, or "lsei" for constrained least squares regression. The fraction of uncharacterized cells ("other") is computed only by the "lsei" method, which estimates cell fractions referred to the total cells in the sample under investigation, allowing intra- and inter-sample comparison. For all the other methods, the cell fractions are referred to immune cells considered in the signature matrix. Default: "lsei."
- All analyses and benchmarkings presented in the quanTIseq publication (Finotello et al., 2019) were run with the "lsei" algorithm.
- *mRNAscale*: specifies whether cell fractions must be scaled to account for cell-type-specific mRNA content. Default: TRUE.

- *totalcells*: path to a tab-separated text file containing the total cell densities (e.g., cells/mm$^2$) estimated from images of histological slides (e.g., H&E or IHC). This information will be used to scale the deconvoluted cell fractions, for the 10 immune cells and "other" cells, to cell densities (e.g., cells/mm$^2$). This parameter is optional and, if not set, only the cell fractions are returned.
- *rmgenes*: specifies which genes must be removed from the signature matrix before running deconvolution. This parameter can be equal to:
  - "none": no genes are removed
  - "default": A set of 17 genes with variable expression in RNA-seq data is removed from the TIL10 signature matrix (Finotello et al., 2019): *CD36, CSTA, NRGN, C5AR2, CEP19, CYP4F3, DOCK5, HAL, LRRK2, LY96, NINJ2, PPP1R3B, TECPR2, TLR1, TLR4, TMEM154* and *CD248*.
  - A path to a text file containing a list of genes to be removed. The text file must contain one gene symbol per line. This option can be used to specify noisy genes that might bias deconvolution results (e.g., immune genes with very high expression in the samples of interest).
  - Default: "default." (but set to "none" when **--arrays = TRUE**).
- *rawcounts***:** specifies whether a file containing raw read counts per gene should be generated in addition to TPM. This file can be used for downstream analyses with third-party tools, like differential gene expression (e.g. with the R-packages edgeR (Robinson, McCarthy, & Smyth, 2010) or DESeq2 (Love, Huber, & Anders, 2014)). Default: FALSE.
- *prefix*: prefix of the output files. Default:"quanTIseq."
- *threads*: number of threads to be used. Default: 1.

  Kallisto results (gene counts and TPM) for paired end reads can differ slightly when using more than one thread. The estimate for the fragment length is computed from about 10,000 reads, which reads are used becomes nondeterministic in multithreading mode.
- *phred*: encoding of the RNA-seq quality scores for read preprocessing with Trimmomatic: "33" for Phred-33 or "64" for Phred-64. Default: 33.
- *adapterSeed*: maximum number of seed mismatches for the identification of adapter sequences by Trimmomatic. Default: 2.
- *palindromeClip*: threshold for palindrome clipping of adapter sequences by Trimmomatic. Default: 30.
- *simpleClip*: threshold for simple clipping of adapter sequences by Trimmomatic. Default: 10.

- *trimLead*: minimum Phred quality required by Trimmomatic to keep a base at the start of a read. Bases with lower quality are trimmed. Default: 20.
- *trimTrail*: minimum Phred quality required by Trimmomatic to keep a base at the end of a read. Bases with lower quality are trimmed. Default: 20.
- *minLen*: minimum read length required by Trimmomatic to keep a read. Reads shorter than this threshold are discarded. Default: 36.
- *crop*: maximum read length required by Trimmomatic. Longer reads are trimmed to this maximum length by removing bases at the end of the read. Default: 10000.
- *avgFragLen*: estimated average fragment length required by Kallisto for single-end data. Default: 50.
- *sdFragLen*: estimated standard deviation of fragment length required by Kallisto for single-end data. Default: 20.

## 4.4 Result files

Once the analysis is complete the output files will be stored in the specified output directory. According to the options specified, up to three text files will be created:

- *prefix_cell_fractions.txt*: a tab-delimited text file containing the cell fractions estimated by quanTIseq deconvolution module. The first column lists the sample identifiers and the header shows the cell types. The fraction of uncharacterized cells is reported only when `--method=lsei` option is specified.
- *prefix_gene_tpm.txt*: a tab-delimited text file containing gene expression values in TPM. Gene symbols are listed in the first column and the header contains the sample identifiers.
- *prefix_gene_count.txt*: a tab-delimited text file containing gene expression values as raw gene counts. The format is the same as for the TPM file. This file is generated only when the `--rawcounts=TRUE` option is specified.

# 5. Examples

The quanTIseq pipeline cannot only be used for tumor samples but can also be applied on RNA-seq data derived from human blood. Here, we will demonstrate two different applications of quanTIseq. The first example shows how to estimate the immune-cell fractions from preprocessed RNA-seq data (in TPM format) generated from blood samples. The second example demonstrates how to run the full pipeline from raw RNA-seq data from bulk-tumor samples collected from melanoma patients.

Besides calculating the immune cell fractions, we will also show how to perform simple evaluations of the results using the R programming language.

## 5.1 Example 1

In this example, we show how quanTIseq can be used to deconvolute blood-derived immune-cell mixtures and compare them with matching flow cytometry data.

The example data set is available online at the Gene Expression Omnibus (GEO: https://www.ncbi.nlm.nih.gov/geo/) with the accession number GSE107572. This data set contains preprocessed RNA-seq data from blood-derived immune-cell mixtures from nine healthy donors. Flow cytometry estimates for the according immune subpopulations, except for macrophages which are not present in blood, are also available for the same samples.

### *5.1.1 Step 1—Download RNA-seq data*

Download the data file of preprocessed RNA-seq data from the "Supplementary files" section on GEO (https://www.ncbi.nlm.nih.gov/geo/query/acc.cgi?acc=GSE107572): "`GSE107572_tpm_PBMC_RNAseq.txt.gz`."

This file contains the preprocessed RNA-seq data in TPM units which can be used to run the quanTIseq pipeline starting at the deconvolution step (`--pipelinestart=decon`).

### *5.1.2 Step 2—Run quanTIseq*

To calculate individual cell fractions the quanTIseq pipeline should be run with the following parameter settings:

```
• bash quanTIseq_pipeline.sh --inputfile=GSE107572_tpm_PBMC_RNAseq.txt\
      --outputdir=Output --prefix=GSE107572 --pipelinestart=decon
```

This command generates a file named "`GSE107572_cell_fractions.txt`" in the folder named "`Output`." The parameter `--prefix`, which specifies the prefix of the output files, is optional and can be changed.

### *5.1.3 Step 3—Compare results with flow cytometry data*

Using the following R script (named "`GSE107572.R`"), the estimated cell fractions from quanTIseq can be compared with the true cell fractions measured by flow cytometry. The script uses the Bioconductor package "GEOquery" to download the flow cytometry data from the Gene

Expression Omnibus database (GEO), generates scatterplots and calculates Pearson's correlation coefficients between flow cytometry and quanTIseq deconvolution cell fraction estimates.

```
# Access flow cytometry cell fractions from GEO

args <- commandArgs(TRUE)
dir <- args[1]

library(GEOquery)

GEOid<-"GSE107572"
gds<-getGEO(GEOid)
GEOinfo<-pData(gds[[1]])

FACSdata<-data.frame(B.cells=GEOinfo$`b cells:ch1`,
  T.cells.CD4=GEOinfo$`cd4+ t cells:ch1`,
  T.cells.CD8=GEOinfo$`cd8+ t cells:ch1`,
  Monocytes=GEOinfo$`monocytes:ch1`,
  Dendritic.cells=GEOinfo$`myeloid dendritic cells:ch1`,
  NK.cells=GEOinfo$`natural killer cells:ch1`,
  Neutrophils=GEOinfo$`neutrophils:ch1`,
  Tregs=GEOinfo$`tregs:ch1`)
rownames(FACSdata)<-gsub("Blood-derived immune-cell mixture from
donor ", "pbmc", GEOinfo$title)

# Load deconvolution data obtained by running quanTIseq
(--pipelinestart=decon)

DCdata<-read.table(paste0(dir,"/GSE107572_cell_fractions.txt"),
header=TRUE, sep="\t", row.names=1)
rownames(DCdata)<-gsub("_.*$", "", sub("_", "", rownames(DCdata)))

ccells<-intersect(colnames(DCdata),colnames(FACSdata))
csbjs<-intersect(rownames(DCdata),rownames(FACSdata))
DCdata<-DCdata[csbjs,ccells]
FACSdata<-FACSdata[csbjs,ccells]

# Compare cell fractions for single cell types and all cell types
together

palette<-c("#451C87", "#b3b300", "#CE0648", "#2363C5", "#AB4CA1",
"#0A839B", "#DD8C24", "#ED6D42")
names(palette)<-c("T.cells.CD4", "Dendritic.cells", "Monocytes",
"T.cells.CD8", "Tregs", "B.cells", "NK.cells", "Neutrophils")

par(mfrow=c(3,3))
colall<-c()
```

```
for (i in 1:(ncol(DCdata)+1)) {

  if (i<=ncol(DCdata)) {

    x<-as.numeric(as.character(FACSdata[,i]))
    y<-DCdata[,i]
    ccell<-colnames(DCdata)[i]
    col<-palette[ccell]

 } else {

    x<-as.numeric(as.vector(as.matrix(FACSdata)))
    y<-as.vector(as.matrix(DCdata))
    ccell<-"All cells"
    col<-colall
 }

  res.cor<-cor.test(y,x)
  R<-round(res.cor$estimate, digits=2)
  p<-format.pval(res.cor$p.value, digits=2)
  RMSE<-round(sqrt(mean((y-x)^2, na.rm=TRUE)), digits=2)
  regl<-lm(y~x)

  ymax<-max(round(max(y),digits=2)*1.3,0.01)
  xmax<-max(round(max(x),digits=2),0.01)

  pdf("GSE107572_correlation.pdf")
  plot(x, y,
       main=gsub("(\\.)", " ", ccell),
       pch=19,
       xlab="Flow cytometry cell fractions",
       ylab="quanTIseq cell fractions",
       col=col,
       cex.main=1.3,
       ylim=c(0,ymax),
       xlim=c(0,xmax),
       las=1)
  abline(regl)

  text(0, ymax*0.98, cex=1, paste0("r = ", R, ", p = ", p), pos=4)
  text(0, ymax*0.9, cex=1, paste0("RMSE = ", RMSE), pos=4)

  colall<-c(colall, rep(col, length(x)))

 dev.off()

 }
```

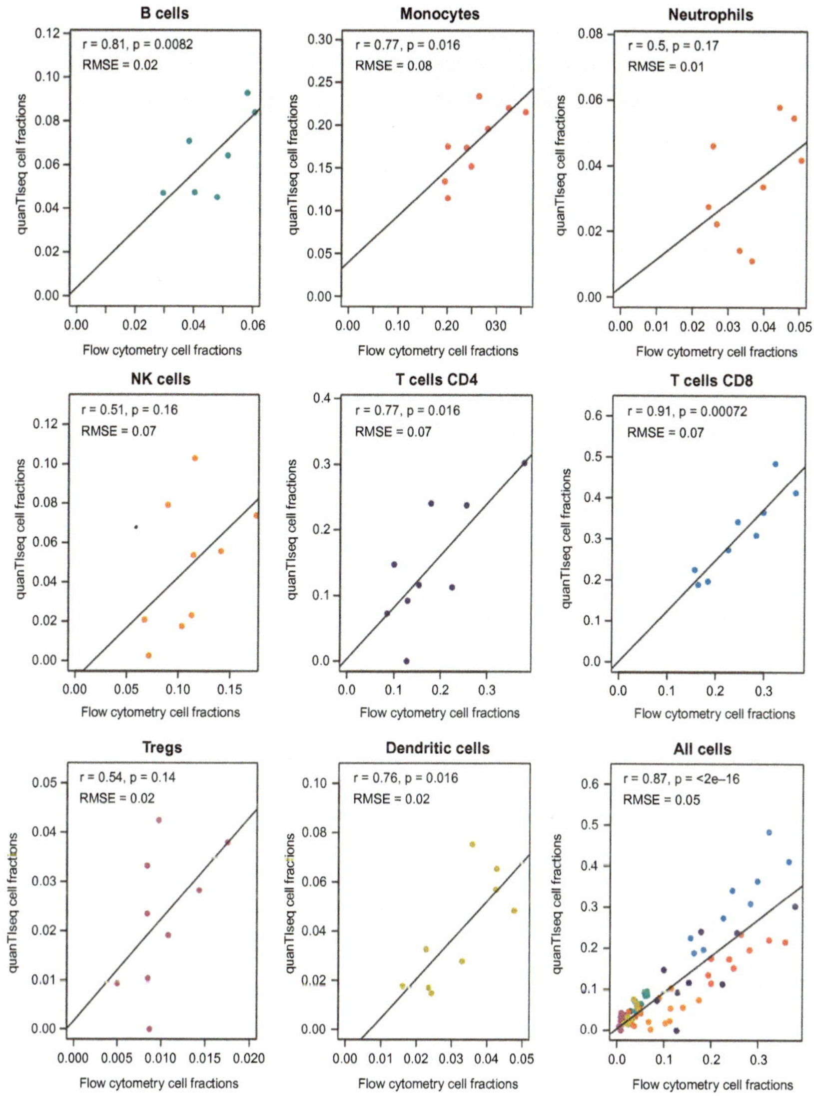


**Fig. 3** Correlation plots comparing the estimated cell fractions from quanTIseq with cell fractions computed by flow cytometry for eight different immune cell types.

Run the R script with following command and specify the path to the quanTIseq output directory (e.g., "`Output`" from Step 2) as argument:

- `Rscript GSE107572.R path/to/directory`

The results of this analytical step are shown in Fig. 3. The overall Pearson's correlation coefficient for the different immune cell types estimated by quanTIseq is 0.87. The highest correlation ($r = 0.91$) is obtained for $CD8^+$ T cells.

The same type of analysis can be run on other datasets where RNA-seq data and matched ground-truth data (e.g., from immunohistochemistry or Coulter counter) are available.

## 5.2 Example 2

The second example analyzes gene expression data from seven melanoma patients who have been treated with kinase inhibitors. The immune cell fractions before and on-treatment with these inhibitors are then assessed and compared.

We first show how to download the publicly available RNAseq data from GEO as SRA files and convert them to FASTQ files. Then we show how to run the full quanTIseq pipeline to quantify the immune cell fractions. The list of samples is shown in Table 5.

**Table 5** List of patients and kinase inhibitors.

| GEO accession | Patient | Sample type | Inhibitor |
|---|---|---|---|
| GSM1949045 | Pt1 baseline | None | None |
| GSM1949046 | Pt1 D85 | On-treatment | BRAFi |
| GSM1949047 | Pt2 baseline | None | None |
| GSM1949048 | Pt2 D726 | On-treatment | BRAFi |
| GSM1949049 | Pt3 baseline | None | None |
| GSM1949050 | Pt3 D22 | On-treatment | BRAFi + MEKi |
| GSM1949051 | Pt4 baseline | None | None |
| GSM1949052 | Pt4 D15 | On-treatment | BRAFi + MEKi |
| GSM1949053 | Pt4 RD261 | On-treatment | BRAFi + MEKi |
| GSM1949054 | Pt6 baseline | None | None |
| GSM1949055 | Pt6 D6 | On-treatment | BRAFi |
| GSM1949056 | Pt6 D12 | On-treatment | BRAFi + MEKi |
| GSM1949057 | Pt6 D15 | On-treatment | BRAFi + MEKi |
| GSM1949058 | Pt7 baseline | None | None |
| GSM1949059 | Pt7 D15A | On-treatment | MEKi |
| GSM1949060 | Pt7 D15B | On-treatment | MEKi |
| GSM1949061 | Pt7 D15C | On-treatment | MEKi |
| GSM1949062 | Pt8 baseline | None | None |
| GSM1949063 | Pt8 D22 | On-treatment | BRAFi + MEKi |

### 5.2.1 Step 1—Get list of files

The RNA-seq data from the seven patients is available on GEO with the accession number GSE75299 (https://www.ncbi.nlm.nih.gov/geo/query/acc.cgi?acc=GSE75299). Open the link in a web browser and in the "Relations" section, follow the link to the sequence read archive (SRA) study with accession number SRP066571. Select the first entry of the resulting SRA page results and select "All runs" in the "Study" section (see Fig. 4).

You will be forwarded to a page with a table showing all runs and samples. All runs with source name "tumor biopsy" should be selected and the "Accession List" table saved as "`SRP066571_Acc_List.txt`". The file now contains the selected "Run" identifiers of the files to be downloaded. In total there should be 19 different identifiers (see Fig. 5).

Additionally, the "RunInfo Table" should also be saved as "`SraRunTable.txt`" which is needed for the annotation in a later step (Fig. 5).

### 5.2.2 Step 2—Download FASTQ files

There are different ways to download the SRA files. In this example we are using the **prefetch** program from the SRA Toolkit (https://www.ncbi.nlm.nih.gov/sra/docs/toolkitsoft/) together with the Aspera Connect Client (https://downloads.asperasoft.com/connect2/), for which three files are required, the ascp executable binary (ascp), the ascp ssh key (asperaweb_id_dsa.openssh) and the file containing your accession numbers from step 1. To download the SRA data files, locate these files on your system and run following bash command:

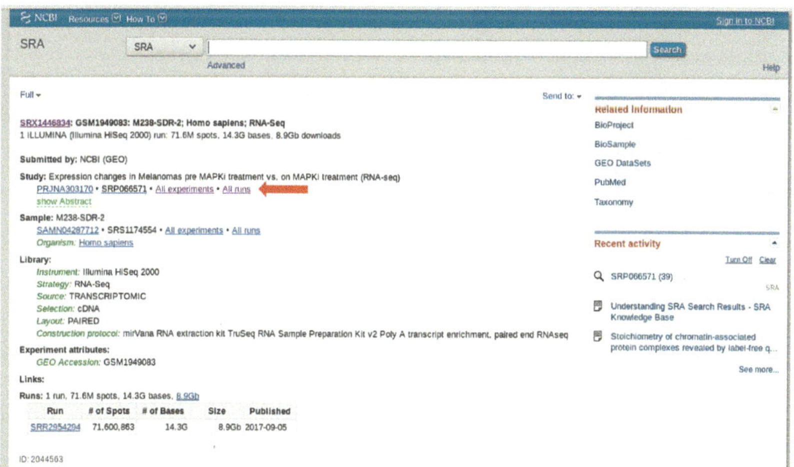


Fig. 4 Screenshot NCBI website to choose "All runs."

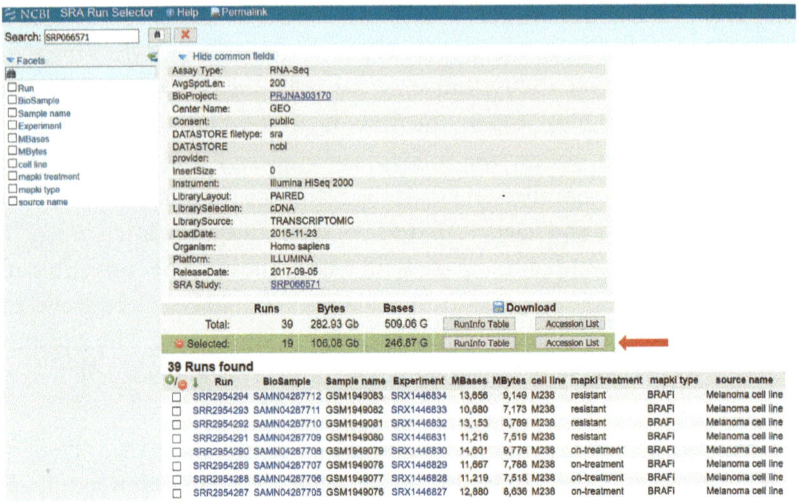

Fig. 5 Screenshot NCBI website to download "Accession List" and "RunInfo Table."

- prefetch --ascp-path "path_to_aspera/connect/bin/ascp|path_to_aspera/connect/etc/asperaweb_id_dsa.openssh" -v --option-file SRP066571_Acc_List.txt

After completion, the SRA files are available locally on the computer in the default folder named "`ncbi`." The **fasterq-dump** tool from the SRA Toolkit can then be used to convert the downloaded SRA files to FASTQ files. Since the sequencing data was generated by paired-end sequencing (see study run info), the reads need to be split into forward and reverse read files by using the `--split-files` option.

Extract and split all paired-end RNA-seq reads from the SRA files with the following bash command:

- for s in \`ls -1 ~/path_to_files/ncbi/sra/*.sra\`; do fasterq-dump $s -p --split-files -O ~/path_to_output/directory; done

To run this example, you need at least 1.7 TB of free disk space.

### *5.2.3 Step 3—Run quanTIseq pipeline*

To run the full quanTIseq pipeline a tab-delimited text file containing the paths to the FASTQ files to deconvolute (see parameter `--inputfile`) needs to be created. Table 6 shows the input file where the first column specifies

**Table 6** SRP066571 input file.

| | | |
|---|---|---|
| SRR2954274 | SRR2954274.sra_1.fastq | SRR2954274.sra_2.fastq |
| SRR2954273 | SRR2954273.sra_1.fastq | SRR2954273.sra_2.fastq |
| SRR2954272 | SRR2954272.sra_1.fastq | SRR2954272.sra_2.fastq |
| SRR2954271 | SRR2954271.sra_1.fastq | SRR2954271.sra_2.fastq |
| SRR2954270 | SRR2954270.sra_1.fastq | SRR2954270.sra_2.fastq |
| SRR2954269 | SRR2954269.sra_1.fastq | SRR2954269.sra_2.fastq |
| SRR2954268 | SRR2954268.sra_1.fastq | SRR2954268.sra_2.fastq |
| SRR2954267 | SRR2954267.sra_1.fastq | SRR2954267.sra_2.fastq |
| SRR2954266 | SRR2954266.sra_1.fastq | SRR2954266.sra_2.fastq |
| SRR2954265 | SRR2954265.sra_1.fastq | SRR2954265.sra_2.fastq |
| SRR2954264 | SRR2954264.sra_1.fastq | SRR2954264.sra_2.fastq |
| SRR2954263 | SRR2954263.sra_1.fastq | SRR2954263.sra_2.fastq |
| SRR2954262 | SRR2954262.sra_1.fastq | SRR2954262.sra_2.fastq |
| SRR2954261 | SRR2954261.sra_1.fastq | SRR2954261.sra_2.fastq |
| SRR2954260 | SRR2954260.sra_1.fastq | SRR2954260.sra_2.fastq |
| SRR2954259 | SRR2954259.sra_1.fastq | SRR2954259.sra_2.fastq |
| SRR2954258 | SRR2954258.sra_1.fastq | SRR2954258.sra_2.fastq |
| SRR2954257 | SRR2954257.sra_1.fastq | SRR2954257.sra_2.fastq |
| SRR2954256 | SRR2954256.sra_1.fastq | SRR2954256.sra_2.fastq |

the sample names and the second and third column contain the paths to the FASTQ files from the paired-end sequencing runs (forward and reverse reads, respectively). In this example the file names are listed without file directory, therefore the pipeline should be run in the same directory where the FASTQ files are located. Alternatively, full file paths can be provided.

This input file should be saved as "`SRP066571_rnaSeqInfoFile.txt`" in the same folder where also the FASTQ files, extracted with fasterq-dump in step 2, are located.

Next, the quanTIseq pipeline can be run with the following command:

```
• bash quanTIseq_pipeline.sh --inputfile=SRP066571_rnaSeqInfoFile.txt\
  --outputdir=Output --prefix=SRP066571 --tumor=TRUE --threads=8
```

In this example, data from melanoma patients are analyzed, therefore the parameter `--tumor=TRUE` needs to be specified, which will remove a set of immune-cell genes with aberrant expression in tumors from the signature matrix.

This command will create two result files, which are located in a folder named "`Output`":

- `SRP066571_cell_fractions.txt`
- `SRP066571_gene_tpm.txt`

The first file contains the cell fractions estimated by quanTIseq, which may be visualized using Excel of R to create stacked bar plots as those shown in Fig. 6. The individual fractions sum up to 1 and reflect their proportion with respect to the total cells in the sample. The second file reports the sample gene expression values in TPM format.

### *5.2.4 Step 4—Analysis/results*

In order to compare the immune cell fractions from patients before and on-treatment with kinase inhibitors, the FASTQ files need to be matched with patient data. An example R script, which is using the Bioconductor package "GEOquery" to download the respective information from the

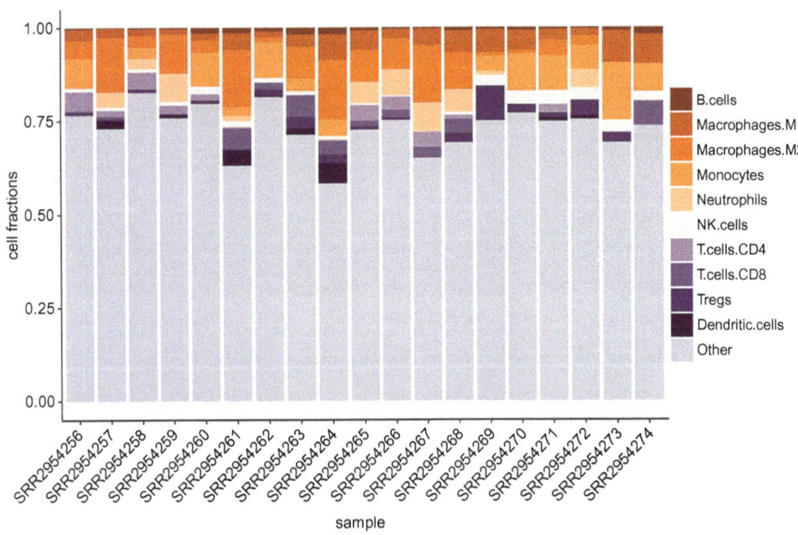


**Fig. 6** Estimated immune cell fractions for all samples.

GEO database, is listed below. This R script (named "`GSE75299.R`") may be used to compare and statistically evaluate the pre- and on-treatment immune cell fractions. It generates boxplots of the cell proportions, uses Wilcoxon test to calculate the *p*-values, and can be run by using the following command:

- `Rscript GSE75299.R path/to/SRP066571_cell_fractions.txt \ path/to/SraRunTable.txt`

where the first argument is the file of cell type fractions obtained from step 2 and the second argument is the SRA run table info file created in step 1.

```
library(GEOquery)
library(ggplot2)
library(reshape2)
library(ggpubr)

args <- commandArgs(TRUE)
path <- args[1]
sra <- args[2]

### Load file with estimated cell fractions by quanTIseq:
cdata <- read.csv(paste0(path,"/SRP066571_cell_fractions.txt"),
header=TRUE, sep="\t", stringsAsFactors = FALSE, row.names=1)

### Annotation
gds<-getGEO("GSE75299")
pdata<-pData(gds[[1]])
sraInfo <- read.csv(paste0(sra,"/SraRunTable.txt"), header=TRUE,
sep="\t", stringsAsFactors = FALSE)

cdata <- cbind(cdata, sraInfo[, "Sample_Name"][match(rownames
(cdata), sraInfo$Run)])
colnames(cdata)[length(cdata)] <- "geo_accession"

pdata<-pdata[grep("^Pt", pdata[,1]),]
cdata <- cbind(cdata, pdata[, "title"][match(cdata$geo_accession,
pdata$geo_accession)])
rownames(cdata) <- cdata[,length(cdata)]

### Remove other cells and the last two columns
cdata<-cdata[,-grep("Other", colnames(cdata))]
cdata<-cdata[,-grep("geo_accession", colnames(cdata))]
```

```
### Transpose
cdata<-t(cdata)

### Boxplots and wilcoxon's test pre/post
groups<-rep("POST", ncol(cdata))
groups[grep("baseline", colnames(cdata))]<-"PRE"
groups<-factor(groups, levels=c("PRE", "POST"))
names(groups)<-colnames(cdata)

dataset <- cdata
colnames(dataset) <- as.vector(groups)
dataset <- melt(dataset)

pdf("GSE75299_qTIs_boxplot.pdf")
ggplot(dataset, aes(Var1, value, fill=Var2))+
  geom_boxplot()+
  ylab("cell fractions") +
  xlab("")+
  theme(axis.text=element_text(size=17),
        axis.title=element_text(size=19))+
  theme(legend.text=element_text(size=17),
        legend.title=element_text(size=19))+
  scale_fill_discrete(name="Treatment")+
  theme(axis.text.x= element_text(angle=45, hjust = 1 ))+
  stat_compare_means(aes(group=Var2),
        symnum.args=list(cutpoints = c(0, 0.001, 0.01, 0.05, 0.1, 1),
        symbols = c("****", "***", "**", "*", "ns")),
        label ="p.signif",
        hide.ns=TRUE, cex=10) #default: Wilcoxon Test
dev.off()
```

The resulting boxplots illustrate the cell-type fraction differences between pre and on-treatment with the kinase inhibitors (PRE vs. POST) for all analyzed samples. For this analytical step all samples were combined. Fig. 7 shows detection of significant increase in M2 macrophage and dendritic cells fractions (Wilcoxon test, $P$ value $< 0.1$) after kinase treatment. All other immune cell types quantified by quanTIseq did not show any significant differences between on-treatment and before treatment samples.

Further analytical steps can be performed with the estimated immune cell fractions, for instance a comparison could be done between the single patients, to show how the immune infiltrates before and on-treatment differs between patients.

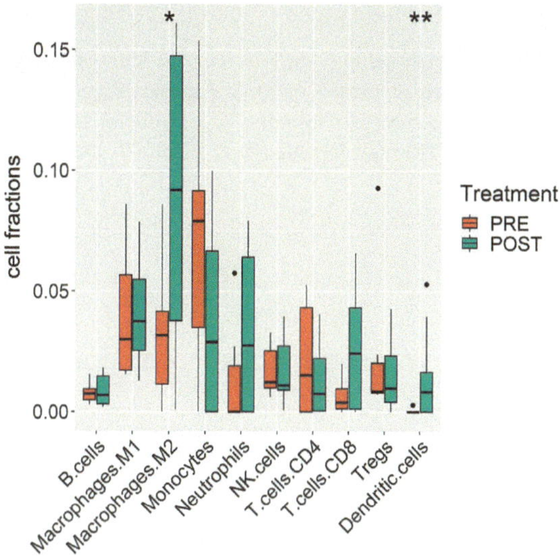


**Fig. 7** Boxplots of the estimated cell fractions (pre vs. on treatment, **P* value < 0.1, ***P* value < 0.05).

## 6. Concluding remarks

quanTIseq is a computational pipeline, developed for the analysis of bulk RNA-seq data that quantifies the fractions and densities of 10 different immune cell types relevant for cancer immunology. quanTIseq is specifically designed for and validated on RNA-seq data, allows full analysis of raw RNA-seq data, from pre-processing to cell fraction deconvolution, and can be used for either tumor or blood samples by simply setting the parameter `--tumor=TRUE/FALSE`.

The containerized pipeline of quanTIseq can estimate immune-cell fractions in a standardized manner across different platforms, enabling reproducible analysis of immune-cell fractions and densities. Hence, it is a useful tool to investigate the immunological effects of cancer drugs (Galluzzi, Buqué, Kepp, Zitvogel, & Kroemer, 2015) that can provide mechanistic insights for the design of combination therapies.

quanTIseq was primarily developed to estimate immune cell fractions relevant for cancer immunology, but it can be helpful also to investigate immune cell fractions and their modulation in other diseases, like autoimmunity, inflammation and infections.

## Acknowledgments

C.P. is a recipient of a DOC Fellowship of the Austrian Academy of Sciences at the Division for Bioinformatics. F.F. was supported by the Austrian Cancer Aid/Tyrol (project No. 17003, "quanTIseq: dissecting the immune contexture of human cancers") and by the Austrian Science Fund (FWF) (project No. T 974-B30).

## References

Balkwill, F. R., Capasso, M., & Hagemann, T. (2012). The tumor microenvironment at a glance. *Journal of Cell Science*, *125*(23), 5591–5596. https://doi.org/10.1242/jcs.116392.

Bolger, A. M., Lohse, M., & Usadel, B. (2014). Trimmomatic: A flexible trimmer for Illumina sequence data. *Bioinformatics*, *30*(15), 2114–2120. https://doi.org/10.1093/bioinformatics/btu170.

Bray, N. L., Pimentel, H., Melsted, P., & Pachter, L. (2016). Near-optimal probabilistic RNA-seq quantification. *Nature Biotechnology*, *34*(5), 525–527. https://doi.org/10.1038/nbt.3519.

Carpenter, A. E., Jones, T. R., Lamprecht, M. R., Clarke, C., Kang, I. H., Friman, O., et al. (2006). Cell profiler: Image analysis software for identifying and quantifying cell phenotypes. *Genome Biology*, 7(10), R100. https://doi.org/10.1186/gb-2006-7-10-r100.

Chen, D. S., & Mellman, I. (2017). Elements of cancer immunity and the cancer–immune set point. *Nature*, *541*(7637), 321–330. https://doi.org/10.1038/nature21349.

Dodt, M., Roehr, J. T., Ahmed, R., & Dieterich, C. (2012). FLEXBAR—Flexible barcode and adapter processing for next-generation sequencing platforms. *Biology*, *1*(3), 895–905. https://doi.org/10.3390/biology1030895.

Finotello, F., & Eduati, F. (2018). Multi-omics profiling of the tumor microenvironment: Paving the way to precision immuno-oncology. *Frontiers in Oncology*, *8*, 430. https://doi.org/10.3389/fonc.2018.00430.

Finotello, F., Mayer, C., Plattner, C., Laschober, G., Rieder, D., Hackl, H., et al. (2019). Molecular and pharmacological modulators of the tumor immune contexture revealed by deconvolution of RNA-seq data. *Genome Medicine*, *11*(1), 34. https://doi.org/10.1186/s13073-019-0638-6.

Finotello, F., & Trajanoski, Z. (2018). Quantifying tumor-infiltrating immune cells from transcriptomics data. *Cancer Immunology, Immunotherapy*, *67*(7), 1031–1040. https://doi.org/10.1007/s00262-018-2150-z.

Fridman, W. H., Pagès, F., Sautès-Fridman, C., & Galon, J. (2012). The immune contexture in human tumours: Impact on clinical outcome. *Nature Reviews. Cancer*, *12*(4), 298–306. https://doi.org/10.1038/nrc3245.

Galluzzi, L., Buqué, A., Kepp, O., Zitvogel, L., & Kroemer, G. (2015). Immunological effects of conventional chemotherapy and targeted anticancer agents. *Cancer Cell*, *28*(6), 690–714. https://doi.org/10.1016/j.ccell.2015.10.012.

Galon, J., Costes, A., Sanchez-Cabo, F., Kirilovsky, A., Mlecnik, B., Lagorce-Pagès, C., et al. (2006). Type, density, and location of immune cells within human colorectal tumors predict clinical outcome. *Science*, *313*(5795), 1960–1964. https://doi.org/10.1126/science.1129139.

Love, M. I., Huber, W., & Anders, S. (2014). Moderated estimation of fold change and dispersion for RNA-seq data with DESeq2. *Genome Biology*, *15*(12). https://doi.org/10.1186/s13059-014-0550-8.

Martin, M. (2011). Cutadapt removes adapter sequences from high-throughput sequencing reads. *EMBnet. Journal*, *17*(1), 10–12. https://doi.org/10.14806/ej.17.1.200.

Newman, A. M., Liu, C. L., Green, M. R., Gentles, A. J., Feng, W., Xu, Y., et al. (2015). Robust enumeration of cell subsets from tissue expression profiles. *Nature Methods*, *12*(5), 453–457. https://doi.org/10.1038/nmeth.3337.

Racle, J., de Jonge, K., Baumgaertner, P., Speiser, D. E., & Gfeller, D. (2017). Simultaneous enumeration of cancer and immune cell types from bulk tumor gene expression data. *eLife*, *6*, e26476. https://doi.org/10.7554/eLife.26476.

Robinson, M. D., McCarthy, D. J., & Smyth, G. K. (2010). edgeR: A bioconductor package for differential expression analysis of digital gene expression data. *Bioinformatics*, *26*(1), 139–140. https://doi.org/10.1093/bioinformatics/btp616.

Sommer, C., Straehle, C., Köthe, U., & Hamprecht, F. A. (2011). Ilastik: Interactive learning and segmentation toolkit. In *2011 IEEE international symposium on biomedical imaging: From nano to macro* (pp. 230–233). https://doi.org/10.1109/ISBI.2011.5872394.

CHAPTER ELEVEN

# Concurrent *in situ* analysis of point mutations and immune infiltrate in FFPE cancers

**Ann-Marie Baker*, Trevor A. Graham***

Barts Cancer Institute, Barts and the London School of Medicine and Dentistry, Queen Mary University of London, London, United Kingdom

*Corresponding authors: e-mail address: a.m.c.baker@qmul.ac.uk; t.graham@qmul.ac.uk

## Contents



## Abstract

Existing methodology for analysis of genetic heterogeneity generally involves digestion of the tumor tissue, followed by bulk DNA extraction or single cell preparation. Such methods destroy the tissue morphology, and therefore opportunities to analyze tumor heterogeneity and clonal architecture within the native spatial context are lost. Thus, there is a clear need for the development of generally applicable methods of *in situ* mutation detection (ISMD), in which tumor cells can be genetically analyzed in the context of their cellular microenvironment, including immune infiltrate. Furthermore, protocols in which ISMD can be combined with immunohistochemical analysis are highly sought after, as the combination of these two techniques enables insight not only into genetic heterogeneity, but is also permissive of genotype-phenotype analysis, whilst preserving tissue morphology and spatial context. Here we describe a novel method for *in situ* point mutation detection using commercially available BaseScope reagents, followed by immunohistochemical detection of immune infiltrate on the same tissue section.

*Methods in Enzymology*, Volume 636
ISSN 0076-6879
https://doi.org/10.1016/bs.mie.2019.05.009

## 1. Introduction

Detection of clinically relevant mutations in human cancers is commonly performed using genomic analysis of "bulk" tumor extracts or flow cytometry-sorted single cells, however, these approaches do not preserve tissue morphology. Consequently, the interrelationship between mutant cells and their local environment cannot be studied. Novel methods that enable genetic analysis and concurrent analysis of protein expression to be carried out on intact tissue sections are therefore highly sought after.

*In situ* mutation detection (ISMD) is a powerful tool for the analysis of genetic heterogeneity that preserves spatial and histopathological context. Broadly, there are two approaches currently used for ISMD in a research setting: first *in situ* PCR-based approaches that rely upon "rolling-circle" amplification of relevant transcripts to generate large fluorescently-labeled DNA coils (Grundberg et al., 2013; Larsson, Grundberg, Soderberg, & Nilsson, 2010), and second "branched" assays such as the commercially available "BaseScope" (from Advanced Cell Diagnostics, ACD), that rely upon sequential binding of complementary DNA oligos to generate a large signal amplification "tree" (Baker et al., 2017).

Assay development for the detection of point mutations *in situ* has traditionally been problematic as it requires probes capable of discrimination between sequences that differ by only one nucleotide base. Thus, existing ISMD protocols often require extensive optimization and technical expertise, and this has limited their adoption by the wider research community. Upon the release of BaseScope technology by ACD, we performed rigorous testing of its capacity for *in situ* detection of point mutations, and we found the BaseScope assay offers excellent sensitivity and specificity, with little or no optimization required by the end user (Baker et al., 2017). The assay relies upon the hybridization of a single pair of DNA oligonucleotide "ZZ" probes across the site of a point mutation in an mRNA transcript. This creates a binding platform for further sequential hybridization of a series of complementary DNA oligos, allowing the formation of a large signal amplification "tree" (Fig. 1). Although BaseScope is a relatively new technology, it has already been successfully applied to the detection of somatic point mutations (Baker et al., 2017) and splice variants (Erben, He, Laeremans, Park, & Buonanno, 2018; Zhu et al., 2018), as well as single nucleotide polymorphisms (SNPs) and insertions/deletions (Baker, unpublished) in FFPE tissue sections.

Multiplexing BaseScope ISMD with the *in situ* analysis of proteins via immunohistochemistry (IHC) is a particularly appealing technique to study

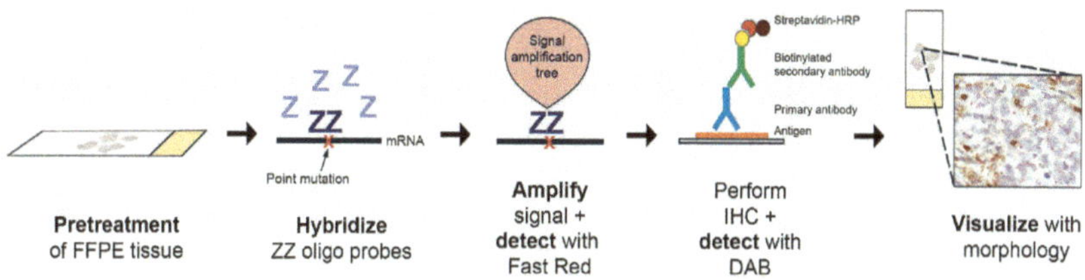


**Fig. 1** Schematic diagram showing the combined ISMD and IHC assay. Following pretreatment of FFPE tissue sections, custom-designed "ZZ" probes are hybridized to target mRNA and this directs binding of a signal amplification tree, allowing visualization of point mutations by Fast Red detection. Then a conventional IHC protocol is performed (sequential binding of primary antibody, biotinylated secondary antibody and streptavidin-horseradish peroxidase), followed by detection with DAB. Staining is visualized using a standard light microscope.

genotype-phenotype/environment relationships. Both ISMD and IHC have the capacity to analyze tumor tissue at single cell level, whilst preserving tissue morphology and spatial context. *In situ* analysis of point mutations and proteins on the same tissue section offers a unique opportunity to study the regulation of protein expression by genetic mutation, cell-of-origin of mutant transcripts and the relationship between mutant clones and their surrounding environment. It has the further benefit of reducing the number of tissue sections required for research, allowing precious archival samples to be used in the most efficient manner.

BaseScope ISMD and conventional IHC lend themselves to multiplexing as the protocols share many of the same steps, for example tissue fixation, target retrieval and signal amplification/detection. However, combining ISH and IHC on the same tissue section is not always straightforward, as the protease treatment that is an essential part of tissue pretreatment for BaseScope probe hybridization could destroy the antigen required for IHC detection. Therefore, careful optimization and validation of protocols is of critical importance. As ISMD with BaseScope is reliant on the preservation of mRNA (which is inherently more unstable than the protein-based antigens required for IHC), we have had greater success with performing the BaseScope assay before IHC (as depicted in Fig. 1).

In the following protocol we describe the concurrent Fast Red-based detection of the KRAS G12V mutation (using ISMD with BaseScope) and DAB-based detection of CD68+ macrophages (using IHC) in FFPE human colorectal cancers. The data generated using this method can be used to gain insight into the immune response elicited by KRAS G12V mutant tumor cells, however with minor modifications the protocol has numerous and broad applications in studying the role of specific somatic mutations in tumor progression.

# 2. ISMD and IHC

## 2.1 Required equipment

- Microtome (*various*)
- Slide holders and staining racks (*various*)
- Hot plate (*various*)
- HybEZ oven (*310010, ACD*)
- HybEZ tray (*310012, ACD*)
- EZ-batch slide holder and wash tray (*optional, 310017, ACD*)
- Dry oven (*various*)

## 2.2 Required materials and reagents

- 10% neutral buffered formalin (*NBF, HT501128-4L, Sigma-Aldrich*)
- Superfrost plus slides (*10149870, Fisher Scientific*)
- RNase free water (*10977049, Fisher Scientific*)
- RNaseZap (*10708345, Fisher Scientific*)
- Xylene (*various*)
- Ethanol (*various*)
- BaseScope v2 reagent kit (*323900, ACD*), containing the following:
  - Ready-to-use Hydrogen peroxide
  - 10 × Target retrieval solution
  - Ready-to-use Protease IV
  - 20 × Wash buffer
  - Detection reagents (AMP1-8 and Fast Red A + B)
- BaseScope 1ZZ positive control probe (*human POLR2A, 701041, ACD*)
- BaseScope 1ZZ negative control probe (*dapB, 701021, ACD*)
- BaseScope probe of interest (*e.g., KRAS G12V, ACD*)
- ImmEdge hydrophobic barrier pen (*310018, ACD*)
- HybEZ humidifying sheets (*310015, ACD*)
- Kimwipes (*various*)
- Phosphate buffered saline (*PBS, various*)
- Tween-20 (*P9416-100ML, Sigma-Aldrich*)
- Normal goat serum (*5425S, New England Biolabs*)
- Bovine serum albumin (*A7906, Sigma-Aldrich*)
- Antibody of interest (*e.g., CD68, M087601-2, Agilent*)
- Biotinylated antibody (*e.g., Rabbit anti-mouse, E035401-2, Agilent*)
- Streptavidin-horseradish peroxidase (*P039701-2, Agilent*)

- DAB (*SK-4100, Vector Labs*)
- Hematoxylin, Gill No. 2 (*GHS232-1L, Sigma-Aldrich*)
- Ammonium hydroxide (*320145, Sigma-Aldrich*)
- Vectamount (*H-5000, Vector Labs*)
- Cover slips (*631-1574, VWR*)

## 2.3 Protocol

### 2.3.1 Sample preparation

1. Rapidly place freshly dissected tissue into a large volume of 10% NBF for 24 h at room temperature.
2. Wash fixed tissue in PBS and embed sample in paraffin using standard histological procedures.
3. Trim the FFPE block using a standard microtome and a fresh blade, and cut the required number of sections at 5 μm thickness.
4. Place sections in a water bath containing RNase-free water at 40 °C.
5. Collect sections onto Superfrost Plus slides and air-dry overnight at room temperature.

### 2.3.2 ISMD for KRAS G12V using BaseScope

1. Bake slides at 60 °C for 1 h in a dry oven.
2. Wash slides twice in xylene (5 min each) with gentle agitation.
3. Wash slides twice in 100% ethanol (2 min each) with gentle agitation.
4. Place slides in a dry oven at 60 °C for 2 min.
5. Allow slides to cool, then apply several drops of hydrogen peroxide (ready-to-use from the BaseScope v2 reagent kit) to completely cover the tissue section, and incubate slides at room temperature for 10 min.
6. Wash slides twice in RNase-free water (1 min each) with gentle agitation.
7. Place slides into 1 × target retrieval solution at 100 °C (on hotplate) for 15 min. Small bubbles should be visible in the solution, however vigorous boiling should be avoided.
8. Wash slides twice in RNase-free water (1 min each) with gentle agitation.
9. Wash slides in 100% ethanol (30 s) with gentle agitation.
10. Place slides in a dry oven at 60 °C for 2 min.
11. Draw a barrier around the section using the ImmEdge hydrophobic barrier pen and allow to dry for 5 min at room temperature.
12. Apply several drops of protease IV (ready-to-use from the BaseScope v2 reagent kit) to completely cover the tissue section and place slides onto the preheated HybEZ tray within the HybEZ oven at 40 °C for 30 min. If using the EZ-Batch system then the EZ-Batch slide

holder can be used at this stage (all subsequent incubations and washes can then be performed within this holder).

13. Wash slides twice in RNase-free water (1 min each) with gentle agitation.
14. Apply several drops of the BaseScope 1ZZ probe (*e.g.*, KRAS G12V, POLR2A or dapB, preheated to 40 °C) to completely cover the tissue section, and place slides onto the preheated HybEZ tray within the HybEZ oven at 40 °C for 2 h.
15. Wash slides twice in 1 × wash buffer (prepared by diluting 20 × Wash buffer in $ddH_2O$, 2 min each).
16. Tilt the slides and use a Kimwipe to remove excess wash buffer, taking care not to touch the tissue section. Apply several drops of AMP 1 reagent to completely cover the tissue section and place slides back onto the preheated HybEZ tray within the HybEZ oven at 40 °C for 30 min.
17. Repeat steps 15 and 16, working through AMP reagents 2–8 with the following conditions: AMP 2 for 30 min at 40 °C, AMP 3 for 15 min at 40 °C, AMP 4 for 30 min at 40 °C, AMP 5 for 30 min at 40 °C, AMP 6 for 15 min at 40 °C, AMP 7 for 30 min at room temperature, AMP 8 for 15 min at room temperature.
18. After the AMP 8 incubation, perform final washes in 1 × wash buffer (twice for 2 min each).
19. Detect the BaseScope signal by adding Fast Red reagent (prepared by adding Fast Red B to Fast Red A at a 1:60 ratio) to completely cover the tissue section.
20. Incubate slides for 10 min at room temperature in the dark.
21. Wash slides in PBS (2 min).
22. Replace with fresh PBS and store slides at 4 °C overnight.

### 2.3.3 Immunohistochemistry for CD68

1. Remove the slides from 4 °C storage and place into room temperature PBST (PBS supplemented with 0.1% Tween-20) for 2 min.
2. Place the slides into a humidified chamber (the HybEZ tray and slide holder can be used) and add blocking buffer (PBS supplemented with 5% normal goat serum and 2% bovine serum albumin) for 1 h at room temperature.
3. Remove blocking reagent by tilting the slides and tapping onto a Kimwipe, then use a Kimwipe to absorb any excess (take care not to touch the tissue section).

4. Add primary antibody diluted in blocking buffer (*e.g.*, CD68 antibody at 1:100) to completely cover the section, and incubate for 1 h at room temperature.
5. Wash slides in PBST three times (5 min each) with gentle agitation.
6. Add the appropriate secondary antibody diluted in blocking buffer (*e.g.*, biotinylated rabbit anti-mouse at 1:400) to completely cover the section, for 30 min at room temperature.
7. Wash slides in PBST three times (5 min each) with gentle agitation.
8. Add streptavidin-HRP diluted in blocking buffer (1:500) to completely cover the section, for 30 min at room temperature.
9. Wash slides in PBST three times (5 min each) with gentle agitation.
10. Prepare DAB reagent according to the manufacturers protocol (Vector Labs), and add several drops to completely cover the tissue section.
11. Incubate for 3 min at room temperature.
12. Wash slides three times in dd$H_2O$ (1 min each).
13. Place slides in hematoxylin (Gill No.2, diluted 1:5 in dd$H_2O$) for 20 s with gentle agitation.
14. Wash slides three times in dd$H_2O$ (1 min each).
15. Place slides in ammonia water (0.02% ammonium hydroxide in dd$H_2O$) for 20 s with gentle agitation.
16. Wash slides three times in dd$H_2O$ (1 min each).
17. Place slides in a dry oven at 60 °C for 15 min.
18. Allow slides to cool to room temperature then add 1–3 drops of Vectamount and carefully lower a cover slip over the tissue. Press down firmly on the cover slip to remove any air bubbles.
19. Allow mountant to completely dry at room temperature before imaging.

## 3. Further considerations

The above protocol describes detection of KRAS G12V and CD68+ macrophages, however, it can be readily adapted to the detection of other mutations and/or proteins. It is advisable to test both the ISMD and IHC protocols in isolation before attempting to combine them (see Fig. 2 for examples). Following successful demonstration of the two individual protocols, there are some important considerations: first various dilutions of the primary antibody should be tested in the ISMD/IHC protocol, as the dilution used in IHC alone will not necessarily produce optimal results in a combined ISMD/IHC protocol. Second, as previously mentioned, the protease pretreatment involved in the ISMD protocol can irreversibly destroy the

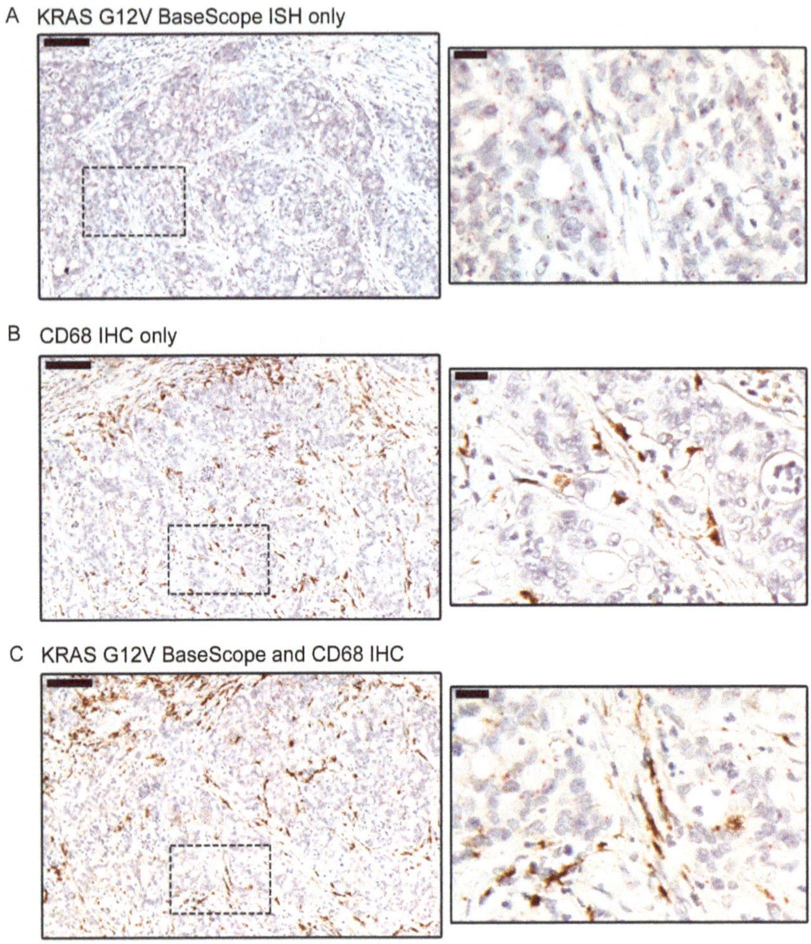


**Fig. 2** Representative images showing ISMD for KRAS G12V (A), IHC for CD68 (B) and concurrent ISMD for KRAS G12V and IHC for CD68 (C) in a human FFPE colorectal cancer. Scale bars in (A–C) represent 100 μm or 20 μm (inset). Images were obtained using a Pannoramic 250 slide scanner (3DHISTECH, Budapest, Hungary).

antigens required for IHC. Therefore optimization of the ISMD protease treatment step may be required in order to preserve the required antigens. However, we have found that more than half of antibodies tested will work with the above ISMD/IHC protocol without requiring further optimization of pretreatment conditions. In Table 1 we have provided information on further antibodies against immune markers that we have tested using the ISMD/IHC protocol.

**Table 1** Description of antibodies against immune markers and their suitability for the ISMD/IHC protocol.

| Antigen | Clone | Supplier | Dilution | Suitable for ISMD/IHC |
|---|---|---|---|---|
| CD68 | PG-M1 | Agilent | 1:100 | Yes |
| CD8 | C8/144B | Dako | 1:50 | Yes |
| CD4 | EPR6855 | Abcam | 1:400 | Yes |
| CD3 | F7.2.38 | Dako | 1:500 | Yes |
| PD-L1 | E1L3N | Cell Signaling Technology | 1:50 | Yes |
| CD20 | L26 | Dako | 1:50 | No |
| CD45RO | UCH-L1 | Abcam | 1:50 | No |
| FOXP3 | 236A/E7 | Abcam | 1:50 | No |

Our protocol describes optimal sample preparation (fixation) of tissue samples for ISMD and IHC; however, it is important to note that archival FFPE samples often have to be analyzed retrospectively. Tissue processing protocols can vary greatly between hospital sites, and "best practice" recommendations have changed over time and indeed are not always adhered to; therefore it is inevitable that a proportion of archival FFPE samples will not have been optimally processed. This may have unpredictable or undesirable effects, particularly on the success of ISMD with BaseScope. A further confounding factor is intra-sample heterogeneity in tissue fixation, which may be particularly evident in large samples which could be poorly fixed in the center, potentially leading to artifactual absence of signal in central areas. Therefore, it is important to always run duplicate sections of all archival samples with positive and negative control probes in order to ensure the validity of results, although we recognize this need must be balanced with the cost of the assay and the limited number of slides that can be processed in one batch. It is notable that we have tested FFPE blocks of up to 20 years old and we have not noted a consistent negative impact of the age of FFPE blocks on the quality of the BaseScope ISMD. In our experience, adherence to the recommended tissue processing guidelines is of more importance to the quality of ISMD results than the age of the FFPE block.

This chapter does not explicitly discuss quantification of ISMD/IHC staining, as the optimal approach is highly dependent on the application and the resources available to the researcher. However, we note that basic analysis can be performed in a manual way by counting the number of "dots" per cell of interest (for ISMD) or the number of positive stained cells (for IHC) (Baker et al., 2017) under a brightfield microscope. The Fast Red precipitate is fluorescent, and furthermore a fluorescent secondary antibody could be used in place of the biotinylated antibody used in the IHC protocol. This would allow visualization and quantification to be performed using fluorescent microscopy. It is notable that the sensitivity of the BaseScope assay does not often result in every mutant cell containing a signal: typically in an archival FFPE block we see approximately 25% of mutant cells containing a positive signal (Baker et al., 2017), although this is highly variable and dependent on the processing of the FFPE tissue, as well as the copy

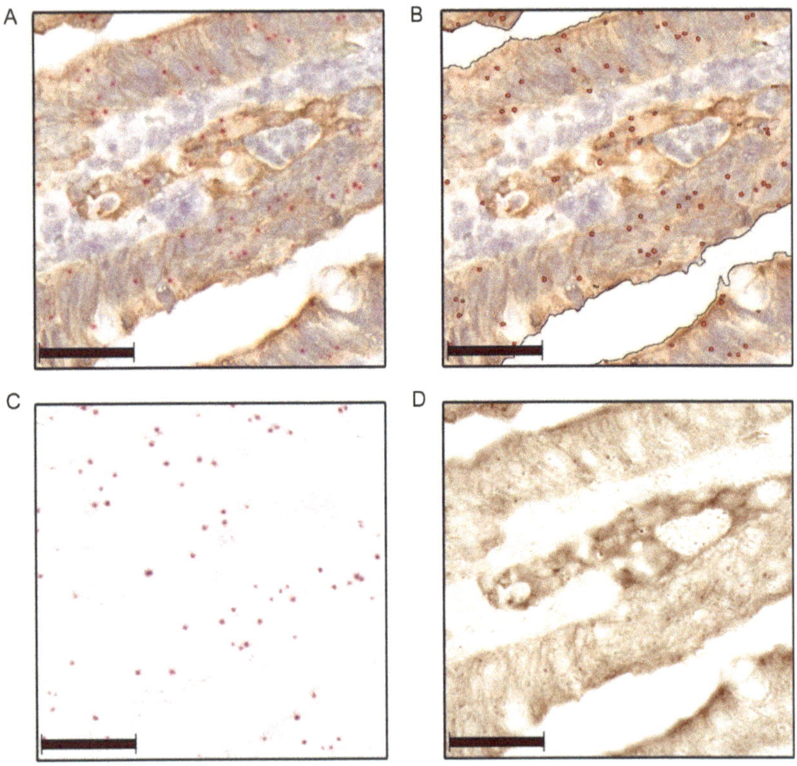


**Fig. 3** (A) Raw unprocessed image showing KRAS G12D signal (BaseScope ISMD, pink spots) and pan-cytokeratin IHC (brown) in a human FFPE colorectal cancer. (B) HALO processed image showing tissue boundaries (black line) and BaseScope ISMD positive signals highlighted in red. These signals are then quantified by the software. (C) HALO stain separation of the BaseScope ISMD positive signals. (D) HALO stain separation of the pan-cytokeratin IHC stain. Scale bars in (A–D) represent 50 μm.

number and half-life of the target mRNA. The quantification of ISMD (and indeed IHC) can also be performed in an automated way, and we have had good experience with the HALO (Indica Labs, Fig. 3) and Visiopharm packages.

## 4. Concluding remarks

This chapter has described a general protocol that can be applied to the analysis of point mutations and protein expression concurrently on the same FFPE tissue section, focusing specifically on the KRAS G12V mutation and CD68+ macrophage infiltrate in colorectal cancers. Clearly there is much scope to adapt this protocol to a variety of applications, and we have described the caveats that may be encountered when doing so and highlighted the steps at which optimization may be required. In summary, concurrent ISMD and IHC is a powerful technique that enables the analysis of tumor heterogeneity and genotype-phenotype correlations whilst preserving tissue morphology and spatial context.

## Acknowledgments

The authors thank Advanced Cell Diagnostics for providing ongoing technical support for the BaseScope assay, and the Barts Cancer Institute histopathology team for expert tissue processing. The authors are supported by Cancer Research UK (grant number A19771).

## References

Baker, A. M., Huang, W., Wang, X. M., Jansen, M., Ma, X. J., Kim, J., et al. (2017). Robust RNA-based in situ mutation detection delineates colorectal cancer subclonal evolution. *Nature Communications*, *8*(1), 1998. https://doi.org/10.1038/s41467-017-02295-5.

Erben, L., He, M. X., Laeremans, A., Park, E., & Buonanno, A. (2018). A novel ultrasensitive in situ hybridization approach to detect short sequences and splice variants with cellular resolution. *Molecular Neurobiology*, *55*(7), 6169–6181. https://doi.org/10.1007/s12035-017-0834-6.

Grundberg, I., Kiflemariam, S., Mignardi, M., Imgenberg-Kreuz, J., Edlund, K., Micke, P., et al. (2013). In situ mutation detection and visualization of intratumor heterogeneity for cancer research and diagnostics. *Oncotarget*, *4*(12), 2407–2418. https://doi.org/10.18632/oncotarget.1527.

Larsson, C., Grundberg, I., Soderberg, O., & Nilsson, M. (2010). In situ detection and genotyping of individual mRNA molecules. *Nature Methods*, *7*(5), 395–397. https://doi.org/10.1038/nmeth.1448.

Zhu, Y., Sharp, A., Anderson, C. M., Silberstein, J. L., Taylor, M., Lu, C., et al. (2018). Novel junction-specific and quantifiable in situ detection of AR-V7 and its clinical correlates in metastatic castration-resistant prostate cancer. *European Urology*, *73*(5), 727–735. https://doi.org/10.1016/j.eururo.2017.08.009.

CHAPTER TWELVE

# RNA interference screening methods to identify proliferation determinants and mechanisms of resistance to immune attack

**Yong-Wei Zhang[a], Rochelle E. Nasto[a,b,c], Sandra A. Jablonski[a], Ilya G. Serebriiskii[b,d], Rishi Surana[a], Joseph Murray[a], Michael Johnson[a], Rebecca B. Riggins[a], Robert Clarke[a], Erica A. Golemis[b], Louis M. Weiner[a,*]**

[a]Lombardi Comprehensive Cancer Center, Georgetown University, Washington, DC, United States
[b]Program in Molecular Therapeutics, Fox Chase Cancer Center, Philadelphia, PA, United States
[c]School of Biomedical Engineering, Science and Health Systems, Drexel University, Philadelphia, PA, United States
[d]Kazan Federal University, Kazan, Russia
*Corresponding author: e-mail address: weinerl@georgetown.edu

## Contents



## Abstract

We have used RNA interference (RNAi) screening technology to reveal unknown components of biological signaling pathways including survival mechanisms of

*Methods in Enzymology*, Volume 636
ISSN 0076-6879
https://doi.org/10.1016/bs.mie.2019.06.002

estrogen-independent breast cancer cell growth and cancer cell resistance to immune attack. In this chapter, a detailed protocol describing the use of RNAi screening to identify factors important for the proliferation of estrogen-independent MCF7 breast cancer cells will be described. Resistance to therapies that target the estrogen pathway remains a challenge in the treatment of estrogen receptor-positive breast cancer. To address this challenge, small interfering-RNA (siRNA)-based libraries targeting an estrogen receptor (ER)- and aromatase-centered network, including 631 genes relevant to estrogen signaling, was designed and constructed for RNAi screening. This protocol will include the following parts: (1) selection of RNAi transfection reagent for specific cells; (2) optimization of RNAi screening conditions using Z′-factor; (3) procedure of ER-network gene siRNA library screening using automated machines under optimized experimental conditions; and (4) method of analysis for RNAi screening data to identify specific determinants important for cell proliferation. 46 genes were found to be selectively required for the survival of estrogen-independent MCF7-derived cells.

## 1. Introduction

RNA interference (RNAi) is a biological process that can be exploited to inhibit gene expression by causing the destruction of specific mRNA molecules. Knockdown of specific genes by RNAi technology is often associated with phenotypic changes, which has made RNAi widely used in life science research. Two systems are utilized for high-throughput RNAi screening, one is a lentivirus-based short hairpin RNA (shRNA) library method for screening; the other is chemically-synthesized small interference RNA (siRNA)-based screening (Boutros & Ahringer, 2008). shRNA-based infection induces stable gene knockdown in cells. siRNA-based transfection induces transient gene knockdown. Lentiviral pooled shRNA libraries consist of lentiviruses containing shRNAs targeting either genomic DNA or a group of genes. Following screening, analysis is required to distinguish target genes, such as chip-based DNA microarray or next generation sequencing (NGS) (Josse et al., 2015; Shuptrine et al., 2017). However, in siRNA-based libraries, siRNAs against each single target gene are individually distributed in the wells of 96-well or 384-well plates. A siRNA library may include many plates depending on the number of targeting genes in this library (Zhang, Jones, Martin, Caplen, & Pommier, 2009; Zhang et al., 2016). The siRNAs in the library are then reverse-transfected into cells to introduce the siRNA into the cells in the same array they are laid out in the 96- or 384-well plates. Thus for siRNA library screening, no further techniques are required to identify targeting genes.

In our studies, we designed the Estrogen receptor (ER)-network around five seed proteins relevant to estrogen signaling: the ER genes ESR1 (ERα) and ESR2 (ERβ), the estrogen-related receptors ESRRA and ESRRG, and CYP19A1 (aromatase). 631 genes were selected for the ER network. Next, we constructed siRNA-based libraries targeting the ER network genes, which were custom-made from QIAGEN (MD, USA) and arrayed into 96-well plates. siRNAs against the selected genes were distributed into 11 96-well plates. Two siRNAs were selected for each gene and mixed in one well (Zhang et al., 2016). The advantage of our method provides high-throughput screening by using automatic machines (Cybio, combi-nl or Wellmate dispenser) to dispense liquid to speed the screening process.

Different types of cancer cell lines had been used for RNAi screening with our methods to identify proliferation determinants and mechanisms of resistance to immune attack (Astsaturov et al., 2010; Murray et al., 2014; Zhang et al., 2009, 2016), such as estrogen positive breast cancer MCF7cells, estrogen-independent MCF7 cells (LCC1 and LCC9), triple negative breast cancer MDA-MB-231 cells, epidermoid cancer A431 cells and human fibroblast HFF1 cells. For each cell line, the optimal transfection reagent was determined before RNAi library screening. Z′-factor was taken as a quantitative parameter to control the experimental quality for various cell lines and corresponding transfection reagents. In this example, we utilize ER-network RNAi screening in MCF7 cells to describe the protocol (Zhang et al., 2016). It also can be used with other cell lines or other gene network RNAi libraries with minor modifications, such as type of transfection reagent, cell plating density, Cell Titer blue incubation time or RNAi library scale (total number of siRNA library plates), which will be noted. In this example, the protocol will be described in three parts (Fig. 1): (1) selection of transfection reagents; (2) Z′-factor determination; (3) screening an RNAi library.

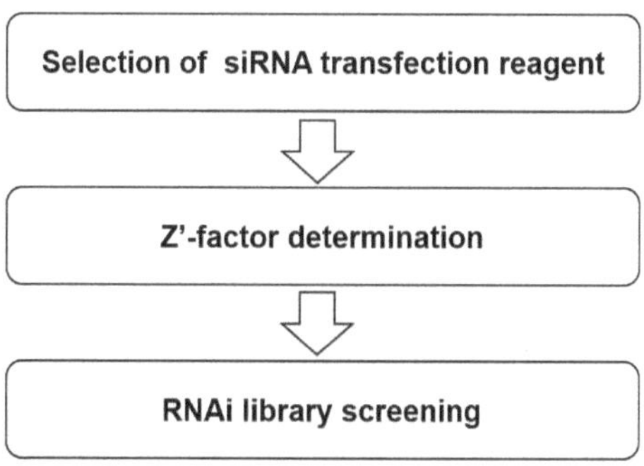


**Fig. 1** Outline of the protocol.

## 2. Materials and reagents

1. ER network siRNA library plates (Customized from QIAGEN) siRNA-based library targeting Estrogen receptor (ER)- and aromatase-centered network was designed around five seed proteins relevant to estrogen signaling: the ER genes ESR1 (ERα) and ESR2 (ERβ), the estrogen-related receptors ESRRA and ESRRG, and CYP19A1 (aromatase). 631 genes were selected as ER network, which was described in Supplementary Material in the online version at https://doi.org/10.1016/bs.mie.2019.06.002. siRNA against each gene was ordered from QIAGEN (Germantown, MD).
2. Pipette tips for CyBio-Well Vario 96 channel simultaneous Pipettor (Thermo Fisher Scientific, Thermo Scientific™, catalog number: 5587)
3. V-bottom 96-well plates (Corning, catalog number: 3357)
4. Flat-bottom 96-well plates (Corning, catalog number: 3595)
5. 50 mL conical tube
6. Corning 0.22 μm vacuum filter system (Corning, catalog number: 431098)
7. T75 flasks (Corning, Costar)
8. Labels with Barcode
9. MCF7 cells (Tissue Culture Shared Resource, Lombardi Cancer Center, Georgetown Univ.)
10. AllStars Negative Control siRNA (QIAGEN, catalog number: 1027281)
11. AllStars Hs Cell Death siRNA (QIAGEN, catalog number: 1027299)
12. AP2A1 siRNA (QIAGEN, catalog number: SI04371283)
13. GRB14 siRNA (QIAGEN, catalog number: SI00430703)
14. Opti-MEM reduced serum medium (Thermo Fisher Scientific, Gibco™, catalog number: 31985070)
15. IMEM medium (Mediatech, catalog number: 10-024-CV)
16. Trypsin-EDTA (0.5%), no phenol red (Thermo Fisher Scientific, Gibco™, catalog number: 15400054)
17. Charcoal-stripped bovine calf serum (CCS) (Gemini Bio-Products, catalog number: 100-213)
18. Estradiol (Sigma-Aldrich, catalog number: E8875)
19. Cell Titer Blue (Promega, catalog number: G8082)
20. Hank's balanced salt solution (HBSS) without calcium, magnesium, phenol red (GE Healthcare, Hyclone, catalog number: SH30588.01)
21. siRNA suspension buffer (QIAGEN)

22. Lipofectamine RNAiMAX transfection reagent (Thermo Fisher Scientific, Invitrogen™, catalog number: 13778500)
23. HiPerfect (QIAGEN, catalog number: 301704)
24. Dharmafect 1–4 transfection reagent (GE Dharmacon, catalog numbers: T-2001, T-2002, T-2003, T-2004)
25. RNAiFect (QIAGEN)
26. 70% (v/v) ethanol (filtered via Corning 0.22 μm vacuum filter system)
27. 0.22 μm filtered $ddH_2O$

## 3. Equipment

1. CyBi-Well Vario 96 channel simultaneous Pipettor (CyBio)

2. Multidrop Combi nL reagent dispenser (Thermo Fisher Scientific, catalog number: 5840400)

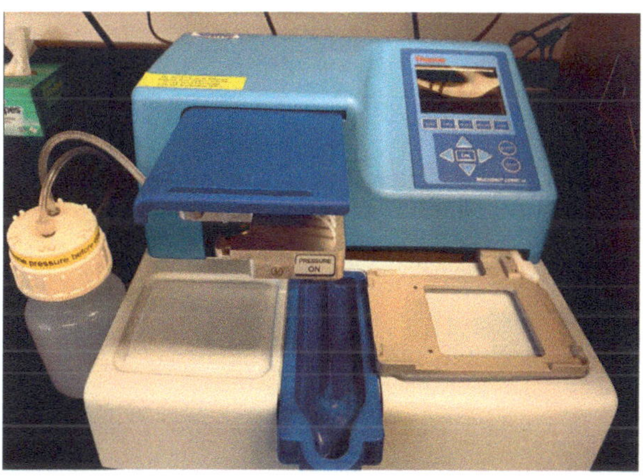

3. WellMate microplate dispenser (Thermo Scientific Matrix)

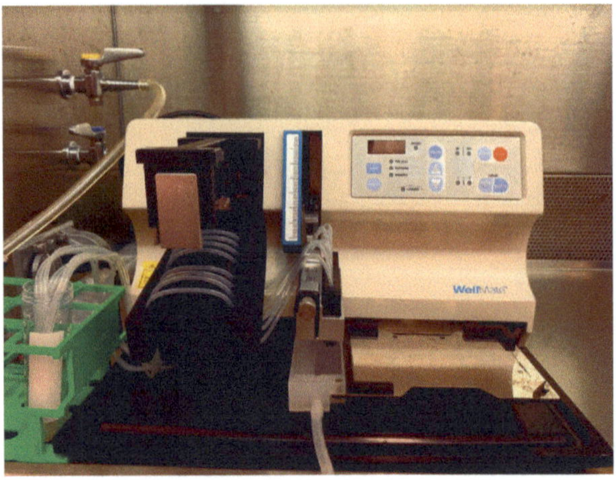

4. Envision multi-label plate reader with $560_{Ex}/590_{Em}$ filter set (PerkinElmer, catalog number: 2104-0010)
5. 3R centrifuge with Ch.003741 rotor (Thermo Fisher Scientific, catalog number: 4393) and swing rectangular buckets with adapters (Thermo Fisher Scientific, catalog number: 75006449)
6. Magnetic stirrer (Thermo Fisher Scientific)
7. 500 mL glass bottle (Corning, Costar)

## 4. Part I. Selection of transfection reagents

Because siRNA transfection is the most critical step for RNAi library screening, it is important to select the suitable transfection reagent, which yields high transfection efficiency for a certain cell line. Multiple commercial transfection reagents will be tested by measuring cell viability in cells transfected with negative control siRNAs (siNEG), death control siRNAs (siDEATH), or AP2A1 siRNA (AP2A1 knockdown will yield moderate cell death). The siRNA is reverse-transfected into the tissue culture cells (lipid tranfection reagent and siRNA are mixed first in the plate, and cells are added to the mixture). The procedure for the selection of transfection reagents in breast cancer MCF7 cells is described.

### 4.1 Procedure

Transfection with multiple transfection reagents (lipid) in breast cancer MCF7 cells:

1. Transfect cells in 96-well plate in 7 blocks, for each block (12 wells; see Fig. 2 for plate layout):
   3 wells: lipid + Opti-MEM medium.
   3 wells: 20 nM AllStars negative control siRNA (QIAGEN, MD) + lipid + Opti-MEM.
   3 wells: 20 nM AllStars Death control siRNA (QIAGEN, MD) + lipid + Opti-MEM.
   3 wells: 20 nM AP2A1 siRNA (QIAGEN, MD) + lipid + Opti-MEM.
2. For lipid: diluted lipid (recipe as in Table 1) will be added (15 μL/well) after appropriate dilution in Opti-MEM (Invitrogen, MD). Diluted lipid will be aliquoted into 12 wells.
3. For siRNA: 1 μM siRNA is diluted in Opti-MEM (1:3) and 7 μL diluted siRNA is added to each corresponding well. Each siRNA is dispensed into 21 wells, therefore, 147 μL are needed. To account for pipetting loss, we made 165 μL diluted siRNA: 55 μL 1 μM siRNA + 110 μL Opti-MEM.
4. Split cells, count and calculate dilution for 8000 cells per well in 100 μL IMEM + 5% charcoal-stripped bovine calf serum (CCS) + 1 nM estradiol (Sigma-Aldrich, MI).
5. Set up a 96-well plate:
   a. Pipette siRNA (Opti-MEM for lipid control wells: A1–12, E1–9), 7 μL/well to following wells: siNEG, wells B1–12 and F1–9; siAP2A1, wells C1–12 and G1–9; siDEATH, wells D1–12 and H1–9.
   b. Pipette lipid mixture (Opti-MEM for medium control wells: E10–12, F10–12, G10–12 and H10–12), 15 μL/well to following wells: A1–12, B1–12, C1–12, D1–12, E1–9, F1–9, G1–9 and H1–9 (see Fig. 2 for plate layout).
   c. Incubate for 20 min at room temperature, next add 100 μL MCF7 cells/well with WellMate Microplate dispenser (Thermo Scientific Matrix, USA), then incubate at 37 °C, 5% $CO_2$ for 5 days.
   d. Five days later, add 20 μL of 1:1 mixture of Cell Titer Blue:HBSS to each well and incubate at 37 °C to allow cells to convert resazurin to resorufin. The fluorescent signal is measured by Envision multi-label plate reader (with excitation wavelength 560 nm/emission wavelength 590 nm) every hour up to 4 h. For this experiment, 2 h is typically the optimal time point to read out with 4 h nearing the maximum signal of the assay where dynamic range is not compromised.

| | 1 | 2 | 3 | 4 | 5 | 6 | 7 | 8 | 9 | 10 | 11 | 12 |
|---|---|---|---|---|---|---|---|---|---|---|---|---|
| A | HiPerfect alone | HiPerfect alone | HiPerfect alone | RNAiFect alone | RNAiFect alone | RNAiFect alone | DharmaFect1 alone | DharmaFect1 alone | DharmaFect1 alone | DharmaFect2 alone | DharmaFect2 alone | DharmaFect2 alone |
| B | HiPerfect negative control | HiPerfect negative control | HiPerfect negative control | RNAiFect negative control | RNAiFect negative control | RNAiFect negative control | DharmaFect1 negative control | DharmaFect1 negative control | DharmaFect1 negative control | DharmaFect2 negative control | DharmaFect2 negative control | DharmaFect2 negative control |
| C | HiPerfect AP2A1 | HiPerfect AP2A1 | HiPerfect AP2A1 | RNAiFect AP2A1 | RNAiFect AP2A1 | RNAiFect AP2A1 | DharmaFect1 AP2A1 | DharmaFect1 AP2A1 | DharmaFect1 AP2A1 | DharmaFect2 AP2A1 | DharmaFect2 AP2A1 | DharmaFect2 AP2A1 |
| D | HiPerfect Death | HiPerfect Death | HiPerfect Death | RNAiFect Death | RNAiFect Death | RNAiFect Death | DharmaFect1 Death | DharmaFect1 Death | DharmaFect1 Death | DharmaFect2 Death | DharmaFect2 Death | DharmaFect2 Death |
| E | DharmaFect 3 alone | DharmaFect 3 alone | DharmaFect 3 alone | DharmaFect4 alone | DharmaFect4 alone | DharmaFect4 alone | RNAiMax alone | RNAiMax alone | RNAiMax alone | Medium | Medium | Medium |
| F | DharmaFect 3 negative control | DharmaFect 3 negative control | DharmaFect 3 negative control | DharmaFect4 negative control | DharmaFect4 negative control | DharmaFect4 negative control | RNAiMax negative control | RNAiMax negative control | RNAiMax negative control | Medium | Medium | Medium |
| G | DharmaFect 3 AP2A1 | DharmaFect 3 AP2A1 | DharmaFect 3 AP2A1 | DharmaFect4 AP2A1 | DharmaFect4 AP2A1 | DharmaFect4 AP2A1 | RNAiMax AP2A1 | RNAiMax AP2A1 | RNAiMax AP2A1 | Medium | Medium | Medium |
| H | DharmaFect 3 Death | DharmaFect 3 Death | DharmaFect 3 Death | DharmaFect4 Death | DharmaFect4 Death | DharmaFect4 Death | RNAiMax Death | RNAiMax Death | RNAiMax Death | Medium | Medium | Medium |

**Fig. 2** Layout of transfection reagent selection plate.

**Table 1** Recipe of diluted lipid. To account for pipetting loss, 187.5 μL (12.5 × 15 μL) diluted lipid will be made in Opti-MEM.

| | Lipid (μL) | Opti-MEM (μL) |
|---|---|---|
| Block 1: HiPerfect | 9.4 | 178.1 |
| Block 2: RNAiFect | 11.25 | 176.3 |
| Block 3: DharmaFect 1 | 6.25 | 181.3 |
| Block 4: DharmaFect 2 | 6.25 | 181.3 |
| Block 5: DharmaFect 3 | 6.25 | 181.3 |
| Block 6: DharmaFect 4 | 6.25 | 181.3 |
| Block 7: RNAiMax | 3.75 | 183.8 |

### 4.2 Data analysis

Normalize all cell growth to control cells (in wells E, F, G and H10–11). Next, assess the growth inhibition induced by negative control siRNAs (siNEG, in wells B and F), death control siRNAs (siDEATH, in wells D and H), and AP2A1 siRNA (in wells C and G). Then, compare the effect of various transfection reagents. Viability of AP2A1 siRNA should be within the middle of the dynamic range between siNEG and siDEATH. Select the lipid mixture containing the transfection reagent that provides not only the highest viability with siNEG, but also the lowest viability with siDEATH.

## 5. Part II. Z′-factor determination

Before performing the high-throughput RNAi library screening, the experimental conditions should be optimized. Z′-factor, a measure of dynamic range and signal-to-noise, was assessed to determine whether the assay conditions were suitable for screening (Birmingham et al., 2009; Zhang, Chung, & Oldenburg, 1999). A Z′-factor siRNA plate including negative control siRNAs (siNEG) and death control siRNAs (siDEATH) was used to calculate Z′-factor and test the potential experimental conditions. The procedures of Z′-factor determination was described here.

### 5.1 Procedure

The workflow of high throughput screening using multiple automated machines is shown in Fig. 3. Briefly, prepare siRNA plates with

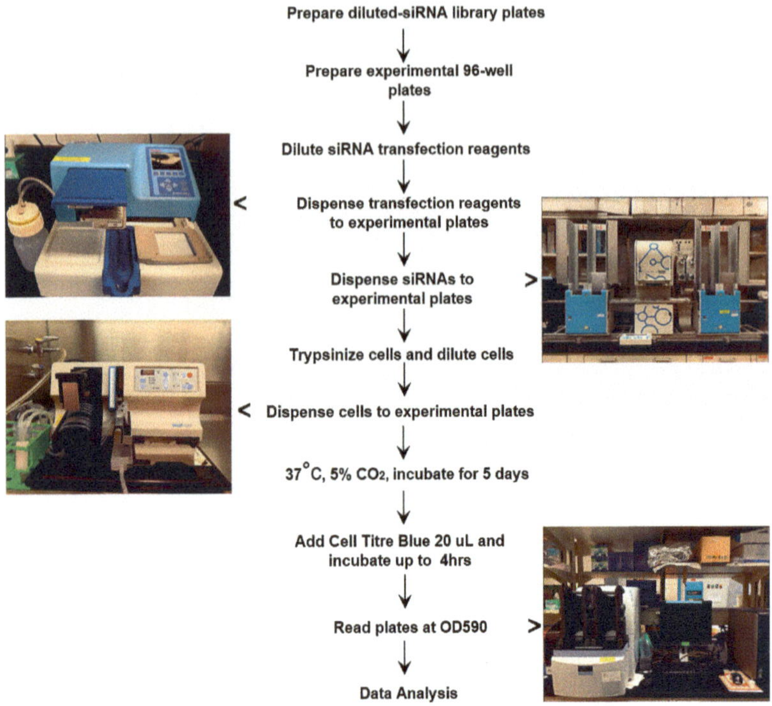


**Fig. 3** Workflow of high throughput-RNAi screening.

concentration of siRNA at 0.24 μM. Next, dispense 10.5 μL of diluted lipid transfection reagent to experimental plates (using Combi-nl), mix with 10 μL of diluted ER library siRNA (using CyBio) and 8000 cells per well (using WellMate). Cells were incubated for 5 days. Viability was measured by Cell Titer Blue methods. In this procedure of Z′-factor determination, Z′-factor plates are used in the place of the siRNA library plates.

### *5.1.1 Procedure of day 1*

Set up a siRNA Z′-factor plate in V-bottom 96-well plate containing 0.24 μM siNEG and siDEATH (using layout shown in Fig. 4). To make 0.24 μM siRNA solution: 72 μL 20 μM siRNA is mixed in 5928 μL siRNA suspension buffer; or 1440 μL 1 μM siRNA is mixed in 4560 μL siRNA suspension buffer. One-hundred μL of diluted siNEG or siDEATH (0.24 μM) is dispensed into each well (using layout shown in Fig. 4). Then siRNA Z′-factor is frozen at −20 °C for later use.

| | 1 | 2 | 3 | 4 | 5 | 6 | 7 | 8 | 9 | 10 | 11 | 12 |
|---|---|---|---|---|---|---|---|---|---|---|---|---|
| A | DEATH | NEG | NEG | DEATH | NEG | NEG | DEATH | NEG | NEG | NEG | NEG | NEG |
| B | NEG | DEATH | DEATH | NEG | DEATH | DEATH | NEG | DEATH | DEATH | NEG | DEATH | DEATH |
| C | DEATH | NEG | NEG | DEATH | NEG | NEG | DEATH | NEG | NEG | DEATH | NEG | NEG |
| D | NEG | DEATH | NEG | NEG | DEATH | NEG | NEG | NEG | DEATH | NEG | NEG | DEATH |
| E | DEATH | NEG | DEATH | DEATH | NEG | DEATH | DEATH | NEG | DEATH | DEATH | DEATH | NEG |
| F | NEG | DEATH | NEG | NEG | DEATH | NEG | NEG | DEATH | NEG | NEG | DEATH | NEG |
| G | DEATH | NEG | DEATH | DEATH | NEG | DEATH | DEATH | NEG | DEATH | DEATH | NEG | DEATH |
| H | NEG | DEATH | NEG | DEATH | DEATH | NEG | NEG | DEATH | DEATH | NEG | DEATH | DEATH |

**Fig. 4** Layout of siRNA Z′-factor plate.

### 5.1.2 Procedure of day 2

Each experimental plate (96-well plate, Costar, Corning, USA) must have 10.5 μL of diluted lipid transfection reagent, 10 μL of siRNA and 8000 cells per well. Two replicate plates are run at one experiment. The Z′-factor siRNA plate is thawed at room temperature. For two plates (192 wells), 2016 μL of lipid transfection reagents are required. Due to the loss of machine priming, 3 mL of total transfection reagent is loaded into Combi-nl machine that then distributes 10.5 μL to each well of two experimental plates. The plates are moved to the CyBio machine, and it is used to pipette 10 μL from the siRNA Z′-factor plate into each experimental replicate plate containing the wells with 10.5 μL of diluted transfection. The final concentration of siRNA in cells is 20 nM. Next, cells will be added into experimental plates for culture by Wellmate microplate dispenser.

1. Clean WellMate microplate dispenser with 15 mL of 70% ethanol (pre-filtered by 0.22 μm vacuum filter), then 15 mL of dd$H_2O$ (pre-filtered by 0.22 μm vacuum filter), and lastly 15 mL of IMEM (no serum).
2. Split cells and count. Dilute 8000 cells in 100 μL for one well (For one 96-well plate, 80,000 cells/mL, need 16 mL, prepare 40 mL in a 50 mL conical tube).
3. Dispense 15 mL Opti-MEM in a 50 mL conical tube.
4. Clean Combi-nL with 7 mL of filtered (0.22 μm) 70% ethanol, 7 mL distilled water (0.22 μm filtered), 7 mL Opti-MEM.
5. Take siRNA Z′-factor plate out from −20 °C and thaw plate at room temperature. Next, spin plate in centrifuge to get liquid in center of wells at 3500 rpm (1935 ×*g*), 5 min, room temperature, take plate out immediately to keep condensation from forming.
6. Dilute transfection reagent: 144 μL in 2.86 mL Opti-MEM in a 50 mL conical tube.
7. Dispense 10.5 μL/well of diluted transfection reagent to each Costar 96-well plate (experimental plate) with Combi-nL machine.
8. Start Cybio machine, load plates and distribute siRNAs from siRNA Z′-factor plate to experimental plate (pre-loaded with lipid from step 2g). The loading position for plates: (Fig. 5)
   - **i.** siRNA Z′-factor plate—loaded on stack A (left arm) on CyBio.
   - **ii.** Experimental plate—loaded on stack A (right arm, with diluted lipid) on CyBio.
9. Run the program (Fig. 6) to dispense siRNAs to experimental plates.

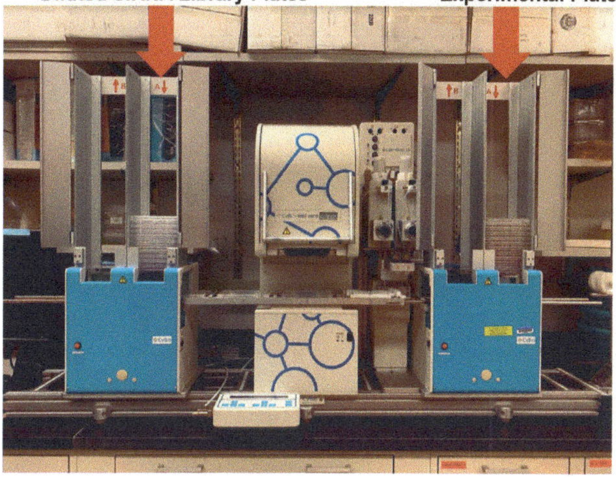


**Fig. 5** Loading of siRNA plates and experimental plates.

**10.** Wait for 10 min at room temperature for siRNA-lipid complexes to form in experimental plate. While waiting, set up dispensing program on the WellMate dispenser.

**11.** Use the WellMate microplate dispenser to dispense 100 μL cells/well into each experimental plate, incubate experimental plate at 37 °C, 5% $CO_2$ for 5 days.

**12.** Clean WellMate dispenser with 15 mL filtered dd$H_2O$, then 15 mL filtered 70% ethanol, switch off machine.

**13.** Clean the Combi-nL dispenser with 7 mL distilled water, and 7 mL of filtered (0.45 μm) 70% ethanol, switch off machine.

### *5.1.3 Procedure of day 7*

Add 20 μL of 1:1 mixture of Cell Titer Blue:HBSS to each well and read out every hour up to 4 h (as described above).

## 5.2 Data analysis

**1.** Based on Cell-Titer Blue read out, calculate the average viability value of siNEG and siDEATH. Calculate Z′-factor (Birmingham et al., 2009; Zhang et al., 1999):

$$Z' - \text{factor} = 1 - \frac{(3 \times \text{S.D.of siDEATH} + 3 \times \text{S.D.of siNEG})}{(\text{Average viability reading of siNEG-Average viability reading of siDEATH})}$$

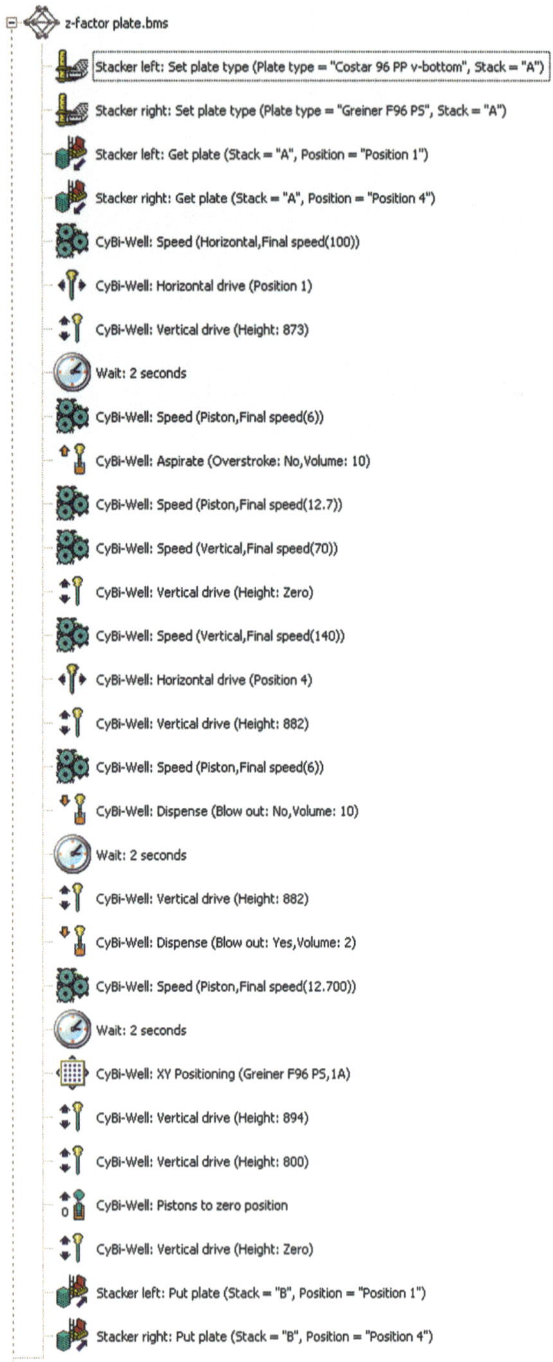


**Fig. 6** Cybio program to dispense siRNA into experimental plates.

An assay is considered suitable for screening if Z′-factor is in the range of 0.7–1. If Z′-factor is smaller than 0.7, modify assay parameters (such as cell numbers, transfection reagent and siRNA final concentration) and repeat the Z′-factor experiment. Once Z′-factor is within range, continue with large scale RNAi library screening.

2. A representative viability measurement (Fig. 7A) and dot plots (Fig. 7B) from one Z′-factor test were shown here. In this experiment, average viability reading of siNEG, siDEATH and corresponding standard deviation values (S.D.) were obtained from Fig. 7A. Then Z′-factor was calculated as 0.81, therefore, the prescribed conditions used in this experiment were optimal for following siRNA library screening.

# 6. Part III. Screening an RNAi library

## 6.1 Procedure for siRNA control preparation

1. In this article, we utilize an ER-network RNAi library screening as an example to describe the protocol. The ER network siRNA library contains siRNAs against 631 genes, which were custom-made from

A

| | 1 | 2 | 3 | 4 | 5 | 6 | 7 | 8 | 9 | 10 | 11 | 12 |
|---|---|---|---|---|---|---|---|---|---|---|---|---|
| A | 512821 | 2732333 | 3153999 | 462617 | 2904101 | 2908701 | 565297 | 2951710 | 3164610 | 3582064 | 3549669 | 3021951 |
| B | 3212173 | 462494 | 551609 | 3340972 | 560306 | 463258 | 3379221 | 517948 | 497057 | 3280402 | 506548 | 530900 |
| C | 439970 | 2297361 | 3082661 | 422603 | 2811268 | 2958623 | 466030 | 2956318 | 2963699 | 434236 | 3018473 | 2799434 |
| D | 2736371 | 416437 | 2947868 | 2885317 | 415219 | 3020473 | 3242222 | 3101489 | 484255 | 3212510 | 3322089 | 571912 |
| E | 434035 | 2636991 | 441501 | 471890 | 2982055 | 479466 | 606192 | 2550881 | 472511 | 406943 | 508755 | 2910021 |
| F | 3028686 | 473975 | 3097356 | 3019020 | 437968 | 2973422 | 2899782 | 541957 | 3338163 | 3147214 | 422435 | 3084216 |
| G | 453870 | 3484683 | 427296 | 411977 | 3217215 | 606284 | 479326 | 3315395 | 445224 | 458876 | 3291320 | 459777 |
| H | 3320426 | 481893 | 3198396 | 492183 | 438258 | 3325481 | 3355647 | 566108 | 486995 | 3533333 | 496251 | 590437 |

B

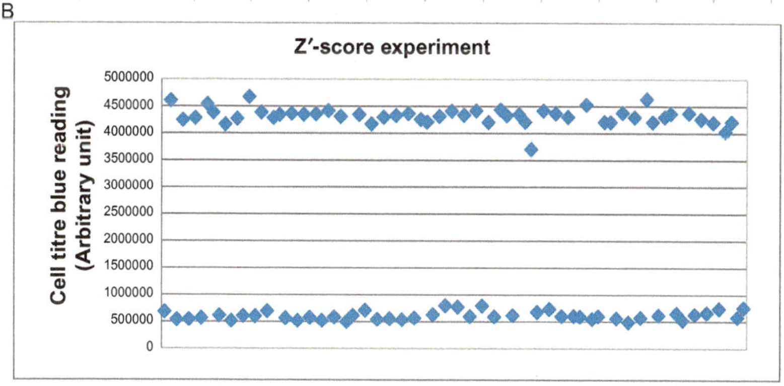


**Fig. 7** Representative data of Z′-factor test. (A) Viability measurement by Cell Titer Blue reading. Purple wells: siDEATH-transfected cells; white wells: siNEG-transfected cells. (B) Dot plots of viability measurements. Each dot indicated the value of cell tire blue reading in each well. Upper line of dots indicated those wells containing siNEG-transfected cells; lower line of dots indicated those wells with siDEATH-transfected cells.

QIAGEN (MD, USA). siRNAs against those genes were distributed into 11 96-well plates. Two siRNAs were selected for each gene and mixed in one well.

*Notes: For other siRNA library screening, the total number of plates will be determined based on the number of targeted genes and number of targeting siRNAs per gene as well. In summary, the total number and final layout of wells containing siRNA in an RNAi library will modify the screening protocol.*

2. Calculate the amount of various control siRNA (0.24 μM) needed to make 11 plates (100 μL each well):
   a. Preparation of negative control siRNA (siNEG)

$$\frac{14\,\text{wells}}{\text{plate}} \times \frac{100\,\mu\text{L}}{\text{well}} \times 11\,\text{plates} = 15{,}400\,\mu\text{L of siNEG}\,(0.24\,\mu\text{M})$$

   Accounting for pipetting error, 18 mL of 0.24 μM siNEG will be prepared as followed: 4.32 mL siNEG (1 μM) + 13.68 mL siRNA suspension buffer.

   b. Preparation of DEATH, AP2A1(X) and GRB14(Y) siRNA

$$\frac{6\,\text{wells}}{\text{plate}} \times \frac{100\,\mu\text{L}}{\text{well}} \times 11\,\text{plates} = 6{,}600\,\mu\text{L of Death, X or Y siRNA at }0.24\,\mu\text{M}$$

   Accounting for pipetting error, 8 mL of 0.24 μM DEATH, X and Y siRNA will be prepared as followed: 1.92 mL DEATH, X or Y siRNA (1 μM) + 6.08 mL siRNA suspension buffer.

   c. Preparation of ER network library siRNAs

   For each siRNA mix in one well, prepare 100 μL 0.24 μM siRNA as followed: 24 μL 1 μM stock siRNA +76 μL siRNA suspension buffer.

## 6.2 Estrogen receptor siRNA library screening using RNAiMAX transfection reagent in MCF7 cells

### *6.2.1 Procedure of day 1*

Autoclave 500 mL glass bottles and magnetic stir bars.

### *6.2.2 Procedure of day 2*

#### 6.2.2.1 Introduction to siRNA library screening procedure

1. Set up siRNA Library plates (stock concentration: 0.24 μM, in V-bottom 96-well plates). Layouts are shown in the following 96-well plates (Plates #1–10, Fig. 8; Plates #11, Fig. 9). The total

| | 1 | 2 | 3 | 4 | 5 | 6 | 7 | 8 | 9 | 10 | 11 | 12 |
|---|---|---|---|---|---|---|---|---|---|---|---|---|
| A | NEG | AP2A1 | NEG | GRB14 | MOCK | DEATH | NEG | AP2A1 | NEG | GRB14 | MOCK | NEG |
| B | DEATH | | | | | | | | | | | DEATH |
| C | MOCK | | | | | | | | | | | GRB14 |
| D | AP2A1 | | | | | | | | NEG | | | NEG |
| E | NEG | | | | | | | | NEG | | | DEATH |
| F | GRB14 | | | | | | | | | | | MOCK |
| G | DEATH | | | | | | | | | | | AP2A1 |
| H | NEG | MOCK | GRB14 | NEG | AP2A1 | NEG | DEATH | MOCK | AP2A1 | NEG | GRB14 | NEG |

**Fig. 8** Layout of ER network siRNA library plate #1–10.

| | 1 | 2 | 3 | 4 | 5 | 6 | 7 | 8 | 9 | 10 | 11 | 12 |
|---|---|---|---|---|---|---|---|---|---|---|---|---|
| A | NEG | AP2A1 | NEG | GRB14 | MOCK | DEATH | NEG | AP2A1 | NEG | GRB14 | MOCK | NEG |
| B | DEATH | | | | | | | | | | MOCK | DEATH |
| C | MOCK | | | | | | | | | | MOCK | GRB14 |
| D | AP2A1 | | | | | | | | NEG | | MOCK | NEG |
| E | NEG | | | | | | | | NEG | | MOCK | DEATH |
| F | GRB14 | | | | | | | | | | MOCK | MOCK |
| G | DEATH | | | | | | | | | MOCK | MOCK | AP2A1 |
| H | NEG | MOCK | GRB14 | NEG | AP2A1 | NEG | DEATH | MOCK | AP2A1 | NEG | GRB14 | NEG |

**Fig. 9** Layout of ER network siRNA library plate #11

number of siRNA library plates is 11. Unlabeled wells are for siRNAs against ER network target genes. Labeled wells are for various controls (NEG, DEATH, AP2A1 and GRB14 and MOCK siRNA suspension buffer). 100 μL diluted siRNA was dispensed into each well, stock in −20 °C.

2. Next, prepare 11 flat-bottom 96-well plates as experimental plates. Barcodes for these 11 plates need to be printed and pasted on plates (front and middle). After labeling, these plates are stored in a cell culture hood overnight. Meanwhile, confirm that all materials, including screening media, transfection reagent, plates, transfer pipette tips, and cell cultures are ready.
3. The Cybio program (Fig. 10) will pipette 10 μL from 0.24 μM siRNA ER library plate in V-bottom plate, then mixed into Costar 96-well plate that already has 10.5 μL of diluted RNAimax transfection reagent. The final concentration of siRNA in cells was 20 nM.
4. Each experimental plate (Costar 96-well plate) must have 10.5 μL of diluted lipid transfection reagent (0.5 μL RNAimax + 10 μL Opti-MEM medium) added to each

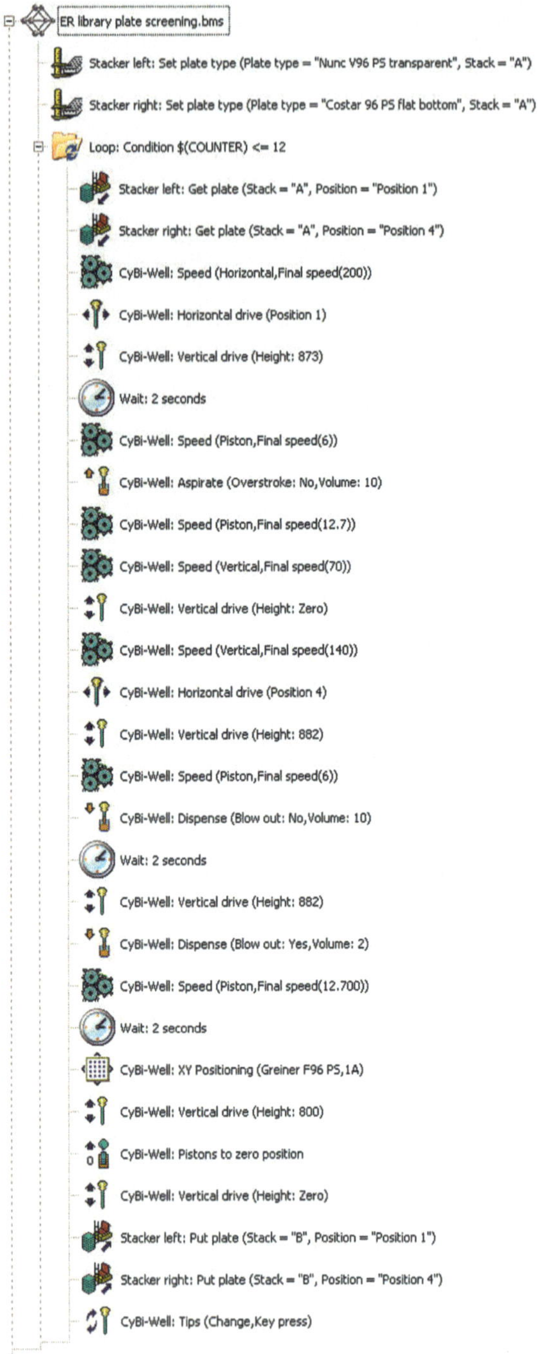


**Fig. 10** Cybio program to dispense siRNA into RNAi screening experimental plates.

5. Well with Combi-nl dispensing machine. 96 wells × 10.5 μL = 1008 μL, 11 plates 11,088 μL. Accounting for loss from machine priming, make 15 mL for 11 plates. Then load these siRNA and experimental plates on Cybio machine as following (indicated as in Fig. 5):
   i. siRNA ER library plates (0.24 μM, 11 plates)—loaded on Left Arm, Stack A, plates ascending from the bottom with lids removed before loading;
   ii. Experimental plates (11 plates)—loaded on Right Arm, Stack A; plates with ascending from the bottom with lids removed before loading.

#### 6.2.2.2 siRNA library screening procedure and timeline

a. Thaw siRNA library plates (stock concentration: 0.24 μM) at room temperature.
b. Clean WellMate microplate dispenser machine with 15 mL of filtered 70% ethanol, then 15 mL of filtered $ddH_2O$, and 15 mL IMEM medium (without serum).
c. Split MCF7 cells and count. Dilute 8000 cells in 100 μL (80,000 cells/mL). For one plate: 16 mL of diluted cells are needed. Therefore, for 11 plates: 250 mL diluted cells in 500 mL sterile bottle, with magnet stir bar on the magnetic stirrers, low speed. Plate extra cells in new flasks if necessary.
d. Clean Combi-nl dispensing machine with 7 mL of filtered 70% ethanol, 7 mL filtered $ddH_2O$, and 7 mL Opti-MEM medium.
e. When the siRNA library plates have thawed, centrifuge them at 3500 rpm (1935 × *g*) for 5 min at room temperature. Take plate out immediately to keep condensation from forming.
f. Dispense 15 mL Opti-MEM in a 50 mL conical tube.
g. Dilute transfection reagent: 720 μL RNAimax in 14.3 mL Opti-MEM in a 50 mL conical tube.
h. Dispense 10.5 μL/well of diluted transfection reagent to 11 experimental plates (Costar 96-well plates).
i. Load experimental plates and siRNA plates on designated stacks of Cybio-machine. siRNA plates: Left Arm, Stack A; experimental plates: Right Arm, Stack A. Lids were removed before loading, and plates ascending from the bottom.
j. Start Cybio machine, run program (Fig. 10) to distribute 11 plates of library siRNAs as shown in attached plate layout (Figs. 8 and 9).

Set up a timer, record the time of each plate being processed. Proceed to the next step while completing this step.

*Note: In this example, the time for running one plate and changing tips was 1 min and 15 s. The total machine running process will take about 15 min.*

**k.** After the first four experimental plates containing diluted siRNA-lipid mixture are completed, transfer these plates to the cell culture hood. A second person should continue running the remainder of plates on Cybio machine.

**l.** After waiting 10–15 min from the time of siRNA-lipid mixture, use the WellMate microplate dispenser machine to dispense 100 μL of cells/well to the first four experimental plates, and then transfer to 37 °C $CO_2$ incubator. Wait for another 5 min, then prime the WellMate machine with 5–7 mL cell suspension, dispense cells for the next four plates and transfer them to the 37 °C $CO_2$ incubator; then wait for another 5 min, prime machine with 5–7 mL cell suspension, dispense cells for the last three plates and transfer them to the 37 °C incubator. Document the time each plate received cells.

**Notes**

**a.** *Four plates as a group to be added with cells.*

**b.** *Put WellMate dispenser probe into diluted cells only before adding cells into experimental plates, not earlier, and then prime machine with 5–7 mL cell suspension for each group.*

**m.** Clean WellMate dispenser with 15 mL filtered dd$H_2O$, then 15 mL filtered 70% ethanol, and switch off the machine.

**n.** Clean the Combi-nL dispenser with 7 mL distilled water, and 7 mL of filtered (0.45 μm) 70% ethanol, switch off machine.

### 6.2.3 Procedure of day 7

Five days after incubation at 37 °C, 5% $CO_2$, add 20 μL of 1:1 mixture of Cell Titer Blue:HBSS to each well and read out every hour up to 4 h (as described above).

## 6.3 Data analysis

1. Calculate median viability values (Cell Titer blue readout, arbitrary unit) of each control from the data of corresponding wells, mock (Plate 1–10: A5, A11, C1, F12, H2 and H8; Plate 11, A5, A-G11, C1, F12, H2, H8 and G10); siNEG (all plates: A1, A3, A7, A9, A12, E1, E9, D9, D12, H1, H4, H6, H10 and H12); siDEATH (all plates: B1, B12, A6, E12, G1 and H7); siAP2A1 (all plates: A2, A8, D1, G12, H5 and H9); siGRB14 (all

plates: A4, A10, C12, F1, H3 and H11). Determine if AP2A1 or GRB14 siRNA yields median killing of cells, and if death control siRNA kills >90% of cells. If so, continue with following data processing and analysis. If not, go back to trouble shoot with technique.

2. Calculate the viability index (VI) of each siRNA-transfected cell normalized to the

$$VI = \frac{\text{viability of siRNA-transfected cells}}{\text{median viability of negative control-transfected cell}}$$

3. Three independent experiments will be needed with average VI of each gene-knocked down-cells is obtained. An arbitrary threshold of VI <0.5 can be utilized. These groups of genes whose knockdown induced a loss of 50% viability or more were identified as genes of interest, which were considered as reflecting a robust biological effect and then continued with additional validation.
4. siRNA Validation

   Hits identified by a loss of 50% viability of more following siRNA knockdown (VI $\leq$ 0.5) underwent validation studies. For each hit identified, four different siRNAs (QIAGEN, MD) targeting the same gene were tested in individual wells. Two out of the four siRNAs were the same target sequences as the siRNAs in the screen, when available. The other two siRNAs were new sequences to test. Cells were screened as described above. If at least two out of four of the siRNAs tested reduced viability by at least 50%, the candidate passed validation as a putative hit.

## 6.4 Notes

1. Each siRNA library screening experiment is repeated three times. In order to reproduce the data efficiently, cell viability of various control siRNAs (NEG, DEATH, AP2A1 and GRB14 and MOCK siRNA suspension buffer) is used to monitor if there is any experimental artifacts or batch-effects. If AP2A1 or GRB14 siRNA yields median killing of cells, and if death control siRNA kills >90% of cells, continue with data processing and analysis. If not, the data from a certain batch would not be included for statistical analysis. This will ensure the reproducibility and robustness of this method and analysis.
2. When dealing with multiple plates in one experiment, keep in mind the lipid-siRNA complex formation time should be limited to 10–15 min,

record the lipid adding time (CyBio), and calculate the cells dispensing time 10–15 min after (WellMate dispenser). Shorter or longer waiting time can cause variability with transfection efficiency and baseline cell viability, which can be detected using the relevant controls included on each plate.

3. As far as setting up the threshold of VI for identifying hits and validation, we chose 0.5 (VI) in our studies. Those genes resulted in a loss of 50% viability or more following gene knockdown were identified as hits. The criteria of choosing an appropriate threshold is that the threshold must reflect a robust biological effect. The value can be adjusted according to projects.
4. Time recording

| Plate number | Cybio time | WellMate time |
|---|---|---|
| 1 | | |
| 2 | | |
| 3 | | |
| 4 | | |
| 5 | | |
| 6 | | |
| 7 | | |
| 8 | | |
| 9 | | |
| 10 | | |
| 11 | | |

5. Screening check list

| | Complete |
|---|---|
| Costar plates | |
| V-bottom plates | |
| Media (IMEM) | |
| CCS | |
| Opti-MEM | |

| | Complete |
|---|---|
| Trypsin | |
| Cells | |
| T75 flasks | |
| Filtered Mono-Q water | |
| Filtered 70% Ethanol | |
| Cell Titer Blue | |
| Labels w/Barcode | |
| Control siRNAs | |
| Autoclaved 500 mL bottle (1) with small stir bar inside | |
| Calculations | |

## Acknowledgments

This protocol was adapted from previous work Zhang et al. (2016). We thank Wei Xu, Alan Zwart, David Goldstein and Annie Zuo for their technical assistance. The authors were supported by R01CA050633, CA51880, U54 CA149147 (to L.M.W.), R01CA63366 and R21CA181287 (to E.A.G.).

## References

Astsaturov, I., Ratushny, V., Sukhanova, A., Einarson, M. B., Bagnyukova, T., Zhou, Y., et al. (2010). Synthetic lethal screen of an EGFR-centered network to improve targeted therapies. *Science Signaling*, *3*(140), ra67. https://doi.org/10.1126/scisignal.2001083.

Birmingham, A., Selfors, L. M., Forster, T., Wrobel, D., Kennedy, C. J., Shanks, E., et al. (2009). Statistical methods for analysis of high-throughput RNA interference screens. *Nature Methods*, *6*(8), 569–575. https://doi.org/10.1038/nmeth.1351.

Boutros, M., & Ahringer, J. (2008). The art and design of genetic screens: RNA interference. *Nature Reviews. Genetics*, *9*(7), 554–566. https://doi.org/10.1038/nrg2364.

Josse, R., Zhang, Y. W., Giroux, V., Ghosh, A. K., Luo, J., & Pommier, Y. (2015). Activation of RAF1 (c-RAF) by the marine alkaloid lasonolide a induces rapid premature chromosome condensation. *Marine Drugs*, *13*(6), 3625–3639. https://doi.org/10.3390/md13063625.

Murray, J. C., Aldeghaither, D., Wang, S., Nasto, R. E., Jablonski, S. A., Tang, Y., et al. (2014). c-Abl modulates tumor cell sensitivity to antibody-dependent cellular cytotoxicity. *Cancer Immunology Research*, *2*(12), 1186–1198. https://doi.org/10.1158/2326-6066.CIR-14-0083.

Shuptrine, C. W., Ajina, R., Fertig, E. J., Jablonski, S. A., Kim Lyerly, H., Hartman, Z. C., et al. (2017). An unbiased in vivo functional genomics screening approach in mice identifies novel tumor cell-based regulators of immune rejection. *Cancer Immunology, Immunotherapy*, *66*(12), 1529–1544. https://doi.org/10.1007/s00262-017-2047-2.

Zhang, J. H., Chung, T. D., & Oldenburg, K. R. (1999). A simple statistical parameter for use in evaluation and validation of high throughput screening assays. *Journal of Biomolecular Screening*, *4*(2), 67–73. https://doi.org/10.1177/108705719900400206.

Zhang, Y. W., Jones, T. L., Martin, S. E., Caplen, N. J., & Pommier, Y. (2009). Implication of checkpoint kinase-dependent up-regulation of ribonucleotide reductase R2 in DNA damage response. *The Journal of Biological Chemistry*, *284*(27), 18085–18095. https://doi.org/10.1074/jbc.M109.003020.

Zhang, Y. W., Nasto, R. E., Varghese, R., Jablonski, S. A., Serebriiskii, I. G., Surana, R., et al. (2016). Acquisition of estrogen independence induces TOB1-related mechanisms supporting breast cancer cell proliferation. *Oncogene*, *35*(13), 1643–1656. https://doi.org/10.1038/onc.2015.226.

CHAPTER THIRTEEN

# An altered miTRAP method for miRNA affinity purification with its pros and cons

**Simon Jasinski-Bergner*, Christoforos Vaxevanis, Nadine Heimer, Maria-Filothei Lazaridou, Michael Friedrich, Barbara Seliger**

Institute for Medical Immunology, Faculty of Medicine, Martin Luther University Halle-Wittenberg, Halle, Germany

*Corresponding author: e-mail address: simon.jasinski@uk-halle.de

## Contents



## Abstract

The major mechanisms of posttranscriptional gene regulation involve microRNAs (miRs) and RNA-binding proteins (RBPs). Recent studies not only identified functionally and characterized such factors, but rather investigated their use as biomarkers and suitability as biopharmaceuticals. Indeed, some miR-based drugs are currently tested in clinical studies as potential anti-viral and as anti-cancer agents. For the chemical application, a profound knowledge of the binding affinities of miRs and RBPs to their target RNA is essential.

The authors recently identified several miRs regulating the non-classical human leukocyte antigen (HLA)-G, and characterized their binding affinity to the 3′ untranslated

*Methods in Enzymology*, Volume 636
ISSN 0076-6879
https://doi.org/10.1016/bs.mie.2019.05.016

region (UTR) of HLA-G. These miRs identified by miTRAP were classified into high affinity and low affinity miRs, which were either key regulators or fine tuners of HLA-G.

While the miTRAP method has been described in detail, a novel modified miTRAP technique has been established, which completely consists of commercially available components and uses a simplified cloning strategy. This technique allows the identification and characterization of miRs and RBPs for any RNA sequence of interest.

## Abbreviations

**AGO** argonaute
**APM** antigen processing and presentation machinery
**e.g.** exempli gratia
**HLA** human leukocyte antigen
**mAb** monoclonal antibody
**miR** microRNA
**miTRAP** miRNA trapping by *in vitro* RNA affinity purification
**n.d.** not detectable
**nt** nucleotide
**oligo** oligonucleotide
**PBS** phosphate buffered saline
**PCR** polymerase chain reaction
**RBP** RNA-binding protein
**RT** room temperature
**UTR** untranslated region

## 1. Introduction

The microRNA (miR) mediated post-transcriptional gene regulation is for various reasons of growing interest. These small ~ 21 nt long non-coding RNAs, bind sequence specific to their target mRNAs preferentially, but not exclusively to the 3′ unstranslated region (UTR) leading to translational inhibition and mRNA decay (Jasinski-Bergner et al., 2015; Nielsen, Gloggnitzer, & Martinez, 2009). Recently, several miR based biopharmaceuticals reached preclinical and even clinical phase III trials, like the miR-122 miravirsen (Santaris Pharma, Denmark). In addition, drugs composed of miRs and small interfering RNAs (siRNAs) are currently patented by leading RNA-therapeutic companies (Chakraborty, Sharma, Sharma, Doss, & Lee, 2017) for the further establishment of the first clinical trials (van Zandwijk et al., 2017).

It is generally accepted that one miR regulates more than one single mRNA, and *vice versa* one single mRNA can be regulated by many different miRs depending on the presence of binding motifs. Using the immunologic and tumor biologic relevant non-classical human leukocyte antigen (HLA) class Ib molecule HLA-G, as a model the authors identified many different miRs binding and negatively regulating the target mRNA, but with distinct affinities leading to the classification of these miRs as key regulators and fine tuners of the HLA-G expression. Furthermore, miRs of different miR families show diverse binding affinities resulting as well in the distribution of key regulators and fine tuners (Jasinski-Bergner et al., 2016). Even within a single miR family, whose members share the same target mRNA relevant seed region, ranging from the second to the seventh nucleotide (nt) from the 5′ end of the miR, different binding affinities for the same target binding site were observed.

In the recent years the function of human miRs was characterized and tumor suppressive or oncogenic miRs were described depending on their target mRNAs. This is an important issue for tumor immunology, since miRs regulate many immunologic relevant molecules including, e.g., NKG2D ligands, antigen processing and presentation machinery (APM) components, HLA molecules and programmed death ligand 1 (PD-L1) (Eichmüller, Osen, Mandelboim, & Seliger, 2017; Jasinski-Bergner, Mandelboim, & Seliger, 2014).

A broad array of publications describes novel miR-mRNA-interactions in different cell systems, but without any comparability regarding the different miR affinities for the same target mRNA. The different methods applied for miR overexpression, such as transient mimic transfection or stable miR plasmid transfection, result in a weak comparability. In addition, miRs can exert a tissue specific expression profile suggesting that miRs could have an altered affinity in a different cellular context. However, for miR based drug design it is of immense importance to distinguish between miRs acting as key regulators with a high affinity for the target mRNA and these with a low affinity for the target mRNA in the cell system of interest.

The authors here present a modified miTRAP method consisting of fully commercial available components allowing the affinity based purification of miRs from a cell lysate with a RNA sequence of choice as bait. This approach also allows competitive experiments in analogy to the principles of a competitive ELISA whether miRs exert "cross reactivity" for other mRNA sequences labeled to beads and used as competitive baits next to the actual bait for miR purification.

A fully detailed instruction of the whole procedure is provided in this manuscript from cloning of the RNA sequence of choice to the analysis strategies of eluates. Furthermore, the advantages and disadvantages of the modified miTRAP method will be discussed in comparison to the conventional miTRAP technique (Braun, Misiak, Busch, Krohn, & Huttelmaier, 2014).

## 2. Cloning the constructs

Any desired sequence derived from every organism and even artificial synthesized sequences can be applied as potential bait for miR affinity purification. For practical reasons longer sequences should be divided into overlapping fragments. The authors suggest sequences with a final length ranging from 1000 to 1500 nt.

**2.1.** The Taq-DNA polymerase (Thermo-Fisher Scientific, Waltham, USA) was applied to amplify the sequences of interest. In Table 1 the respective primers for the amplification of the HLA-G 3′-UTR are listed. This Taq-DNA polymerase has a non-template-dependent terminal transferase activity adding a single deoxyadenosine (A) to the 3′ ends of PCR products. The resulting PCR products were gel purified and cloned into a linearized pCR™2.1-TOPO® vector (Thermo-Fisher Scientific) by TA cloning (Fig. 1). The ligation was performed over night at 4 °C.

**2.2.** The ligation products were transformed into competent DH5α bacterial cells (NEB, Ipswich, USA) by heat shock at 42 °C for 30 s in a water bath. The transformed cells were incubated on ampicillin plates over night at 37 °C. Positive clones were identified by colony PCR. From these positive clones overnight cultures were inoculated (4 mL LB-medium (Carl Roth GmbH + Co. KG, Karlsruhe, Germany) containing ampicillin) for plasmid amplification and further plasmid extraction by usage of the NucleoSpin® Plasmid kit (MACHEREY-NAGEL GmbH & Co. KG, Düren, Germany). The plasmids were sequenced by Eurofins Genomics Germany GmbH (Ebersberg, Germany).

**2.3.** Plasmids, which contain the correct sequence, were used as template for an additional PCR reaction (Fig. 2) in combination with special and long PCR oligonucleotides (Table 1). For that 2nd PCR reaction a proof-reading DNA polymerase should be employed, which generates blunt ends (important for later in vitro transcription), like the Pfu DNA polymerase (Promega, Madison, WI, USA).

**Table 1** List of the applied oligonucleotides.

| Primer | Application | Sequence (5′ → 3′) | Temp. |
|---|---|---|---|
| 152RT-Rkt | Stem-loop primer | GTCGTATCCAGTGCAGGGTCCGAGGTATTC GCACTGGATACGACCCAAGT | 42 °C |
| 152 PCR fw | qPCR | GCCCTCAGTGCATGACAGA | 60 °C |
| 141RT-Rkt | Stem-loop primer | GTCGTATCCAGTGCAGGGTCCGAGGTATTC GCACTGGATACGACCCATCT | 42 °C |
| 141 PCR fw | qPCR | GCCCTAACACTGTCTGGTAA | 60 °C |
| Stem loop reverse primer | qPCR | GTGCAGGGTCCGAGGT | 60 °C |
| invitroHLA-Gfw | PCR | CCCCCCTAATACGACTCACTATAGGGAAAA AACAGCTGCCCTGTGT | 60 °C |
| invitroHLA-Grev | PCR | TTTTTTTTTTTTTTTTTTTTTTTTTTTTTT TTTTTAAAGTTCTCATGTCTTCCATTT | 60 °C |
| clHLAG3UTRfw | PCR | AAACAGCTGCCCTGTGT | 60 °C |
| clHLAG3UTRrev | PCR | CTCTCAAATTTCAGGAATC | 60 °C |

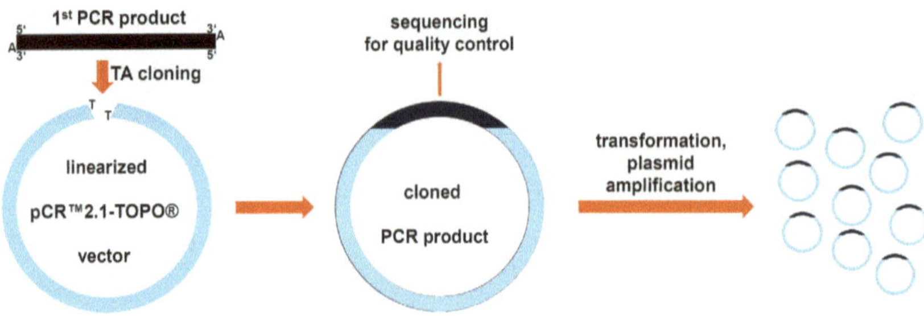


**Fig. 1** Schematic overview about the cloning strategy.

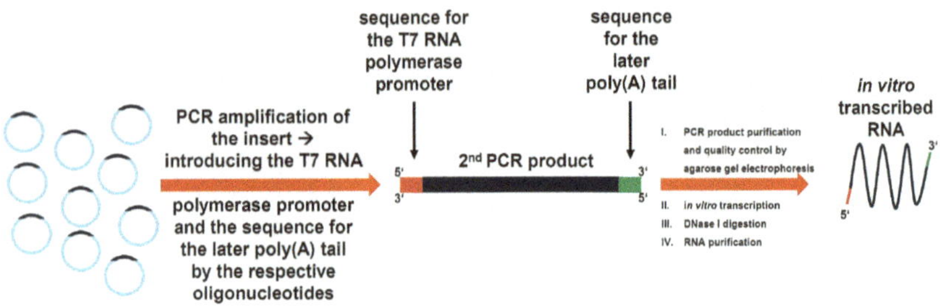


**Fig. 2** Generation of a linear DNA template for further *in vitro* transcription.

The forward oligo contains the T7 promoter sequence for further *in vitro* transcription, while the reverse oligo contains additional 35 thymine residues, which will form the later poly(A) tail after *in vitro* transcription. The design of the oligos is based upon instructions provided by Promega and is explained in more detail in Fig. 3.

For technical reasons the authors performed for every construct 10 identical PCR reactions. The accurate PCR product size was checked by agarose gel electrophoresis. Afterward, the remaining PCR products were pooled directly prior to purification with the NucleoSpin® Gel and PCR Clean-up kit (MACHEREY-NAGEL).

## 3. *In vitro* transcription

The purified PCR products were applied as templates for *in vitro* transcription over night by usage of the T7 RiboMAX™ kit (Promega, Madison, WI, USA). For that purpose, 1 μg of purified linear DNA was employed. After *in vitro* transcription a DNase I (Promega) digestion

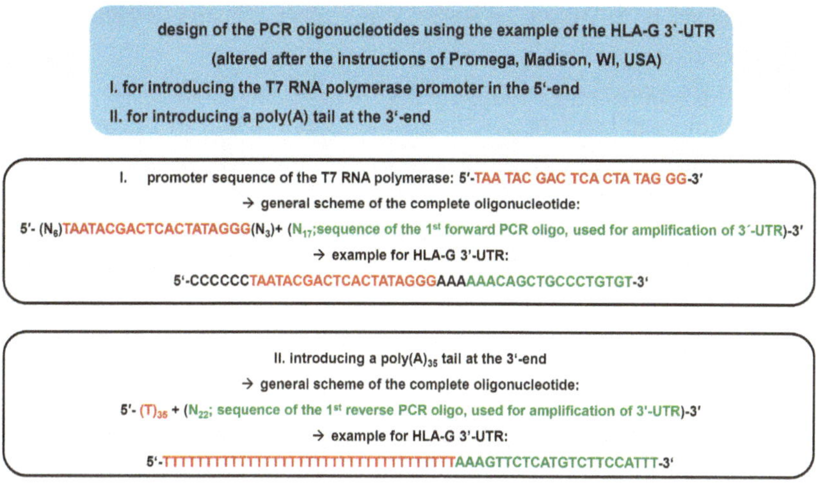


**Fig. 3** Detailed instruction for introducing the T7-RNA polymerase promoter and the sequence for the later poly(A) tail.

followed to destroy contaminating template DNA. To check the accurate size and the quality of the *in vitro* transcribed RNA an agarose gel electrophoresis was performed. The authors suggest using a respective RNA ladder as size marker. The *in vitro* transcribed RNA was then purified with the MEGAclear™ Transcription Clean-Up Kit (Thermo-Fisher Scientific). The RNA concentration was determined and the RNA was stored at −80 °C until further usage.

## 4. Cell culture, cell lysis, treatments, and buffer preparations

### 4.1 Cell culture

The cell lysate containing the factors to be enriched with the *in vitro* transcribed RNA construct as a bait can by derived from any organism. However, the authors only have experience with human cell lines. Therefore, the applied buffers were only tested for human tumor cell lines of distinct origin.

One major prerequisite is that the factors of interest, e.g., miRs or RBPs, are expressed in the respective cell lines. If novel interaction partners will be identified without any knowledge about their expression levels in the analyzed cell lines or tissues, it is possible to combine the cell lysates of different cell lines or different tissues as a mixture. After identification of miRs or

RBPs it could be analyzed, in which of the applied tissues or cell lines the novel interaction partner is expressed and at which levels.

**4.1.1.** In the presented example HLA-G-regulatory miRs were enriched from a cell lysate of $5 \times 10^7$ Kasumi-3 cells (DMSZ, Braunschweig, Germany). The Kasumi-3 cells were grown in Gibco RPMI 1640 Medium (Thermo-Fisher Scientific) supplemented with 10% fetal calf serum (FCS; Merck Millipore, Heidelberg, Germany), 1% penicillin/streptomycin (Sigma Aldrich, St. Louis, USA), 1% sodium pyruvate (Merck Millipore, Heidelberg, Germany), 2 mM L-glutamine (Merck Millipore, Heidelberg, Germany) and detached with trypsin (Sigma Aldrich, St. Louis, USA). The cells were counted with a Neubauer chamber and $5 \times 10^7$ cells were washed with $1 \times$ PBS, before centrifugation at 1500 rpm for 5 min was performed. While the supernatant was discarded, the cell pellet was frozen at −80 °C until use. That freezing step did not yet harm the miR affinity purification assays, but might negatively influence the affinity purification of RBPs due to possible denaturation. This fact should be considered in the experiments.

## 4.2 Cell lysis

For generation of the cell lysates, $5 \times 10^7$ cells were lysed in 5.0 mL lysis/binding buffer (see Section 4.4.). Due to the fact, that the cell lysates also include natural polyadenylated RNA species, these cell lysates were processed prior usage. Hence, all RNA species bigger than 200 nt were excluded by size exclusion columns from the NucleoSpin® miRNA kit (MACHEREY-NAGEL).

**4.2.1.** $5 \times 10^7$ cells were lysed in 1500 μL buffer, vortexed, and incubated for 5 min at RT. Cell lysates were then dispensed on five columns (violet ring), followed by centrifugation for 1 min at 11.000 x *g*. Subsequently, the flow through was collected and pooled, then 250 μL of ethanol 96% (v/v) was added, vortexed, and incubated for 5 min at RT. This lysate was then dispensed on five columns (blue ring), centrifuged for 1 min at 11.000 x *g*. This time the flow through (containing the RNA molecules <200 nt) was collected, pooled and used as cell lysate/input for miTRAP experiment, and was stored on ice until usage.

Important: In the same protocol additional steps are described, including protein precipitation steps. These steps should not be performed regarding the involvement of the Argonaute (Ago) proteins in the process of miR-mRNA binding, which is necessary for the further miTRAP experiment.

## 4.3 Treatments

The miTRAP techniques could be performed with (i) cell lysates of different cell lines or tissues or (ii) lysates of the same cell lines +/− treatment or +/− overexpression/downregulation.

Next to changes of the cell lysates alterations in the applied RNA bait are possible, e.g., mutations/deletions of the binding motifs.

## 4.4 Buffer preparations

All necessary buffers with their recipes were adopted from the Dynabeads® Oligo $(dT)_{25}$ manual (Thermo-Fisher Scientific).

### *4.4.1 Binding buffer*

20 mM Tris-HCl, pH 7.5, 1.0 M LiCl, 2 mM EDTA.

### *4.4.2 Lysis/binding buffer*

100 mM Tris-HCl, pH 7.5, 500 mM LiCl, 10 mM EDTA, 1% LiDS, 5 mM dithiothreitol (DTT).

### *4.4.3 Washing buffer A*

10 mM Tris-HCl, pH 7.5, 0.15 M LiCl, 1 mM EDTA, 0.1% LiDS.

### *4.4.4 Washing buffer B*

10 mM Tris-HCl, pH 7.5, 0.15 M LiCl, 1 mM EDTA 10 mM Tris-HCl, pH 7.5.

# 5. Affinity purification (modified miTRAP)

The complete affinity purification was performed with an altered protocol of the Dynabeads® Oligo (dT)25 (Thermo-Fisher Scientific).

**5.1.** The beads were homogenized by vortexing and then 200 μL of beads were transferred into each tube.

**5.2.** After a brief centrifugation step in a mini centrifuge, the tubes were placed into the magnet (Dynal MPC™; Thermo-Fisher Scientific) for 1 min and the supernatant was removed with a pipette.

**5.3.** In analogy to the previous step: two washing steps with 100 μL binding buffer/tube followed.

**5.4.** The last supernatant was removed. Then 100 μL of the bait RNA with a concentration of $3 \times 10^{13}$ copies/μL diluted in binding buffer and 100 μL of the size purified cell lysate were added to the beads in each tube. Furthermore, the reaction volume was supplemented with 11 μg/mL heparin as RNase inhibitor.

**5.5.** After an incubation step of 120 min at RT on a mixer, the beads were pelleted by a brief centrifugation step and then the tubes were placed in the magnet. Afterward, the supernatant was removed and two washing steps followed.

**5.6.** For the first washing steps 1 mL of washing buffer A was used and for the second washing step 1 mL of washing buffer B was applied.

**5.7.** For RNA extraction a volume of 600 μL Trizol (Thermo-Fisher Scientific) was added directly to the beads. At this stage the samples can be stored at −80 °C or directly used for RNA extraction.

To analyze RBPs other elution techniques should be employed, like RNase digestion in an appropriate buffer system for further analyses of enriched proteins.

**5.8.** Do not forget to use also 100 μL of the size purified cell lysate +600 μL of Trizol for RNA extraction as further input control!

## 6. Sample analyses

Once affinity-purified eluates have been generated, it has to be described how the samples should be analyzed. For investigating enriched proteins a polyacrylamide gel electrophoresis in combination with Coomassie staining and mass spectrometry could lead to the identification of enriched proteins.

For the analysis of enriched miRs by qPCR, a detailed protocol will be provided here. In analogy to that, also unknown miRs could be investigated in combination with qPCR by prior usage of *in silico* predictions. However, more precisely the eluates will be analyzed by Next Generation Sequencing, which will lead to the identification of novel miRs.

### 6.1 RNA extration

**6.1.1.** Addition of 600 μL Trizol (Thermo-Fisher Scientific) to the beads with the coupled baits and bound miRs will lead to denaturation of the beads themselves. Then 200 μL of chloroform (Carl Roth GmbH + Co. KG) are added, followed by a vortex step of 15 s.

**6.1.2.** After a centrifugation at 13.000 rpm for 15 min at 4 °C, the supernatant is carefully removed and transferred into a new RNase-free 1.5 mL reaction tube. The RNA is then precipitated by addition of 1 mL isopropanol, and an incubation step of 10 min at 4 °C. To increase the visualization of even very small RNA pellets 2 μL of RNase-free glycogen (Thermo-Fisher Scientific) may be added.

**6.1.3.** For completion of the RNA precipitation the sample is centrifuged at 13,000 rpm for 10 min at 4 °C. Afterward the supernatant is discarded and the pellet is washed with 1.0 mL of 70% (v/v) ethanol. After an additional centrifugation step the supernatant is again completely removed and the pellet is dried.

**6.1.4.** The pellet is resolved in 15 μL RNase-free water, by various cycles of vortexing and spining down. Usually, the concentration of the enriched miRs (also still including the RNA baits) is very low. Therefore, the authors suggest working comparative as much as possible and treating every sample equally.

At this stage the samples might be stored at −80 °C.

## 6.2 cDNA synthesis

For miR-specific cDNA synthesis, 2 μL of the RNA was reverse transcribed with miR-specific stem loop primers (Chen et al., 2005; Varkonyi-Gasic, Wu, Wood, Walton, & Hellens, 2007) and the RevertAidTM H Minus reverse transcriptase (Thermo-Fisher Scientific, Waltham, USA) in a final reaction volume of 10 μL.

Following program was used the reactions were first incubated for 30 min at 16 °C, 30 min at 42 °C, and 5 min at 85 °C and then held at 4 °C.

The sequences of the stem loop primers for miR-152 and miR-141 are listed in Table 1.

## 6.3 qPCR

The qPCR reactions were prepared according to manufacturers' instructions. For this, 1/10th of the first-strand cDNA reaction was used as template DNA. Following program was applied for the PCR cycler: 1 cycle of 5 min at 95 °C; and 40 cycles of 30 s at 95 °C and 30 s at 60 °C. The required oligonucleotides are listed in the Table 1.

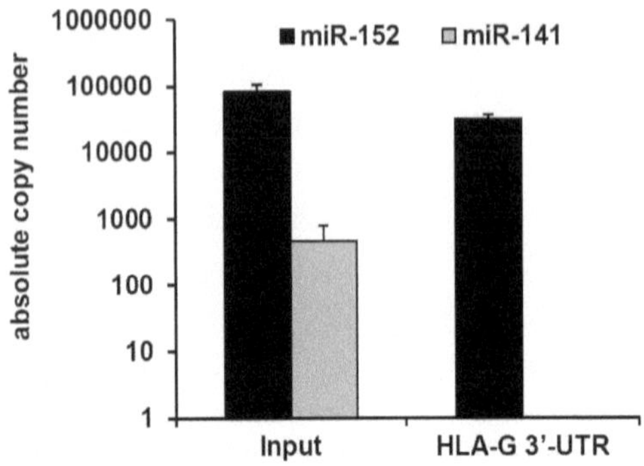


**Fig. 4** qPCR analysis of the internal positive control (miR-152) and internal negative control (miR-141).

The quantification was performed by determination of the absolute copy number by usage of an external plasmid standard, containing the respective PCR product.

In Fig. 4 the necessary internal positive and negative controls are shown using the example of the modified miTRAP with the HLA-G 3′-UTR as bait. As positive control the enrichment of miR-152 is shown. MiR-152 is a reported key regulator of the HLA-G gene expression (Jasinski-Bergner et al., 2016). In contrast, miR-141 is expressed in the respective cell line used for generating the cell lysate (Input), but miR-141 could not be enriched with the HLA-G 3′-UTR as bait. The authors suggest employing exactly such experiment as internal positive and negative control and therefore essential oligonucleotides are listed in Table 1 and Fig. 3.

## 7. Comparison of the modified miTRAP *versus* basic miTRAP

The basic idea of both methods is equal. Both use *in vitro* transcribed RNA constructs as baits to enrich RNA binding molecules out of a cell lysate. However, a modified miTRAP was developed, which uses commercially available components, while for the conventional miTRAP variant the fusion protein of maltose binding protein and MS2 loop binding protein is not commercially available. Recently, the authors also published a detailed protocol for overexpression and purification of that fusion protein for identification of immunomodulatory miRs (Tretbar, Friedrich, Lazaridou, & Seliger, 2019).

Another advantage of the modified miTRAP is the usage of a shorter sequence for binding the bait RNA to the beads. The 35 T-residues are much shorter than the sequence encoding for the MS2 loops reducing the possibility for the enrichment of target-unspecific factors.

Furthermore, the longer sequence encoding for the MS2 loops cannot be introduced by a simple PCR reaction like the 35 T-residues. Therefore, the sequence encoding for the bait RNA has to be cloned into a plasmid, where such MS2 loops are already included at the 3′-end. Such cloning strategy is of higher effort due to the fact that simple TA-cloning is not possible.

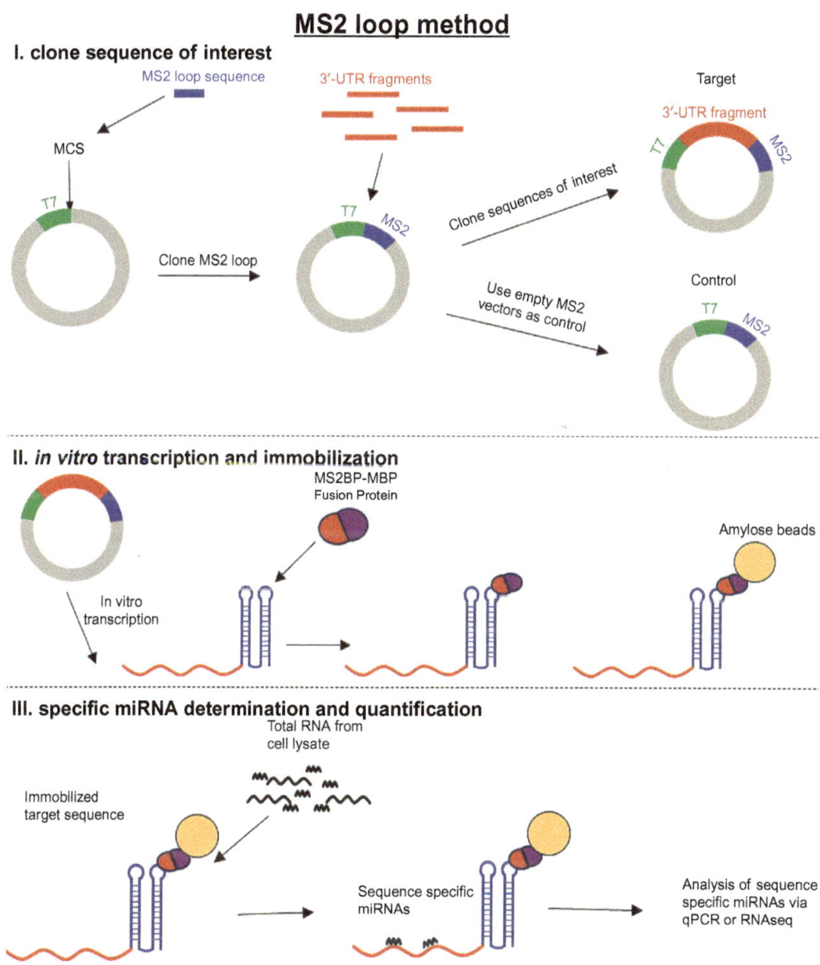


**Fig. 5** Schematic overview of the conventional miTRAP.

A disadvantage of the modified miTRAP is the possible incomplete depletion of polyadenylated RNA species from the applied cell lysate by employing the size exclusion columns. This could cause false enrichments. However, such potential unspecific bindings of contaminating polyadenylated RNA species would occur equally to all oligo-dT beads, also to the internal control reactions using a control RNA as bait, like the suggested HLA-G 3′-UTR next to the actual bait sequence from the gene of interest, e.g., the 3′-UTR from the gene of interest. To minimize such unspecific binding of contaminating polyadenylated RNA species the authors suggest using the RNA baits in saturating excess.

Another disadvantage is the previously mentioned usage of the size exclusion purification of the applied cell lysate *per se*. At this purification step a lot of protein is also lost, which might negatively affect the investigation of RBPs interacting with the bait RNAs.

The complete schematic overview of the basic miTRAP variant is summarized in Fig. 5.

Nevertheless, both miTRAP techniques are easy to perform and enable a lot of possibilities to identify novel binding partners to a RNA sequence of choice.

## Acknowledgments

We want to thank Maria Heise for excellent secretarial help. This study was funded by a grant from the German-Israeli Foundation (GIF; I-37-414.11-2016).

## References

Braun, J., Misiak, D., Busch, B., Krohn, K., & Huttelmaier, S. (2014). Rapid identification of regulatory microRNAs by miTRAP (miRNA trapping by RNA in vitro affinity purification). *Nucleic Acids Research*, *42*(8), e66.

Chakraborty, C., Sharma, A. R., Sharma, G., Doss, C. G. P., & Lee, S. S. (2017). Therapeutic miRNA and siRNA: Moving from bench to clinic as next generation medicine. *Molecular Therapy–Nucleic Acids*, *8*, 132–143.

Chen, C., Ridzon, D. A., Broomer, A. J., Zhou, Z., Lee, D. H., Nguyen, J. T., et al. (2005). Real-time quantification of microRNAs by stem-loop RT-PCR. *Nucleic Acids Research*, *33*(20), e179.

Eichmüller, S. B., Osen, W., Mandelboim, O., & Seliger, B. (2017). Immune modulatory microRNAs involved in tumor attack and tumor immune escape. *Journal of the National Cancer Institute*, *109*(10).

Jasinski-Bergner, S., Mandelboim, O., & Seliger, B. (2014). The role of microRNAs in the control of innate immune response in cancer. *Journal of the National Cancer Institute*, *106*(10)[pii: dju257].

Jasinski-Bergner, S., Reches, A., Stoehr, C., Massa, C., Gonschorek, E., Huettelmaier, S., et al. (2016). Identification of novel microRNAs regulating HLA-G expression and investigating their clinical relevance in renal cell carcinoma. *Oncotarget*, 7(18), 26866–26878.

Jasinski-Bergner, S., Stoehr, C., Bukur, J., Massa, C., Braun, J., Huttelmaier, S., et al. (2015). Clinical relevance of miR-mediated HLA-G regulation and the associated immune cell infiltration in renal cell carcinoma. *Oncoimmunology*, *4*(6), e1008805.

Nielsen, A. F., Gloggnitzer, J., & Martinez, J. (2009). MicroRNAs cross the line: The battle for mRNA stability enters the coding sequence. *Molecular Cell*, *35*, 139–140.

Tretbar, U. S., Friedrich, M., Lazaridou, M. F., & Seliger, B. (2019). Identification of immune modulatory miRNAs by miRNA enrichment via RNA affinity purification. *Methods in Molecular Biology*, *1913*, 81–101.

van Zandwijk, N., Pavlakis, N., Kao, S. C., Linton, A., Boyer, M. J., & Clarke, S. (2017). Safety and activity of microRNA-loaded minicells in patients with recurrent malignant pleural mesothelioma: A first-in-man, phase 1, open-label, dose-escalation study. *The Lancet Oncology*, *18*(10), 1386–1396.

Varkonyi-Gasic, E., Wu, R., Wood, M., Walton, E. F., & Hellens, R. P. (2007). Protocol: A highly sensitive RT-PCR method for detection and quantification of microRNAs. *Plant Methods*, *3*, 12.

CHAPTER FOURTEEN

# Identification of immunomodulatory RNA-binding proteins in tumors

**Michael Friedrich*, Maria-Filothei Lazaridou, Jette Rahn, Christoforos Vaxevanis, Nadine Heimer, Simon Jasinski-Bergner, Barbara Seliger**

Institute for Medical Immunology, Faculty of Medicine, Martin Luther University Halle-Wittenberg, Halle, Germany

*Corresponding author: e-mail address: michael.friedrich@uk-halle.de

## Contents



## Abstract

By binding RNA in a sequence- and/or structure-dependent manner, RNA-binding proteins (RBPs) and their target RNA form a ribonucleoprotein complex involved in the RNA's fate. In this context, RBPs were shown to act as key players for post-transcriptional gene regulation by controlling RNA editing, splicing, polyadenylation, translocation, and stability. So far, over 1900 RBPs were identified and their deregulation has been associated with the development and progression of various disorders including cancer.

Although a number of sophisticated approaches are available, our knowledge about direct RNA-RBP interactions is, however, quite limited. Here we present a protocol with restricted requirements for equipment and devices to identify RBPs. This approach is based on (i) the purification of biotinylated RNA, (ii) chromatographic separation of co-purified proteins, and (iii) their identification by mass spectrometry.

*Methods in Enzymology*, Volume 636
ISSN 0076-6879
https://doi.org/10.1016/bs.mie.2019.05.042

## Abbreviations

**APM** antigen processing and presenting machinery
**HLA** human leukocyte antigen
**PVC** packed cell volume
**RBP** RNA-binding proteins
**UTR** untranslated region

# 1. Introduction

Up to now, systematic approaches identified >1900 RNA-binding proteins (RBPs) for *Homo sapiens* (Hentze, Castello, Schwarzl, & Preiss, 2018). These proteins are characterized by their capacity to bind RNA in a sequence- and/or structure-dependent manner. Such an interaction results in the formation of a ribonucleoprotein complex that is involved in gene expression (Dreyfuss, Kim, & Kataoka, 2002). RBPs modulate key aspects of post-transcriptional gene regulation, like RNA editing, splicing, polyadenylation, translocation, and control of RNA stability within the complex (Muller-McNicoll & Neugebauer, 2013). Therefore, it was not surprising that deregulation of RBP expression was associated with pathological conditions. Indeed, altered RBP expression has been demonstrated to play a key role in the development and progression of various disorders, such as neurological diseases like the fragile X syndrome and cancer (Hong, 2017). Furthermore, aberrant RBP expression is highly related to the overall survival rate of cancer patients (Busa et al., 2007; Jang, Lee, Kim, Hong, & Lee, 2017).

Despite a growing list of known RBPs and the understanding about their critical functions, very few RBPs have so far been systematically studied. As a consequence, only a limited number of direct RNA-RBP interactions are validated up to date. For example, post-transcriptional regulation of the essential antigen processing and presenting machinery (APM) by RBPs is only poorly analyzed. Although considered as major factor of immune escape mechanisms, only four RBPs were shown to modulate human leukocyte antigen (HLA) class I expression. In detail, the RBPs mex3b, mex3c and syncrip specifically target HLA-A (Cano et al., 2012; Huang et al., 2018; Kulkarni et al., 2017), while hnRNP-R was shown to act as post-transcriptional activator of classical and non-classical HLA class I molecules (Reches et al., 2016). To the best of our knowledge, there is so far no validated interaction of RBPs and components of the antigen processing machinery strengthening the need of further research to understand APM's post-transcriptional regulation (Eichmuller, Osen, Mandelboim, & Seliger, 2017).

This lack of information could be at least partially explained by the fact that several methods to identify RNA-RBP interactions are quite costly, technically challenging, and demand comprehensive equipment. In contrast, the method presented here has proven to be highly reproducible, easy to handle, cost-effective, and expeditious. The workflow is mainly based on publications of Reches et al. and Hammerle et al. with some modifications as shown in Fig. 1 (Hammerle et al., 2013; Reches et al., 2016). First, a sequence of interest has to be cloned into an appropriate vector enabling in vitro transcription. A sequence with comparable G/C content and length and/or an antisense construct of the original target sequence can act as control(s). The RNA synthesis is thereby performed in the presence of biotinylated UTP. Subsequently, the tagged RNA is incubated with a cytoplasmic extract to allow RBPs to bind to the RNA. Biotinylated RNA is immobilized using streptavidin-coated agarose beads and after stringent washing co-purified proteins are precipitated by acetone. Proteins are separated by SDS gel electrophoresis and stained for total protein.

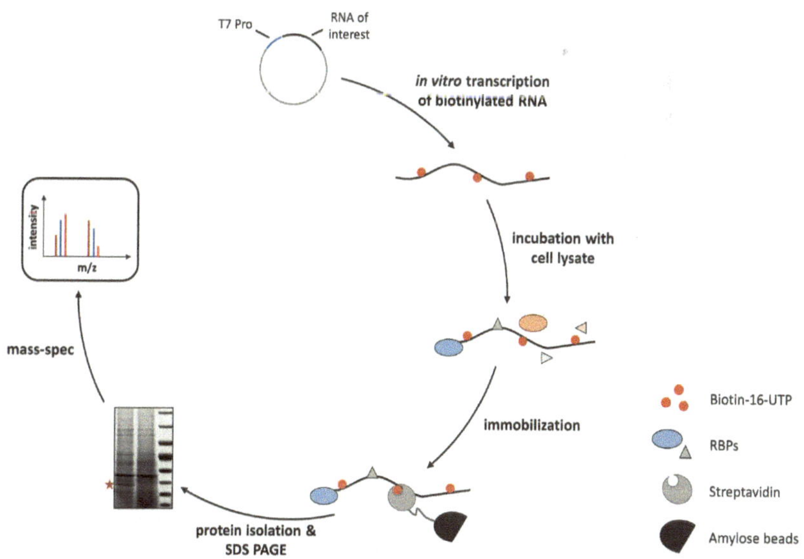


**Fig. 1** Scheme of the workflow for the identification of RNA-binding protein using biotinylated RNA.

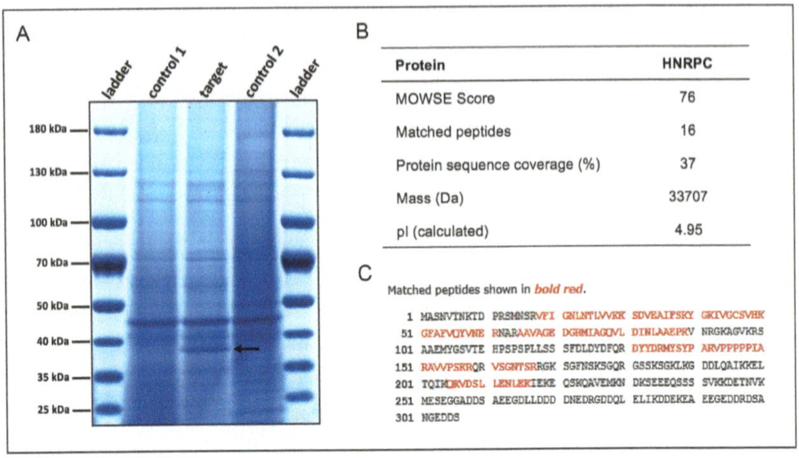


**Fig. 2** Representative SDS-PAGE of co-purified proteins and mass spectrometry-based protein identification. (A) Coomassie-stained SDS-PAGE of co-purified proteins. The arrow indicates the picked protein band for mass spectrometry, which is clearly evident in target sample, but not in control 1 (sequence of comparable length and G/C content) and control 2 (RNA antisense construct). (B) Identification of measured peptides according to the MASCOT database. (C) Mapping of the identified peptides to the hnRNP-C amino acid sequence.

Stained protein bands present in the sample of interest, but not in the control are extracted and identified by mass spectrometry (Fig. 2). After successful identification and verification of the RBP binding to the RNA, the method is also useful to map the interacting RNA sequence. This would require the analysis of the RBP binding capacity to progressively smaller target sequences by Western blotting as recently proposed by Panda and coworkers (Panda, Martindale, & Gorospe, 2016). In the following the different work steps will be described in detail.

## 2. Cell culture and cytoplasmic extract preparation

1. Depending on the cell line, it is recommended to harvest approximately fifteen 10 cm plates of ~90% confluent cells (1–5 × $10^8$ cells) or as many cells in T75 flasks as needed in order to have a cell pellet (=packed cell volume, pcv) of at least 700 μL (see Notes 1 and 2). Upon trypsin/EDTA detachment, cells are collected in 50 mL tubes and washed twice with 1 × PBS.

2. The cell pellet can be snap-frozen and stored at −80 °C until use.
3. For cytoplasmic extract preparation, resuspend cells in 5 pcv hypotonic buffer (10 mM HEPES pH 7.9, 1.5 mM $MgCl_2$, 10 mM KCl) and pellet the cells by spinning for 5 min with 350 g at 4 °C. Discard the supernatant. Resuspend the pellet in 3 pcv hypotonic buffer and place the cells on ice for 30 min in order to swell (see Note 3).
4. Use a glass dounce homogenizer to disrupt cell membranes (about 50strokes with loose pestle, see Note 4). Spin the cells for 15 min with 3300 g at 4 °C in a 15 mL tube to pellet nuclei. Transfer carefully the supernatant into a new 15 mL tube and repeat centrifugation until the supernatant is clear.
5. Supplement the cytoplasmic extract by adding 0.11 volumes of 10 × cytoplasmic buffer (0.3 M HEPES pH 7.9, 1.4 M KCl, 30 mM $MgCl_2$) and determine protein concentration via BCA or Bradford assay. It is recommended to use at least 3 mg of protein per reaction.
6. Add RNase inhibitor (100 units/mL of extract).
7. Keep the cytoplasmic extract on ice for further use in the assay or freeze the extracts in aliquots at −80 °C (see Note 5).

## 3. Cloning and RNA preparation

1. The sequence of interest is PCR amplified and cloned downstream of a T7 or SP6 promoter sequence (e.g. pJET1.2 blunt end vector, ThermoFisher; see Notes 6 and 7). Target sequences can be the 5′-untranslated region (UTR) of a gene, the coding region or the 3′-UTR. A sequence with comparable G/C content and length and/or the antisense construct of the original target sequence can act as control(s).
2. DNA template for in vitro transcription is either prepared by linearization of the plasmid DNA (see Note 8) or by generation of a PCR product (see Note 9).
3. A transcription system (e.g. RiboMAX™ Large Scale RNA Production Systems; Promega) enables in vitro synthesis of RNA. Ribonucleotides are mixed with biotinylated UTP (B-16-UTP, Jena Biosciences) in order to achieve 5–6 B-16-UTPs per transcript (see Note 10). Following the RNA synthesis, template DNA is removed by DNase I treatment and RNA is purified using an RNA transcription clean up kit (e.g. MEGAclear™ Transcription Clean-Up, ThermoFisher; see Note 11).

4. RNA integrity can be analyzed by denaturing agarose gel electrophoresis (see Note 12) and spectral photometric analysis. The molarity of the RNA is calculated using the following equation:

$$\text{RNA}\,[\text{pmol}/\mu L] = \frac{\text{RNA}\,[\text{ng}/\mu L] \times 1000}{\text{MW}(\text{RNA})}$$

MW (RNA) = number of nucleotides × 325 g/mol (=average MW of nucleotides).

## 4. RNA immunoprecipitation and co-purification of RBPs

1. Use 50 μL of streptavidin-coated agarose beads (e.g. GE Healthcare; 17-5113-01) per reaction and ~200 μL for pre-clearing of the cytoplasmic extract. Wash the beads twice with 500 μL of LS-washing buffer (20 mM HEPES pH 7.9, 0.1 M KCl, 10 mM $MgCl_2$, 0.01% NP40, 1 mM DTT; see Note 13) and spin each time at 1500 g for 1 min at room temperature (Note 14).
2. Remove washing buffer completely and incubate the beads with 500 μL blocking buffer (20 mM HEPES pH 7.9, 0.1 M KCl, 10 mM $MgCl_2$, 0.01% NP40, 1 mM DTT, 1 mg/mL BSA, 0.09 mg/mL glycogen, 20 μg/mL yeast tRNA) per sample for at least 2.5 h (or overnight) at 4 °C on a turning wheel.
3. Spin the beads at 1500 g for 1 min, discard the supernatant and wash the beads three times with 500 μL HS300-washing buffer (20 mM HEPES pH 7.9, 0.3 M KCl, 10 mM $MgCl_2$, 0.01% NP40, 1 mM DTT) for 5 min at room temperature on a turning wheel, spin each time at 1500 g for 1 min.
4. Pre-clear the cytoplasmic extract with the washed beads in order to minimize unspecific binding to the streptavidin-coated agarose beads. For RNA coupling, add 250 pmol of biotinylated RNA and 250 μL HS300-washing buffer per sample. Both, coupling and pre-clearing are performed for 4–5 h (or overnight) at 4 °C on a turning wheel.
5. Centrifuge the cytoplasmic extract at 1500 g for 3 min at 4 °C and collect the supernatant. Wash the beads coupled to RNA four times with 1 mL HS400-washing buffer (20 mM HEPES pH 7.9, 0.4 M KCl, 10 mM $MgCl_2$, 0.01% NP40, 1 mM DTT) and spin each time at 1500 g for 1 min at room temperature. Place the supernatant of the

pre-cleared cytoplasmic extract (respective to 3 mg of protein per reaction) on the beads that are coupled to the RNA and incubate overnight at 4 °C on a turning wheel.

6. Wash the beads seven times with 1 mL HS400-washing buffer for 5 min at 4 °C on a turning wheel, spin each time at 1500 g for 1 min (4 °C) and discard the supernatant. Elute proteins from the beads with 100 μL 6 M urea, 0.01% NP40, 1 mM DTT by shaking at 900 rpm on a thermomixer at room temperature for 1 h.
7. Spin the beads at 1500 g for 1 min at room temperature and transfer the supernatant to a new 1.5 mL Eppendorf tube. Add 5 volumes (500 μL) ice cold acetone, mix well and precipitate proteins at −20 °C overnight.
8. Pellet the proteins by spinning at the maximum speed and room temperature for 30 min. Wash protein pellets twice with 1 mL 80% acetone, spin down at maximum speed for 10 min after each wash and discard the supernatant.
9. Heat the 2 × Laemmli buffer (120 mM Tris-HCl, pH 6.8; 4% SDS; 20% glycerol) at 65 °C for 5 min and resuspend the pellets in ~30 μL hot 2 × Laemmli buffer. Separate co-purified proteins on a 4–20% gradient SDS-PAA gel (e.g. 4–20% Mini-PROTEAN® TGX™ Precast Protein Gel) to obtain fine resolution.
10. Upon completion of SDS-PAGE, the gel is rinsed twice with deionized water (see Note 15) and stained overnight in Coomassie blue (2.55% phosphoric acid; 150 mM aluminum sulfate, 0.5 mM Coomassie Brilliant Blue G-250, 5 M methanol). Gel can be destained by several washes with deionized water (see Note 16) until protein bands become clearly visible.

## 5. Protein band picking and sample preparation

1. Gel pieces of appropriate protein bands are picked semi-manually with a spot picker (e.g. GelPal spot picker, Genetix; see Note 17) using a needle with a diameter of 0.6 mm. The picking procedure consist of three steps:
   a. Excising a piece of gel with 30 μL suction volume.
   b. Placing it along with 50 μL deionized water into a 96-well reaction plate.
   c. Purging the needle twice with 500 μL sterile $H_2O$, before excising the next gel piece.

2. After the picking is completed, the 96-well reaction plate is placed into an automated protein digesting and sample preparation system (e.g. DigestPro MSi, Intavis; see Note 18).
3. In Table 1 the automated destaining and in-gel digestion steps of the gel pieces are outlined.
4. After completion of the digestion process the DigestPro MSi is switched into the *spotting mode* for desalting and concentrating peptides (e.g. ZipTips®, Millipore) and subsequent spotting on the MALDI (matrix-assisted laser desorption/ionization) target plate (e.g. MTP 384 ground steel, Bruker).
5. The processed loop of the sample preparation is shown in Table 2. The samples are thereby spotted using the dried droplet preparation with HCCA (α-Cyano-4-hydroxycinnamic acid, Bruker).

**Table 1** Steps of the automatic destaining and digestion.

| Step | Process | Time; Temp. | Buffer | Volume (μL) | Repetitions |
|---|---|---|---|---|---|
| 1 | Incubation | 10 min; RT | 50% ABC/ACN | 60 | 1 |
| 2 | Incubation | 10 min; RT | 50 mM ABC | 60 | 1 |
| 3 | Incubation | 10 min; RT | 100% ACN | 60 | 1 |
| 4 | Incubation | 20 min; 60 °C | 10 mM DTT | 30 | 1 |
| 5 | Incubation | 15 min; RT | 50 mM IAM | 30 | 1 |
| 6 | Incubation | 10 min; RT | 50 mM ABC | 60 | 1 |
| 7 | Incubation | 10 min; RT | 25 mM ABC | 60 | 1 |
| 8 | Incubation | 10 min; RT | 100% ACN | 60 | 2 |
| 9 | Wait | 10 min; RT | – | 0 | 1 |
| 10 | Incubation/Wait | 3 h; 37 °C | Trypsin[a] | 15 | 1 |
| 11 | Wait | 3 h; 37 °C | $ddH_2O$ | 10 | 1 |
| 12 | Aliquot | RT | 1% TFA | 10 | 1 |
| 13 | Incubation | 20 min; RT | 25 mM ABC | 20 | 1 |
| 14 | Incubation | 20 min; RT | 100% ACN | 20 | 1 |
| 15 | Wait | 10 min; RT | – | 0 | 1 |
| 16 | Collect to elution plate (8 °C) | | | | |

[a]0.008 μg/μL (25 mM ABC, 400 μM $CaCl_2$); ABC, ammonium bicarbonate; ACN, acetonitrile. DTT, dithiothreitol; IAM, iodoacetamide; RT, room temperature; TFA, trifluoroacetic acid.

**Table 2** Steps of the automated spotting on the MALDI target plate.

| Loop | Process | Buffer | Volume (μL) |
|---|---|---|---|
| L1 | TipWash | 0.1% TFA; 100% ACN | 20; 20 |
| L2 | TipLoad | Sample | 50 |
| L3 | TipElute on target | 100% ACN; Matrix[a] (50% ACN; 0.1% TFA) | 20; 1 |
| L4 | Acetone wash | Acetone | 10 |
| L5 | Rinse Needle | 0.1% TFA | 50; 250[b] |

[a]HCCA, α-cyano-4-hydroxycinnamic acid.
[b]Inside; outside; ACN, acetonitrile; TFA, trifluoroacetic acid.

6. When the sample spots are dry, standards for the external calibration (e.g. calibration standard II, Bruker) of the mass spectrometer in the mass range between 700 and 3500 Da are spotted manually. Therefore, 1 μL of the dissolved peptide calibration standard II is mixed with 1 μL of the HCCA matrix solution and afterwards 1 μL of the mixture is spotted on the MALDI target plate.

## 6. MALDI-TOF mass spectrometry analysis

1. Analyses are performed on an appropriate mass spectrometer (e.g. ultrafleXtreme™, Bruker) in positive reflector mode using an accelerating voltage of 25 kV.
2. Before each measurement the instrument is externally calibrated, and the peptide mass fingerprint dataset analyses are done using the MASCOT search engine (Matrix Science) with following parameter settings:
   a. Database: Swiss-Prot
   b. *Homo sapiens* sequences
   c. Fixed modifications: carbamidomethylation of cysteines
   d. Variable modifications: oxidation of methionine
   e. Cleavage enzyme: trypsin
   f. One missed cleavage is permitted
   g. Monoisotopic mass tolerance: 50 ppm
3. When the MOWES score exceeded 55 according to the MASCOT-defined significance threshold for false-positive events at $P < 0.05$, the proteins are assigned to be identified.

## 7. Concluding remarks

The identification of RNA-RBP interactions is a crucial step toward understanding the post-transcriptional regulation of genes. Particularly in connection with the development and progression of cancer, deregulation of RBPs was shown to be of key importance (Hong, 2017). Unfortunately, only a very limited number of validated RNA-RBP interactions have been identified up to now, especially for immune modulatory genes including HLA class I APM components. With minimal requirements, the presented protocol here provides guidance for the identification of proteins bound to a target RNA of choice. This method has been successfully employed to identify RBPs (e.g. Hammerle et al., 2013; Kuwano et al., 2008; Reches et al., 2016). However, we are aware that the presented method here has its limitations, in particular regarding the high amount of protein (reflecting a large number of cells) needed. Most of all, this could be problematic for cells which cannot be easily propagated. Another restriction of the method is the eye-based selection of candidate RBPs via a Coomassie-stained gel demanding for abundant proteins and high enrichment ratios. Alternatively, quantitative mass spectrometry (Bantscheff, Lemeer, Savitski, & Kuster, 2012; Bantscheff, Schirle, Sweetman, Rick, & Kuster, 2007) using the total eluate (instead of distinct protein bands) could avoid this issue. It is noteworthy that mass spectrometry-based evaluation of RNA pull-down assays usually identifies a number of potential RBPs, from which only very few directly bind to the target RNA. To exclude indirect and, even more importantly, non-specific interactions, validation (e.g. by overexpression and/or knockdown of the respective RBP) is an essential step in RNA affinity purification studies. In this context, a slight modification of the presented method using progressive target sequence deletions would allow a further verification and simultaneous identification of the RBP-bound RNA sequence (Panda et al., 2016).

## 8. Notes

1. Depending on the cell line of choice, optimal growth conditions vary. Recommendations are given by the American Type Culture Collection (ATCC, Manassas, USA).
2. Transfer of the cell pellet in a 15 mL falcon tube enables the estimation of the packed cell volume.

3. Incubation for up to 90 min can improve the swelling and enhance cytoplasmic protein yield.
4. Perform the homogenization slowly, especially down strokes.
5. It is recommended to prepare the cytoplasmic extract freshly.
6. Use a proofreading polymerase for amplification of the target sequence to prevent mismatched nucleotides. Sequence of bases should be verified by sequencing.
7. If a PCR product is used for template generation, virtually all cloning vectors are suitable irrelevant of an existing promoter.
8. An appropriate restriction enzyme for plasmid linearization cuts directly (or in short distance) 3′ to the target sequence.
9. The cloned sequence can also serve as a template in a PCR. Considering a T7 polymerase based in vitro transcription, this would require the adding of the T7 promoter sequence (5′-GCGGAGATCTAATAC GACTCACTATAGG-3′) to the 5′ end of the forward primer.
10. Exact quantity of B-16-UTP depends on the sequence length and uridine frequency.
11. Alternative non-kit-based purification methods are available (e.g. lithium chloride precipitation).
12. Denaturing agarose gel electrophoresis using guanidinium thiocyanate has shown to be suitable to estimate the success of the transcription reaction and to check for RNA integrity.
13. Due to its low stability DTT is always added just before use.
14. Vacuum pump apparatus can be used to aspirate off wash solutions.
15. Cautious boiling of the gel in a microwave to remove the SDS may improve quality of Coomassie stain.
16. Alternatively, a Coomassie destaining solution (2.55% phosphoric acid; 5 M methanol) can be used.
17. Further options of picking protein bands are possible (e.g. manual cut out using a scalpel).
18. There is also the possibility of manual sample preparation.

## References

Bantscheff, M., Lemeer, S., Savitski, M. M., & Kuster, B. (2012). Quantitative mass spectrometry in proteomics: Critical review update from 2007 to the present. *Analytical and Bioanalytical Chemistry*, *404*, 939–965.

Bantscheff, M., Schirle, M., Sweetman, G., Rick, J., & Kuster, B. (2007). Quantitative mass spectrometry in proteomics: A critical review. *Analytical and Bioanalytical Chemistry*, *389*, 1017–1031.

Busa, R., Paronetto, M. P., Farini, D., Pierantozzi, E., Botti, F., Angelini, D. F., et al. (2007). The RNA-binding protein Sam68 contributes to proliferation and survival of human prostate cancer cells. *Oncogene*, *26*, 4372–4382.

Cano, F., Bye, H., Duncan, L. M., Buchet-Poyau, K., Billaud, M., Wills, M. R., et al. (2012). The RNA-binding E3 ubiquitin ligase MEX-3C links ubiquitination with MHC-I mRNA degradation. *EMBO Journal*, *31*, 3596–3606.

Dreyfuss, G., Kim, V. N., & Kataoka, N. (2002). Messenger-RNA-binding proteins and the messages they carry. *Nature Reviews Molecular Cell Biology*, *3*, 195–205.

Eichmuller, S. B., Osen, W., Mandelboim, O., & Seliger, B. (2017). Immune modulatory microRNAs involved in tumor attack and tumor immune escape. *Journal of the National Cancer Institute*, *109*, 1–14.

Hammerle, M., Gutschner, T., Uckelmann, H., Ozgur, S., Fiskin, E., Gross, M., et al. (2013). Posttranscriptional destabilization of the liver-specific long noncoding RNA HULC by the IGF2 mRNA-binding protein 1 (IGF2BP1). *Hepatology*, *58*, 1703–1712.

Hentze, M. W., Castello, A., Schwarzl, T., & Preiss, T. (2018). A brave new world of RNA-binding proteins. *Nature Reviews Molecular Cell Biology*, *19*, 327–341.

Hong, S. (2017). RNA binding protein as an emerging therapeutic target for cancer prevention and treatment. *Journal of Cancer Prevention*, *22*, 203–210.

Huang, L., Malu, S., McKenzie, J. A., Andrews, M. C., Talukder, A. H., Tieu, T., et al. (2018). The RNA-binding protein MEX3B mediates resistance to cancer immunotherapy by downregulating HLA-A expression. *Clinical Cancer Research*, *24*, 3366–3376.

Jang, H. H., Lee, H. N., Kim, S. Y., Hong, S., & Lee, W. S. (2017). Expression of RNA-binding motif protein 3 (RBM3) and cold-inducible RNA-binding protein (CIRP) is associated with improved clinical outcome in patients with colon cancer. *Anticancer Research*, *37*, 1779–1785.

Kulkarni, S., Ramsuran, V., Rucevic, M., Singh, S., Lied, A., Kulkarni, V., et al. (2017). Posttranscriptional regulation of HLA-A protein expression by alternative polyadenylation signals involving the RNA-binding protein syncrip. *The Journal of Immunology*, *199*, 3892–3899.

Kuwano, Y., Kim, H. H., Abdelmohsen, K., Pullmann, R., Jr., Martindale, J. L., Yang, X., et al. (2008). MKP-1 mRNA stabilization and translational control by RNA-binding proteins HuR and NF90. *Molecular and Cellular Biology*, *28*, 4562–4575.

Muller-McNicoll, M., & Neugebauer, K. M. (2013). How cells get the message: Dynamic assembly and function of mRNA-protein complexes. *Nature Review Genetics*, *14*, 275–287.

Panda, A. C., Martindale, J. L., & Gorospe, M. (2016). Affinity pulldown of biotinylated RNA for detection of protein-RNA complexes. *Bioprotocol*, *6*, 1–10.

Reches, A., Nachmani, D., Berhani, O., Duev-Cohen, A., Shreibman, D., Ophir, Y., et al. (2016). HNRNPR regulates the expression of classical and nonclassical MHC class I proteins. *The Journal of Immunology*, *196*, 4967–4976.

CHAPTER FIFTEEN

# Techniques for the generation of humanized mouse models for immuno-oncology

**Chun I. Yu*, Florentina Marches, Te-Chia Wu, Jan Martinek, Karolina Palucka***

The Jackson Laboratory for Genomic Medicine, Farmington, CT, United States
*Corresponding authors: e-mail address: chun.yu@jax.org; karolina.palucka@jax.org

## Contents



*Methods in Enzymology*, Volume 636
ISSN 0076-6879
https://doi.org/10.1016/bs.mie.2019.06.003

## Abstract

Mouse models of human cancer have been used extensively to circumvent the complexity in human patients. However, murine models often inadequately recapitulate the human cancer-immune interface partly due to important differences between mouse and human immune systems. Immunodeficient mice when transplanted with $CD34^+$ hematopoietic progenitor cells (HPCs) develop multilineage human immune cells. While there remain limitations, efforts have been made to improve the function of human immune system. Thus, humanized mice, defined as mice with human immune system, have become an emerging model to study human cancers. Humanized mouse models have been used for various areas of cancer research including adoptive transfer of chimeric antigen receptor (CAR)-modified T cells, neoantigen vaccination to increase T cell repertoire and reprograming tumor microenvironment. Here, we describe the essential techniques to generate humanized mouse models for immuno-oncology studies.

## 1. Introduction

The treatment landscape of human cancer has changed dramatically over the past decade with improved survival for patients treated with blocking antibodies targeting immune regulatory checkpoints (Hamid et al., 2013; Hodi et al., 2010). Thus far, there has been approvals for six antibodies targeting cytotoxic T-lymphocyte-associated protein (CTLA)-4, programmed death (PD)-1 or programmed death ligand (PDL)-1 on various cancer types including melanoma and non-small cell lung carcinoma (Ribas & Wolchok, 2018). Correlative studies suggest that the presence of tumor-reactive T cells that recognize neoantigen arisen from tumor-specific mutations is among key mechanisms underlying clinical response to checkpoint inhibitor therapy, so-called hot tumors (Le et al., 2015; Schumacher & Schreiber, 2015). Thus, the challenge is how to transform "cold tumors" in patients with low preexisting levels of T cells and low frequency of somatic mutations to "hot" tumors that could be controlled by checkpoint blockade (Alexandrov et al., 2013). To this end, other potential approaches under investigation include adoptive transfer of chimeric antigen receptor (CAR)-modified T cells (June, O'Connor, Kawalekar, Ghassemi, & Milone, 2018; Maude et al., 2018); cancer vaccines to increase T cell repertoire (Banchereau & Palucka, 2017; Bulik-Sullivan et al., 2018; Keskin et al., 2019) and reprograming tumor microenvironment (Binnewies et al., 2018; Palucka & Coussens, 2016; Wu et al., 2014, 2018).

Mouse models of human cancer have enabled proof-of-concept studies and the identification of causative links (Zitvogel, Pitt, Daillère, Smyth, & Kroemer, 2016). Nevertheless, murine models often inadequately recapitulate the human cancer-immune interface partly due to important differences between mouse and human immune systems (Hagai et al., 2018; Kanazawa, 2007; Mestas & Hughes, 2004; Williams, Flavell, & Eisenbarth, 2010). Thus, humanized mice, defined as mice with human immune system, could be an attractive alternative (Shultz, Brehm, Garcia-Martinez, & Greiner, 2012; Theocharides, Rongvaux, Fritsch, Flavell, & Manz, 2016; Victor Garcia, 2016; Zhang & Su, 2012). To this end immunodeficient mice lacking common gamma chain (γc) like NOD-SCID-IL2$\gamma c^{-/-}$ (NSG), or BALB/c-Rag$2^{-/-}$-$\gamma c^{-/-}$ (BRG) (Matsumura et al., 2003; Traggiai et al., 2004) can be humanized by transplantation of human CD34$^{+}$ hematopoietic progenitor cells (HPCs). Based on the sources of T cells, the model can be further categorized into two types: (1) model in which mature T cells are isolated from the donor of HPCs and adoptively transferred (Aspord et al., 2007; Pedroza-Gonzalez et al., 2011; Wu et al., 2014, 2018; Yu et al., 2008); in this case the T cells have been selected in human thymus; and (2) model in which endogenous T cells are de novo generated from human CD34$^{+}$ HPCs (Matsumura et al., 2003; Traggiai et al., 2004); in which case human T cells are selected in mouse thymus. The biggest limitation of the first model is graft-versus-host disease; the biggest limitation of the second model is a limited number of T cells able to recognize human major histocompatibility complex (MHC). The current areas of investigation aiming to improve humanized mouse models include: (1) the long-term engraftment of human CD34$^{+}$ HPCs (Cosgun et al., 2014); (2) the development of myeloid cells (Billerbeck et al., 2011; Rongvaux et al., 2014); (3) the development of secondary lymphoid organs (Li et al., 2018; Takahashi et al., 2017); (4) and last but not least maturation of human adaptive immunity in the context of human MHC (Billerbeck et al., 2013; Danner et al., 2011; Najima et al., 2016).

Our group has pioneered using humanized mice to study tumor microenvironment. We have demonstrated the origin and role of type 2 T cells (Th2) in breast tumor progression (Aspord et al., 2007; Pedroza-Gonzalez et al., 2011; Wu et al., 2018) and how cancer cells corrupt dendritic cells (DCs) to drive Th2 inflammation via pathogenic loops involving cancer cell-derived TSLP and TGF-β, and myeloid cell-derived IL-1β (Wu et al., 2018). In melanoma studies utilizing MISTRG mice (Rongvaux et al., 2014), we found that tumor progression is dependent on human CSF-1-driven macrophages and can be partly controlled by VEGFA blockade (Rongvaux et al., 2014). We have also demonstrated that humanized mice can be used to study therapeutic response to checkpoint inhibition (Wang et al., 2018). Indeed, anti-PD1 antibody could control both cell-line derived and patient-derived xenografts (PDX) from numerous types of solid tumors (Capasso et al., 2019; Wang et al., 2018). While the treatment efficacy varied between donors, tumor regression was dependent on T cells as documented in CD8 depletion experiments (Wang et al., 2018). Recent studies also used humanized mice to decode the mechanisms of CAR T cell toxicity (Gulati et al., 2018; Kim et al., 2018). Finally, specific $CD8^+$ T cells developed in mice after vaccination with tumor-associated antigen-loaded DCs demonstrating the proof-of-concept for the preclinical testing of cancer vaccination in humanized mouse models (Najima et al., 2016). Here, we describe the techniques to generate humanized mouse models for immuno-oncology studies from the type of mice for HPC transplant to tumor implantation and immune monitoring.

## 2. The construction of humanized mice

### 2.1 Model with adoptive T cell transfer

1. NOD.Cg-*Prkdc*$^{scid}$ *B2m*$^{tm1Unc}$/J mice (NOD/SCID/B2m$^{-/-}$; RRID: IMSR_JAX:002570) at the age of 4 weeks are sub-lethally irradiated at 12 cGy per gram of body weight using gamma irradiator (MDS Nordion, Ottawa, Canada) (see Note 1–2).
2. After 24 h, $1–3 \times 10^6$ of $CD34^+$ HPCs from normal donor G-CSF mobilized peripheral blood apheresis are resuspended in 200 μL of 1 × phosphate buffered saline (PBS) and given to previously irradiated mice through tail-vein intravenous (IV) injection (see Note 3–4).
3. At 4–8 weeks after HPC transplant, collect one capillary tube of blood (60 μL) with heparin to monitor the human engraftment in the blood.
4. To obtain T cell competent humanized mice, $1 \times 10^7$ T cells isolated from autologous PBMCs using EasySep Human T cell Enrichment Kit (StemCell Technologies, Vancouver, BC) are injected IV into the mice (see Note 5).

## 2.2 Model with endogenous T cells

1. Newborn NOD.Cg-Prkdc$^{scid}$ *Il2rg*$^{tm1Wjl}$/SzJ (NSG, RRID:IMSR_JAX:005557) mice are sub-lethally irradiated (80–100 cGy irradiation) using gamma irradiation (see Note 6–8).
2. After 6–24 h, $1 \times 10^5$ $CD34^+$ HPCs from fetal liver or full-term cord blood (Advanced Bioscience Resources or Lonza) are resuspended in 20 μL of 1× PBS with 1 μg/mL of anti-CD3 antibody (OKT3, Biolegend) (see Note 9).
3. Restrain pups with one hand and inject HPCs directly into the liver or facial-vein using a 31-g insulin syringe (BD) with the other hand.
4. At 8–20 weeks after HPC transplant, collect one capillary tube of blood with heparin to monitor the human engraftment in the blood (see Note 10).

## 2.3 The analysis of human engraftment

1. Pipette 10 μL of heparinized blood with 40 μL of RBClysis buffer (Biolegend) at room temperature for 5 min to lyse red blood cells (RBC) and count the total number of white blood cells in hemocytometer with 0.4% trypan blue.
2. Pipette the remaining 50 μL of heparinized blood and treat with 10 μg/mL of both human and murine Fc blocker (BD) at 4 °C for 5 min.
3. Stain the blood with antibody cocktail including antibodies to mouse (m)CD45-FITC (30-F11, BD) and human (h)CD45-PerCP-Cy5.5 (HI30, BD), CD33-PE (P67.6, Biolegend), CD14-PE-Cy7 (MqP9, BD), CD19-APC (HIB19, Biolegend), CD3-APC-H7 (SK7, BD) for 30 min at 4 °C (see Note 11).
4. Remove excess antibodies with 1 mL of 1× PBS with 2% fetal calf serum (FCS) and pellet the cells at 500 × *g* for 5 min.
5. Lyse the RBC with RBC lysis buffer (>1 mL) for 5 min and resuspend the pellet with 1× PBS with 2% FCS.
6. Acquire the sample on flow cytometer like BD Canto, LSRII or Symphony A5 (BD) and analyze with FlowJo software (Tree Star).
7. Analyze the blood engraftment by the percentage of human $CD45^+$ leukocytes in total human and mouse $CD45^+$ leukocytes as well as the absolute number of $hCD45^+$ leukocytes in 1 μL of blood. Various cell type including $hCD14^+$ monocytes, $hCD33^+$ myeloid cells, $hCD19^+$ B cells,

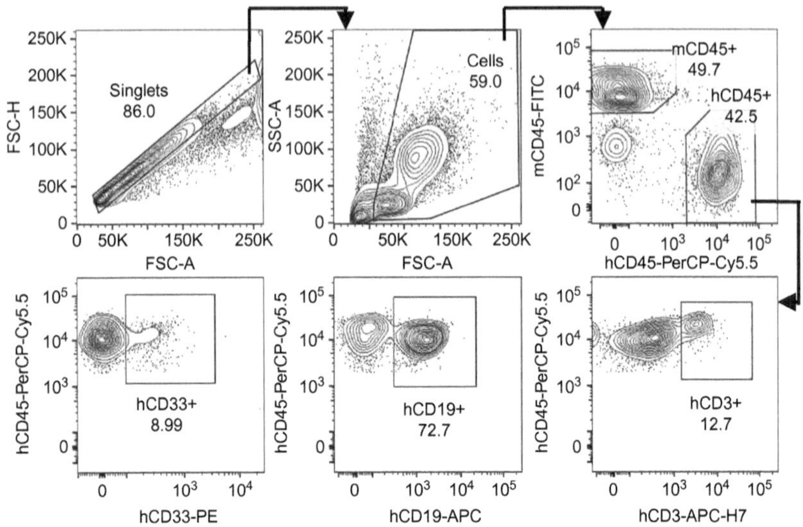


**Fig. 1** Human engraftment in the blood of humanized mice. NSG-SGM3 mice were engrafted with $1 \times 10^5$ cord blood $CD34^+$ HPCs at 4 weeks. Human engraftment in the blood was analyzed at 8-week post-transplant by flow cytometry.

and $hCD3^+$ T cells are enumerated as the percentage in $hCD45^+$ cells and absolute number in 1 μL of blood (see Note 12; Fig. 1).

8. Humanized mice with >20% human engraftment and all subsets of human immune cells are used for the tumor studies.

## 3. Tumor implantation

### 3.1 Cell-line derived xenograft

1. Human triple-negative breast cancer (TNBC) cell line, MDA-MB231 (ATCC, Manassas, Virginia), is cultured in complete RPMI medium (RPMI 1640, 10mM HEPES, 1mM Sodium Pyruvate, 1% non-essential amino acid, 1% penicillin-streptomycin and 2mM L-Glutamine) supplemented with 10% FCS in T150 flasks (see Note 13–14).
2. At day 3–4 after subculture, MDA-MB231 cells reach 90% confluency. MDA-MB231 cells are lifted with 5mL of Trypsin-EDTA solution (ThermoFisher Scientific) and washed once with 1 × PBS.
3. Resuspend the cells in $1–10 \times 10^6$ cells per 100μL of 1 × PBS and inject subcutaneously (SC) into the shaved right lateral flank (see Note 15–16).

## 3.2 Patient-derived xenograft

1. Gently rinse the tumor in 1 × PBS with 1% penicillin-streptomycin and repeat twice.
2. In a Petri dish, remove all hemorrhagic, necrotic parts and tumor capsule from the tumor. Subdivide the tumor specimen into fragments (1–5 $mm^3$) (see Note 17).
3. Preserve two fragments for histological analysis and genomic analysis (see Note 18).
4. Gently rinse the fragments in 1 × PBS with 1% penicillin-streptomycin and repeat twice.
5. Prepare the mice for the surgical implantation of tumor fragments subcutaneously into the right lateral flank by first removing the fur, anesthetizing the mice with isoflurane gas and disinfecting the skin (see Note 19).
6. Using small Iris forceps, load the tumor fragment (30–40 mg) into the 11–13 g tumor implanting trocar needle (Cadence Science) by pulling the plunger back slightly.
7. Using large straight forceps, grab the skin in the center of the right flank and lift up. Gently push the trocar needle through the skin of the mouse and continue to advance the trocar cranially for 1–1.5 cm.
8. Push the trocar plunger to eject the tumor and place the forceps on top of the injection site while slowly pulling the trocar needle out to ensure the fragment stays in the subcutaneous pocket (see Note 20).

## 3.3 Monitor tumor growth

1. Monitor the growth of tumor every 2–7 days.
2. Gently examine the implantation site with fingers. A tumor becomes measurable when a palpable tumor can be moved easily under the skin.
3. Measure the length and width of the tumor using a caliper. The length represents the longest axis of the tumor and the width is the widest point perpendicular to the length axis measured (see Note 21).
4. Tumor volume (ellipsoid) is calculated as follows: $(\text{width})^2 \times \text{length}/2$.

## 3.4 Monitor tumor growth with live imaging

1. An aqueous solution of luciferin (150 mg/kg) is injected intraperitoneally (IP) at 10 min prior to imaging with IVIS Spectrum in vivo imaging system (PerkinElmer) (see Note 22).

2. Mice are anesthetized with Isoflurane gas and placed into the light-tight chamber of the CCD camera system.
3. The photons emitted from the luciferase-expressing cells within the animal are quantified using Living Image software (PerkinElmer).

## 4. Immune monitoring

### 4.1 Processing single cell suspension from the spleen and lymph nodes

1. In a Petri dish, a spleen is injected with 1 mL of enzyme solution containing 50 μg/mL of Liberase TM (Roche) and 24 U/mL of DNase I (Sigma) and incubated for 10 min at 37 °C. Similarly, lymph nodes are incubated with enzyme solution for 30 min.
2. Prepare single cell suspension using two frosted slides to gently disrupt the spleen or lymph nodes and transfer the cell suspension into 15 mL conical tubes.
3. Centrifuge 500 × *g* for 5 min and discard the supernatant.
4. Lyse the RBC with 2 mL of RBC lysis buffer (Biolegend) for 5 min and wash once with 10 mL of 1 × PBS (see Note 23).
5. Resuspend cells with 1 mL of 1 × PBS with 2% FCS and pass through a 70 μm cell strainer.
6. Perform a cell count and proceed to downstream assays. Alternatively, samples can be cryopreserved in FCS with 10% DMSO in a controlled-rate freezing chamber, such as CoolCell (Corning) for 24 h in −80 °C freezer and stored in the air-phase of a liquid nitrogen tank.

### 4.2 Immunophenotypic analysis of T cell activation by multiparameter flow cytometry

1. Resuspend $1 \times 10^6$ cells in 100 μl of PBS/2%FCS and incubate with 1 μg of both mouse and human Fc Blocker (BD) for 5 min to block the Fc receptor.
2. Incubate cells with antibody cocktail including mCD45-BV650 (30-F11, BD), hCD45-BV510 (HI30, BD), CD3-APC-H7 (SK7, BD), CD4-APC (RPA-T4, Biolegend), CD8-FITC (SK1, Biolegend), CD45RA-PerCP-Cy5.5 (HI100, BD), CCR7-PE-Cy7 (3D12, BD), CD25-PE (2A3, BD), PD1-BV421 (EH12.2H7, Biolegend), HLA-DR-Alexa Flour (AF)700 (Bu15, Biolegend), and Brilliant Stain Buffer Plus (BD) for 30 min at 4 °C. (see Note 24).

3. Remove excess antibodies with 1 × PBS with 2% FCS (>1 mL), pellet the cells at 500 × *g* for 5 min and resuspend cells with 0.5 mL of 1 × PBS with 2% FCS.
4. Add 5 μL of 7-AAD viability dye (Biolegend) for 5–10 min, immediately acquire the sample on flow cytometer and analyze with FlowJo software.
5. T cell activation could be defined by the expression of CD45RA, CCR7, CD25, PD1 and HLA-DR.

## 4.3 Measurement of antigen-specific T cell response by tetramer staining

1. Resuspend $1 \times 10^6$ cells in 100 μl of PBS/2%FCS and incubate with 1 μg of both mouse and human Fc Blocker (BD) for 5 min to block the Fc receptor.
2. Incubate cells with antibodies to mCD45-BV650 (30-F11, BD), hCD45-PerCP-Cy5.5 (HI30, BD), CD3-APC-H7 (SK7, BD), CD4-APC (RPA-T4, Biolegend), CD8-FITC (SK1, Biolegend) and peptide-loaded HLA tetramer-PE (MBL) or control peptide-loaded HLA tetramer for 30 min at room temperature (see Note 25).
3. Wash samples with 1 × PBS with 2% FCS, stain with 7-AAD viability dye for 5–10 min and acquire on a flow cytometer.
4. Peptide-specific T cells are measured by the percentage of peptide/MHC-tetramer$^+$ cells in total human CD8$^+$ or CD4$^+$ T cells.

## 4.4 Measurement of atnigen-specific T cell response by intracellular cytokine staining

1. Resuspend $1 \times 10^6$ cells in 200 μL of complete RPMI with 10% human AB serum in 96-well U bottom plate.
2. Stimulate cells with 10 μL/mL of anti-human CD28/CD49d antibodies (BD) with or without 2.5 μg/mL of peptides and culture at 37 °C in a $CO_2$ incubator (see Note 26).
3. After 1 h, add protein transporter inhibitors, 2 μM Monensin and 1 μg/mL Brefeldin (BD), and anti-hCD107-FITC antibody (BD) and incubate for another 5 h at 37 °C in a $CO_2$ incubator (see Note 27).
4. At the end of culture, wash cells with 1 × PBS and stain with Fixable Aqua Dead Cell Stain (ThermoFisher Scientific) for 30 min at 4 °C.
5. Wash cells with 1 × PBS with 2% FCS and proceed with the intracellular staining protocol. First stain cells with surface antibodies to mCD45-BV650 (30-F11, BD), hCD45-PerCP-Cy5.5 (HI30, BD), CD3-APC-H7 (SK7, BD), CD4-V450 (RPA-T4, BD), CD8-ECD (SFC121Thy2D3, Beckman Coulter) for 30 min at 4 °C, followed by

fixation and permeabilization (BD Cytofix/Cytoperm) according to the manufacturer protocol and intracellular staining with antibodies to human IFN-γ-PE (B27, BD), IL-2-APC (MQ1-17H12, BD), TNF-α-PE-Cy7 (Mab11, BD) for 30 min at room temperature.

6. Acquire the sample on flow cytometer and analyze with FlowJo software.
7. Analyze the peptide-specific T cells for their capacity to produce cytokines (IFN-γ, IL-2, TNF-α) and for their degranulation capacity (CD107a) to release granzymes and perforin.

### 4.5 Measurement of tumor-infiltrating immune cells

1. Fresh tumors are embeded in OCT (Sakura Finetek USA) and snap frozen with liquid nitrogen.
2. Prepare 6 μm frozen sections, air dry on Superfrost plus slides, and fix with cold acetone for 5 min.
3. Tissue sections are first treated with 0.03% Hyaluronidase for 15 min, followed by Fc Receptor Block (Innovex Bioscience) for 30 min and Background Buster (Innovex Bioscience) for 40 min.
4. The sections are then stained with antibodies to hCD45 (HI30, mouse IgG1 isotype, Biolegend), mCD45 (30-F11, rat IgG2b isotype, BD) and cytokeratin 19 (A53-B/A2, mouse IgG2a isotype, Abcam) diluted in PBS/5%BSA/0.1% Saponin for 1 h at room temperature. Respective isotype antibodies are used as the control (see Note 28).
5. After rinse twice with PBS/0.1% saponin for 5 min each, tissue sections are stained with isotype-specific secondary antibodies including anti-mouse IgG1-AF488, anti-mouse IgG2a-AF568, anti-rat IgG2b-AF647 (ThermoFisher Scientific).
6. Finally, sections are counterstained with 1 μg/mL of 4′,6-diamidino-2-phenylindole (DAPI), mounted with Fluoromount-G (Thermo Fisher Scientific), and visualized using a confocal microscope like Leica SP 8 with Leica LAS software or a fluorescence microscope like Zeiss Axio with ZEN software.
7. Analyze $hCD45^+$ immune cells present in the tumor.

## 5. Immunological intervention

### 5.1 Specific T cell response elicited by DC vaccination

1. NOD/SCID/$B2m^{-/-}$ mice at 4-week of age are transplanted with $1–3 \times 10^6$ G-CSF mobilized $CD34^+$ HPCs as described in Section 2.1.

2. At 4–8 weeks post-transplant, mice are bled to confirm the human immune cell engraftment as described in Section 2.3.
3. Autologous T cells are purified from G-CSF mobilized PBMCs using total T cell enrichment kit (Stemcell). $1 \times 10^7$ total human T cells are transplanted IV through tail-vein injections into the mice.
4. On the next day, mice receive $1 \times 10^6$ autologous monocyte-derived DCs loaded with tumor antigens SC (see Note 29).
5. At day 10–14 after vaccination, mice are bled or harvested to evaluate specific $CD8^+$ T cell response as described in Sections 4.3 and 4.4.
6. Peptide-specific T cells are measured in the lymph nodes, spleen and blood to evaluate the efficacy of DC vaccination (Fig. 2).

## 5.2 Neutralizing IL-1B blocks breast cancer progression

1. Female NSG-SGM3 (RRID:IMSR_JAX:013062) mice, 4-weeks old, are irradiated with 150 cGy and transplanted with $1 \times 10^5$ $CD34^+$ HPCs from cord blood as described in Sction 2.2.
2. At 8 weeks after transplant, mice are bled to evaluate the engraftment in the blood with >20% $hCD45^+$ engraftment as described in Section 2.3.
3. At 8–12 weeks after transplant, $1 \times 10^7$ MDA-MB231 cells are implanted SC as described in Section 3.1.
4. IL-1 receptor antagonist, Anakinra (Kineret) is injected IP daily at the dose of 10 mg/kg body weight after tumor implantation. The control group is treated with 1 × PBS.
5. The tumor volume is measured every 3–4 days as described in Section 3.3.
6. The efficacy is determined by the regression of tumor mass in the Anakinra group compared to the control group.

## 5.3 Cancer immunotherapy with antibodies to immune checkpoint

1. Female NSG mice are transplanted at 4-week of age with human $CD34^+$ HPCs from cord blood as described in Section 2.2.
2. At 12 weeks post-transplant, mice are bled to confirm the human immune cell engraftment as described in Section 2.3.
3. Human breast cancer cell line, MDA-MB231, is implanted into mammary fat pads of mice at $5 \times 10^6$ cells in 50 μL.
4. The tumor volume is measured every 3–4 days.

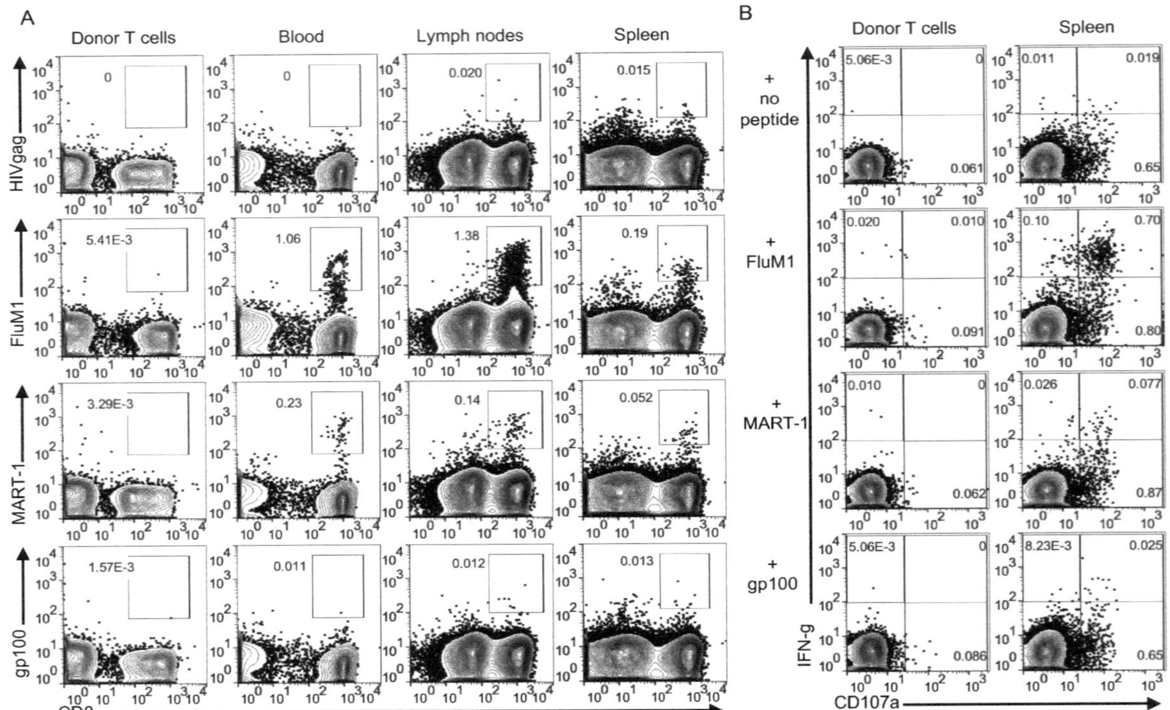


**Fig. 2** Specific $CD8^+$ T cell response elicited by DC vaccine in humanized mice. NOD/SCID/*B2m*$^{-/-}$ mice were transplanted with $2 \times 10^6$ G-CSF-mobilized, HLA-A0201$^+$ $CD34^+$ HPCs. At 8-weeks post-transplant, mice received $1 \times 10^7$ donor T cells IV and were vaccinated the next day SC with $1 \times 10^6$ monocyte-derived DCs loaded with HLA-A0201$^+$ restricted FluM1, MART-1 and gp100 peptides. (A) The frequency of specific $CD8^+$ T cells binding HIVgag-tetramer, FluM1-tetramer, MART-1-tetramer, or gp100-tetramer in donor T cells, and tissues of humanized mice at day 12 after DC vaccination by flow cytometry. (B) Donor T cells and single cell suspension from the spleen of humanized mice were stimulated for 6h with peptides and antibodies against CD28 and CD49d. The frequency of IFN-$\gamma$ and CD107a expression on $CD8^+$ T cells from donor T cells and the spleen of humanized mice was analyzed by flow cytometry.

5. Once tumors reach 50–120 mm$^3$, mice are randomized to receive either anti-PD1 antibody (Pembrolizumab, Merck) or vehicle control saline (Hospira) every 5 days IP for a total of 5 doses. Anti-PD1 antibody is administered at 10 mg/kg for the first dose and 5 mg/kg for the rest.
6. The efficacy is determined by the regression of tumor mass in the anti-PD1 group compared to the control group.

## 6. Concluding remarks

The methods presented here provide a rapid and reliable approach to generate humanized mice for immune-oncological studies.

## 7. Notes

1. The approval from the institutional review board for the use of human materials and institutional animal care and use committee for the use of mice are mandatory before launching the experiments.
2. Several other mouse strains lacking murine major histocompatibility complex (MHC) class I with reduced xenogeneic graft-versus-host disease response maybe suitable substitutes, such as NOD.Cg-*B2m*$^{tm1Unc}$ *Prkdc*$^{scid}$ *Il2rg*$^{tm1Wjl}$/SzJ (RRID:IMSR_JAX:010636); NOD.Cg-*Prkdc*$^{scid}$ *H2-K1*$^{tm1Bpe}$ *H2-D1*$^{tm1Bpe}$ *Il2rg*$^{tm1Wjl}$/SzJ (RRID: IMSR_JAX:023848); NOD.Cg-*Prkdc*$^{scid}$ *H2-Ab1*$^{em1Mvw}$ *H2-K1*$^{tm1Bpe}$ *H2-D1*$^{tm1Bpe}$ *Il2rg*$^{tm1Wjl}$/SzJ (RRID:IMSR_JAX:025216).
3. Adult CD34$^+$ HPCs from G-CSF mobilized peripheral blood or bone marrow may be obtained from Allcells (Alameda, CA, USA) or Lonza (Basel, Switzerland). Cord blood CD34$^+$ HPCs may be used at a lower number like $1 \times 10^5$ cells per mouse.
4. HLA matching is needed if tumor cells are obtained from allogenic source.
5. At 4–8 weeks after transplant, monocytes, DCs and B cells are differentiated and found in the blood. Autologous T cells are usually reconstituted 1 day prior to the tumor implantation to obtain T cell competent mice. Genetic engineered T cells, like CAR T cells may be used before or after the tumor is established.
6. Other mouse strains lacking common gamma chain maybe suitable substitutes, e.g., *Il2rg*$^{tm1Wjl}$ Tg(CMV-IL3,CSF2,KITLG)1Eav/MloySzJ (NSG-SGM3; RRID:IMSR_JAX:013062). If the study outcome is T cell response, it might be important to match HLA type with HLA

transgenic variants, such as NOD.Cg-*Prkdc*$^{scid}$ *Il2rg*$^{tm1Wjl}$ Tg(HLA-A/H2-D/B2M)1Dvs/SzJ (RRID:IMSR_JAX:014570).

7. To increase the survival rate of newborn pups, 1–2-day old newborn mice with milk spot should be used. At this time point, the mice are the most efficient for human HPC engraftment.
8. Human CD34$^+$ HPCs can be transplanted into 4-week old mice sublethally irradiated at 12 cGy per gram of body weight using gamma irradiator.
9. Anti-CD3 antibody (OKT3) can eliminate potential T cell contamination from mature T cells.
10. At 8–12 weeks after transplant, DCs, monocytes, B cells and T cells are differentiated and found in the blood. Depending on the types and donors of CD34$^+$ HPCs, the percentage of each cell type and the timing of T cell development may be variable. Bleeding at 3–4 weekly intervals is adequate to evaluate the engraftment.
11. Antibodies should be tailored to the specification of flow cytometer.
12. It's important to report the absolute number of hCD45$^+$ cells and the subtypes. It reflects the output of human HPCs and allows the comparison between different research groups.
13. All cell lines need to be free of mycoplasma and verified for the identity by short tandem repeat (STR) analysis. Service for mycoplasma testing and STR analysis are available through ATCC.
14. Different cancer cell lines may be used to make cell-line derived xenograft tumor. The recommended culture media and conditions vary.
15. Different cell numbers and cell lines other than MDA-MB231 maybe use in the study. However, the growth curve in the mice maybe very different. It's necessary to titrate the number of cells ($1 \times 10^5$–$1 \times 10^7$) to obtain optimal tumor growth curve.
16. Tumor cells may be implanted into orthotopic site like mammary fat pad for breast cancer.
17. Tumor specimen can be either fresh or frozen from patients or donor mice. Fresh tumor fragments can be frozen at the density of five pieces per 1 mL vial of freezing medium (DMEM, 1% penicillin-streptomycin and 2 mM L-Glutamine, 10 U/mL Heparin, 10% FCS, 10% DMSO) in a controlled-rate freezing chamber, such as CoolCell (Corning) for 24 h in −80 °C freezer, and stored in the air-phase of a liquid nitrogen tank. Frozen tumor fragments need to be thawed at 37 °C and rinse with 1 × PBS with 1% penicillin-streptomycin prior to implantation.

**18.** For histological analysis, a tumor fragment can be snap frozen in OCT to evaluate proliferation (Ki67 staining) and the expression of cancer-associated antigens, etc. For genomic analysis, a tumor fragment can be snap frozen and processed at a later time point.

**19.** Surgical procedure is conducted in accordance with the institutional guideline for survival rodent surgery.

**20.** Wipe trocar needle, plunger and large forceps with chlorhexidine/alcohol pads between animals. Between cages, rinse trocar needle with sterile saline, wipe with chlorhexidine/alcohol pads and sterilize 10–15 s with hot dry bead sterilizer.

**21.** A tumor is three-dimensional. The axis measured (length and width) can be rotated around the tumor to find the longest and widest points as long as the length and width are perpendicular to each other.

**22.** The tumor growth in cell line-derived xenograft tumor model can be monitored with live imaging when the cell line is labeled with luciferase using Lentiviral vector.

**23.** The cell debris from tissue processing or mouse cells may affect some downstream assays. In such case, human cells can be enriched with Ficoll-Plaque density-based centrifugation or the EasySep Mouse/Human Chimera Isolation Kit (Stemcell).

**24.** PD1 marker measured by flow cytometry with the antibody clone EH12.2H7 is affected after anti-PD1 antibody (Pembrolizumab) treatment due to binding of antibodies to the same site.

**25.** Tetramer staining is an easy way to enumerate antigen-specific T cells with TCRs binding to a tetramer complexes that is composed of MHC class I or class II molecules and peptide fragments. However, the usage is limited by the knowledge of peptide sequences restricted to each HLA allele. For example, only HLA-A*0201$^+$ melanoma patients may develop CD8$^+$ T cells to gp100 and that can be detected by HLA-A*0201/gp100 154-162 (KTWGQYWQV) tetramer.

**26.** An activation control which uses superantigen staphylococcal enterotoxin B (SEB, 1 μg/mL, Sigma) or Phorbol 12-myristate 13-acetate (PMA, 10 ng/mL, Sigma) and ionomycin (1 μg/mL, Sigma) is useful to setup the flow cytometer.

**27.** CD107a is a lysosomal associated membrane protein (LAMP-1) on the membrane of cytolytic granules within the cytoplasm. During activation-induced granule exocytosis, CD107a is expressed on the cell surface transiently and rapidly internalized. Therefore, T-cell cytolytic potential is measured with surface CD107a antibody staining during cell stimulation.

**28.** Cytokeratin 19 is often used to identify human breast cancer cells.

**29.** Other vaccines designed to elicit specific T cell response maybe tested here. However, a higher number of T cells ($3 \times 10^7$) maybe needed due to the low frequency of specific T cells.

## Acknowledgments

We thank patients and healthy donors for participation in our studies over the years. We thank Dr. Jim Keck and Dr. Lenny Shultz for collaboration and Vanessa K. P. Oliveira for help with experiments. Support over the years came from Baylor Institute for Immunology Research and Baylor University Medical Center Foundation. Current funding comes from US National Institutes of Health (CA195712, CA219880), US Department of Defense (W81XWH-17-1-0010) and The Jackson Laboratory.

## References

Alexandrov, L. B., Nik-Zainal, S., Wedge, D. C., Aparicio, S. A., Behjati, S., Biankin, A. V., et al. (2013). Signatures of mutational processes in human cancer. *Nature*, *500*(7463), 415–421. https://doi.org/10.1038/nature12477.

Aspord, C., Pedroza-Gonzalez, A., Gallegos, M., Tindle, S., Burton, E. C., Su, D., et al. (2007). Breast cancer instructs dendritic cells to prime interleukin 13-secreting $CD^{4+}$ T cells that facilitate tumor development. *The Journal of Experimental Medicine*, *204*(5), 1037–1047. https://doi.org/10.1084/jem.20061120.

Banchereau, J., & Palucka, K. (2017). Cancer vaccines on the move. *Nature Reviews. Clinical Oncology*, *15*, 9. https://doi.org/10.1038/nrclinonc.2017.149.

Billerbeck, E., Barry, W. T., Mu, K., Dorner, M., Rice, C. M., & Ploss, A. (2011). Development of human CD4+FoxP3+ regulatory T cells in human stem cell factor-, granulocyte-macrophage colony-stimulating factor-, and interleukin-3-expressing NOD-SCID IL2Rgamma(null) humanized mice. *Blood*, *117*(11), 3076–3086. https://doi.org/10.1182/blood-2010-08-301507.

Billerbeck, E., Horwitz, J. A., Labitt, R. N., Donovan, B. M., Vega, K., Budell, W. C., et al. (2013). Characterization of human antiviral adaptive immune responses during hepatotropic virus infection in HLA-transgenic human immune system mice. *Journal of Immunology*, *191*(4), 1753–1764. https://doi.org/10.4049/jimmunol.1201518.

Binnewies, M., Roberts, E. W., Kersten, K., Chan, V., Fearon, D. F., Merad, M., et al. (2018). Understanding the tumor immune microenvironment (TIME) for effective therapy. *Nature Medicine*, *24*(5), 541–550. https://doi.org/10.1038/s41591-018-0014-x.

Bulik-Sullivan, B., Busby, J., Palmer, C. D., Davis, M. J., Murphy, T., Clark, A., et al. (2018). Deep learning using tumor HLA peptide mass spectrometry datasets improves neoantigen identification. *Nature Biotechnology*. https://doi.org/10.1038/nbt.4313.

Capasso, A., Lang, J., Pitts, T. M., Jordan, K. R., Lieu, C. H., Davis, S. L., et al. (2019). Characterization of immune responses to anti-PD-1 mono and combination immunotherapy in hematopoietic humanized mice implanted with tumor xenografts. *Journal for Immunotherapy of Cancer*, 7(1), 37. https://doi.org/10.1186/s40425-019-0518-z.

Cosgun, K. N., Rahmig, S., Mende, N., Reinke, S., Hauber, I., Schafer, C., et al. (2014). Kit regulates HSC engraftment across the human-mouse species barrier. *Cell Stem Cell*, *15*(2), 227–238. https://doi.org/10.1016/j.stem.2014.06.001.

Danner, R., Chaudhari, S. N., Rosenberger, J., Surls, J., Richie, T. L., Brumeanu, T. D., et al. (2011). Expression of HLA class II molecules in humanized NOD.Rag1KO. IL2RgcKO mice is critical for development and function of human T and B cells. *PLoS One*, *6*(5), e19826. https://doi.org/10.1371/journal.pone.0019826.

Gulati, P., Ruhl, J., Kannan, A., Pircher, M., Schuberth, P., Nytko, K. J., et al. (2018). Aberrant Lck signal via CD28 costimulation augments antigen-specific functionality and tumor control by redirected T cells with PD-1 blockade in humanized mice. *Clinical Cancer Research*, *24*(16), 3981–3993. https://doi.org/10.1158/1078-0432.CCR-17-1788.

Hagai, T., Chen, X., Miragaia, R. J., Rostom, R., Gomes, T., Kunowska, N., et al. (2018). Gene expression variability across cells and species shapes innate immunity. *Nature*, *563*(7730), 197–202. https://doi.org/10.1038/s41586-018-0657-2.

Hamid, O., Robert, C., Daud, A., Hodi, F. S., Hwu, W. J., Kefford, R., et al. (2013). Safety and tumor responses with lambrolizumab (anti-PD-1) in melanoma. *The New England Journal of Medicine*, *369*(2), 134–144. https://doi.org/10.1056/NEJMoa1305133.

Hodi, F. S., O'Day, S. J., McDermott, D. F., Weber, R. W., Sosman, J. A., Haanen, J. B., et al. (2010). Improved survival with ipilimumab in patients with metastatic melanoma. *The New England Journal of Medicine*, *363*(8), 711–723. https://doi.org/10.1056/NEJMoa1003466.

June, C. H., O'Connor, R. S., Kawalekar, O. U., Ghassemi, S., & Milone, M. C. (2018). CAR T cell immunotherapy for human cancer. *Science*, *359*(6382), 1361–1365. https://doi.org/10.1126/science.aar6711.

Kanazawa, N. (2007). Dendritic cell immunoreceptors: C-type lectin receptors for pattern-recognition and signaling on antigen-presenting cells. *Journal of Dermatological Science*, *45*(2), 77–86.

Keskin, D. B., Anandappa, A. J., Sun, J., Tirosh, I., Mathewson, N. D., Li, S., et al. (2019). Neoantigen vaccine generates intratumoral T cell responses in phase Ib glioblastoma trial. *Nature*, *565*(7738), 234–239. https://doi.org/10.1038/s41586-018-0792-9.

Kim, M. Y., Yu, K. R., Kenderian, S. S., Ruella, M., Chen, S., Shin, T. H., et al. (2018). Genetic inactivation of CD33 in hematopoietic stem cells to enable CAR T cell immunotherapy for acute myeloid leukemia. *Cell*, *173*(6), 1439–1453 e1419. https://doi.org/10.1016/j.cell.2018.05.013.

Le, D. T., Uram, J. N., Wang, H., Bartlett, B. R., Kemberling, H., Eyring, A. D., et al. (2015). PD-1 blockade in tumors with mismatch-repair deficiency. *The New England Journal of Medicine*, *372*(26), 2509–2520. https://doi.org/10.1056/NEJMoa1500596.

Li, Y., Masse-Ranson, G., Garcia, Z., Bruel, T., Kok, A., Strick-Marchand, H., et al. (2018). A human immune system mouse model with robust lymph node development. *Nature Methods*, *15*(8), 623–630. https://doi.org/10.1038/s41592-018-0071-6.

Matsumura, T., Kametani, Y., Ando, K., Hirano, Y., Katano, I., Ito, R., et al. (2003). Functional CD5 + B cells develop predominantly in the spleen of NOD/SCID/gammac(null) (NOG) mice transplanted either with human umbilical cord blood, bone marrow, or mobilized peripheral blood CD34 + cells. *Experimental Hematology*, *31*(9), 789–797.

Maude, S. L., Laetsch, T. W., Buechner, J., Rives, S., Boyer, M., Bittencourt, H., et al. (2018). Tisagenlecleucel in children and young adults with B-cell lymphoblastic leukemia. *The New England Journal of Medicine*, *378*(5), 439–448. https://doi.org/10.1056/NEJMoa1709866.

Mestas, J., & Hughes, C. C. (2004). Of mice and not men: Differences between mouse and human immunology. *Journal of Immunology*, *172*(5), 2731–2738.

Najima, Y., Tomizawa-Murasawa, M., Saito, Y., Watanabe, T., Ono, R., Ochi, T., et al. (2016). Induction of WT1-specific human CD8+ T cells from human HSCs in HLA class I Tg NOD/SCID/IL2rgKO mice. *Blood*, *127*(6), 722–734. https://doi.org/10.1182/blood-2014-10-604777.

Palucka, A. K., & Coussens, L. M. (2016). The basis of oncoimmunology. *Cell, 164*(6), 1233–1247. https://doi.org/10.1016/j.cell.2016.01.049.

Pedroza-Gonzalez, A., Xu, K., Wu, T. C., Aspord, C., Tindle, S., Marches, F., et al. (2011). Thymic stromal lymphopoietin fosters human breast tumor growth by promoting type 2 inflammation. *The Journal of Experimental Medicine, 208*(3), 479–490. https://doi.org/10.1084/jem.20102131.

Ribas, A., & Wolchok, J. D. (2018). Cancer immunotherapy using checkpoint blockade. *Science, 359*(6382), 1350–1355. https://doi.org/10.1126/science.aar4060.

Rongvaux, A., Willinger, T., Martinek, J., Strowig, T., Gearty, S. V., Teichmann, L. L., et al. (2014). Development and function of human innate immune cells in a humanized mouse model. *Nature Biotechnology, 32*(4), 364–372. https://doi.org/10.1038/nbt.2858.

Schumacher, T. N., & Schreiber, R. D. (2015). Neoantigens in cancer immunotherapy. *Science, 348*(6230), 69–74. https://doi.org/10.1126/science.aaa4971.

Shultz, L. D., Brehm, M. A., Garcia-Martinez, J. V., & Greiner, D. L. (2012). Humanized mice for immune system investigation: Progress, promise and challenges. *Nature Reviews. Immunology, 12*(11), 786–798. https://doi.org/10.1038/nri3311.

Takahashi, T., Katano, I., Ito, R., Goto, M., Abe, H., Mizuno, S., et al. (2017). Enhanced antibody responses in a novel NOG transgenic mouse with restored lymph node organogenesis. *Frontiers in Immunology, 8*, 2017. https://doi.org/10.3389/fimmu.2017.02017.

Theocharides, A. P., Rongvaux, A., Fritsch, K., Flavell, R. A., & Manz, M. G. (2016). Humanized hemato-lymphoid system mice. *Haematologica, 101*(1), 5–19. https://doi.org/10.3324/haematol.2014.115212.

Traggiai, E., Chicha, L., Mazzucchelli, L., Bronz, L., Piffaretti, J. C., Lanzavecchia, A., et al. (2004). Development of a human adaptive immune system in cord blood cell-transplanted mice. *Science, 304*(5667), 104–107. https://doi.org/10.1126/science.1093933.

Victor Garcia, J. (2016). Humanized mice for HIV and AIDS research. *Current Opinion in Virology, 19*, 56–64. https://doi.org/10.1016/j.coviro.2016.06.010.

Wang, M., Yao, L. C., Cheng, M., Cai, D., Martinek, J., Pan, C. X., et al. (2018). Humanized mice in studying efficacy and mechanisms of PD-1-targeted cancer immunotherapy. *The FASEB Journal, 32*(3), 1537–1549. https://doi.org/10.1096/fj.201700740R.

Williams, A., Flavell, R. A., & Eisenbarth, S. C. (2010). The role of NOD-like receptors in shaping adaptive immunity. *Current Opinion in Immunology, 22*(1), 34–40. https://doi.org/10.1016/j.coi.2010.01.004.

Wu, T. C., Xu, K., Banchereau, R., Marches, F., Yu, C. I., Martinek, J., et al. (2014). Reprogramming tumor-infiltrating dendritic cells for CD103+ CD8+ mucosal T-cell differentiation and breast cancer rejection. *Cancer Immunology Research, 2*(5), 487–500. https://doi.org/10.1158/2326-6066.CIR-13-0217.

Wu, T. C., Xu, K., Martinek, J., Young, R. R., Banchereau, R., George, J., et al. (2018). IL1 receptor antagonist controls transcriptional signature of inflammation in patients with metastatic breast cancer. *Cancer Research, 78*(18), 5243–5258. https://doi.org/10.1158/0008-5472.CAN-18-0413.

Yu, C. I., Gallegos, M., Marches, F., Zurawski, G., Ramilo, O., Garcia-Sastre, A., et al. (2008). Broad influenza-specific CD8+ T-cell responses in humanized mice vaccinated with influenza virus vaccines. *Blood, 112*(9), 3671–3678. https://doi.org/10.1182/blood-2008-05-157016.

Zhang, L., & Su, L. (2012). HIV-1 immunopathogenesis in humanized mouse models. *Cellular & Molecular Immunology, 9*(3), 237–244. https://doi.org/10.1038/cmi.2012.7.

Zitvogel, L., Pitt, J. M., Daillère, R., Smyth, M. J., & Kroemer, G. (2016). Mouse models in oncoimmunology. *Nature Reviews. Cancer, 16*, 759. https://doi.org/10.1038/nrc.2016.91.

FSC
www.fsc.org
MIX
Papier | Fördert gute Waldnutzung
FSC® C083411

Kolibri 360 GmbH
Ferdinand-Jühlke-Straße 7
99095 Erfurt, Deutschland
produktsicherheit@kolibri360.de